MAGNETOSTATIC WAVES IN INHOMOGENEOUS FIELDS

MAGNETOSTATIC WAVES IN INHOMOGENEOUS FIELDS

V.G. Shavrov
V.I. Shcheglov

CISP

CRC Press is an imprint of the
Taylor & Francis Group, an **informa** business

Translated from Russian by V.E. Riecansky

First Edition published 2022
by CRC Press
6000 Broken Sound Parkway NW, Suite 300, Boca Raton, FL 33487-2742

and by CRC Press
2 Park Square, Milton Park, Abingdon, Oxon, OX14 4RN

© 2022 by CISP

CRC Press is an imprint of Taylor & Francis Group, LLC

Reasonable efforts have been made to publish reliable data and information, but the author and publisher cannot assume responsibility for the validity of all materials or the consequences of their use. The authors and publishers have attempted to trace the copyright holders of all material reproduced in this publication and apologize to copyright holders if permission to publish in this form has not been obtained. If any copyright material has not been acknowledged please write and let us know so we may rectify in any future reprint.

Except as permitted under U.S. Copyright Law, no part of this book may be reprinted, reproduced, transmitted, or utilized in any form by any electronic, mechanical, or other means, now known or hereafter invented, including photocopying, microfilming, and recording, or in any information storage or retrieval system, without written permission from the publishers.

For permission to photocopy or use material electronically from this work, access www.copyright.com or contact the Copyright Clearance Center, Inc. (CCC), 222 Rosewood Drive, Danvers, MA 01823, 978-750-8400. For works that are not available on CCC please contact mpkbookspermissions@tandf.co.uk

Trademark notice: Product or corporate names may be trademarks or registered trademarks and are used only for identification and explanation without intent to infringe.

Library of Congress Cataloging-in-Publication Data

ISBN: 978-0-367-49447-6 (hbk)
ISBN: 978-0-367-49449-0 (pbk)
ISBN: 978-1-003-04622-6 (ebk)

Contents

Introduction xi
List of frequently used abbreviations xiv

1. Magnetostatic waves and domain structures in ferrite–garnet films (literature review) — 1

1.1. Oscillations and waves in magnetically ordered media in the approximation of magnetostatics — 1
1.2. Conditions of existence and dispersion of MSWs in magnetic films and structures on their basis — 5
1.3. Spreading of SMSW (surface magnetostic waves) in an arbitrary direction along the film plane — 9
1.4. Distribution of SMSW in ferrite films and structures under the conditions of inhomogeneous magnetization — 13
1.5. Distribution of SMSW in ferrite films and structures with periodic inhomogneities — 16
1.6. Conversion of a magnetostatic wave into electromagnetic on the field inhomogeneity — 21
1.7. Domain structures in ferrite films, FMR and MSW under the conditions of the existence of domain structures — 22
1.8. Features of magnetostatic waves in the long-wave limit — 24
1.9. Use of FMR, MSW and domains in ferritle films for information processing devices — 27
1.10. Basic issues for further explanation — 28
1.11. Some new directions of research of MSW — 30

2. Mathematical apparatus used in calculating the properties of magnetostatic waves — 33

2.1. Landau–Lifshitz equation — 33
2.2. Dynamic sensitivity of a magnetic medium — 39
2.3. Walker's equation — 45
2.3.1. Walker's equation with an arbitrary susceptibility tensor — 45
2.3.2. Walker equation in the Damon–Eshbach problem. — 48
2.4. Dispersion ration for magnetic plate with free surface — 50
2.4.1. Basic equations — 50
2.4.2. Border conditions — 54
2.4.3. Complete problem statement — 55
2.4.4. Solving equations without boundary conditions — 56
2.4.5. Frequency regions of body and surface waves — 63
2.4.6. Derivation of the dispersion relation from the solution and boundary

	conditions	65
2.4.7.	Transition to the polar coordinate system	69
2.4.8.	Potentials	72
2.4.9.	Fields	75
2.4.10.	Magnetization	76
2.4.11.	Cutoff angle for the Damon-Eshbach ratio.	78
2.4.12.	Damon–Eshbach dispersion relation in the Cartesian coordinate system	80
2.5.	Dispersion ratio for metal–dielectric–ferrite–metal (MDFDM) structure and its particular cases	83
2.5.1.	General derivation of the dispersion relation	84
2.5.2.	Dispersion relation for an arbitrary direction of propagation of the phase front	91
2.5.3.	Transition to the polar coordinate system	92
2.5.4.	Passage to the limit for dispersion relations for other structures	95
2.6.	Dispersion ration for metal–dielectric–ferrite–ferrite–dielectric–metal structure (MDFFDM)	98
2.6.1.	General conclusion and character of the dispersion relation	98
2.6.2.	Passage to the limit for dispersion relations for other structures	103
2.7.	Phase and group velocities, phase rise and delay time of wave beams SMSW	105
2.7.1.	Phase and group velocities	106
2.7.2.	Phase run and delay time	112
2.8.	System of equations for the Hamilton–Auld method	113
2.8.1.	General derivation of the Hamilton–Auld equations	114
2.8.2.	Transition to the polar coordinate system	115
2.9.	Derivatives from the dispersion relationship for the ferritic–dielectric–metal structure	123
2.10.	Equivalence of different kinds of equations of dynamics in classical mechanics	127
2.11.	Cauchy's proble in the distribution of SMSW	129
2.12.	Technique for calculating the trajectories of wave beams of MSW in an inhomogeneous field	134

3. Magnetostatic waves in homogenized magnetized ferrite films and structures on their basis   139

3.1.	Conditions of existence and dispersion of SMSW (surface magnetostatic waves) in ferrite films and structures on their basis	139
3.1.1.	Dispersion properties of forward and backward SMSWs in the FDM structure	140
3.1.2.	Experimental study of the dispersion of the SMSW in the structure of the FDM	147
3.1.2.1.	Basic experimental technique	147
3.1.2.2.	Results of an experimental study of the dispersion properties of SMSW	154

3.1.3.	On the possibility of experimental observation of backward waves	157
3.2.	Distribution of SMSW in a two-component environment consists of a free ferrite film and FDM (ferrite–dielectric–metal) structure	158
3.2.1.	Analysis of the refraction of the SMSW using the method of isofrequency curves	159
3.2.1.1.	Formulation of the problem	159
3.2.1.2.	Analysis of orientation dependences by the method of isofrequency curves	162
3.2.1.3.	Strip orientation along the field	162
3.2.1.4.	The orientation of the strip is arbitrary	165
3.2.1.5.	Evaluation of the possibility of manifestation of the effects of dispersive splitting of a wave beam under the conditions of a real experiment	166
3.2.2.	Experimental study of the refraction of the SMSW	167
3.2.2.1.	Strip orientation along the field	169
3.2.2.2.	The orientation of the strip is arbitrary	174
3.2.3.	Reflection coefficient of the SMSW from the interface	176
3.3.	Dispersional properties of SMSW in structures containing two ferrite layers	177
3.3.1.	Ferrite–ferrite (FF) structure	178
3.3.2.	Metal–dielectric–ferrite–ferrite–dielectric–metal structure (MDFFDM)	179
3.3.3.	Experimental study of the variance of SMSW	181

4. Methods of research and analysis of the propagation of SMSW under conditions of magnetization by a longitudinal inhomogeneous field — 185

4.1.	Basic types of inhomogeneities of a magnetizing field	186
4.2.	Spatial configuration of the areas if distribution of the SMSW	187
4.3.	Methods for analysis of SMSW propation under the conditions of inhomogeneous binding (frequency curves and Hamilton–Auld)	189
4.3.1.	Isofrequency curve method	189
4.3.2.	The Hamilton–Auld method	191
4.3.3.	Comparison of methods for analyzing SMSW trajectories	193
4.4.	Distribution of SMSWs in ferrite films with free surfaces	193
4.4.1.	Analysis of SMSW trajectories by the method of isofrequency curves	194
4.4.1.1.	Linearly inhomogeneous field	194
4.4.1.2.	Valley-type field	196
4.4.1.3.	Shaft-type field	197
4.4.2.	Analysis of SMSW trajectories by the Hamilton–Auld method	198
4.4.2.1.	Linearly inhomogeneous field	198
4.4.2.2.	Valley-type field	203
4.4.2.3.	Shaft-type field	203

4.5.	Distribution of SMSW in the ferrite–metal structure	208
4.5.1.	Linearly inhomogeneous field	208
4.5.2.	Valley-type field	210
4.5.3.	Shaft-type field	211
4.5.4.	Channels of the first and second type	211
4.6.	Distribution of SMSWs in the structure of ferrite–dielectric metal	212
4.6.1.	Analysis of SMSW trajectories by the method of isofrequency curves	213
4.6.1.1.	Linearly inhoimogeneous field	213
4.6.1.2.	Valley-type field	216
4.6.1.3.	Shaft-type field	218
4.6.1.4.	General comment	220
4.6.2.	Analysis of SMSW trajectories by the Hamilton–Auld method	220
4.6.2.1.	Linearly inhomogeneous field	221
4.6.2.2.	Valley-type field	223
4.6.2.3.	Shaft-type field	224
4.7.	Phase rise and delay time	227
4.7.1.	Linearly inhomogeneous field	227
4.7.2.	Valley-type field	229
4.7.3.	Shaft-type field	231
4.8.	Experimental study of SMSW trajectories	235
4.8.1.	The main parameters of the experiment	235
4.8.2.	Linearly inhomogeneous field	236
4.8.3.	Valley-type field	237
4.8.4.	Shaft-type field	238
4.8.5.	Change of various parameters of the experiment	240
5.	**Propagation of wave beams of finite width in inhomogeneous magnetized ferrite films**	**248**
5.1.	Spatial transformation of wide beams of SMSW propagating in inhohogensouly magnetized films	248
5.1.1.	Linearly inhomogeneous field	249
5.1.2.	Valley-type field	251
5.1.3.	Shaft-type field	252
5.2.	Method for analysis of amplitude-frequency and phase-frequency characteristics of transmission lines of SMSW	253
5.2.1.	General scheme of the method for calculating the frequency phase responses	253
5.2.2.	Frequency response diagram	256
5.2.3.	PFC construction scheme	257
5.3.	Amplitude-frequency characteristics of transmision lines on ferrite films magnetized by fields of different configurations	260
5.3.1.	Homogeneous field	260
5.3.2.	Linearly inhomogeneous field	265

5.3.3.	Valley-type field	267
5.3.4.	Shaft-type field	270
5.4.	Ampliture–frequency characteristics of waveguard channel for SMSW formed by inhomogeneous 'shaft'-type field	277
5.4.1.	Changing the length of the channel	278
5.4.2.	Changing the channel excitation conditions	279
5.4.2.1.	Symmetrical arousal	280
5.4.2.2.	Asymmetrical excitement	281
5.4.2.3.	Transverse shift of the emitting transducer	282
5.5.	Amplitude–frequency characteristics of the transmission line to the SMSW at an arbitrary orientation of the magnetizing field	284
5.5.1.	The general geometry of two variants of the location of the transducers: mutually opposite and mutually shifted	284
5.5.2.	Filtration of the first type, mutually opposite geometry	286
5.5.3.	Filtering of the second type, mutually shifted geometry	289
5.6.	Experimental study of SMSW beams of finite width and amplitude–frequency characteristics	291
5.6.1.	Linearly inhomogeneous field	292
5.6.2.	Valley type field	293
5.6.3.	Shaft-type field	294

6.1. Amplitude–frequency properties of trasmission lines on magnetostatic waves taking into account the phase run 299

6.1.	General characteristics of typical transmission lines to SMSW	299
6.2.	General case of waves in a magnetic medium	300
6.3.	The case of surface magnetostatic waves (SMSW)	306
6.4.	Amplitude transmission line characteristics and its different geometric parameters	307
6.4.1.	Dependence of the amplitude of the transmitted signal on frequency when changing the relative orientation of the transducers	308
6.4.2.	Dependence of the amplitude of the transmitted signal on the frequency with a change in the width of the wave beam	310
6.4.3.	Dependence of the amplitude of the transmitted signal on the relative orientation of the transducers at a fixed signal frequency	311
6.4.4.	Dependence of the phase of the transmitted signal on frequency when changing the relative orientation of the transducers	315
6.5.	Effect of the phase run on AFC	317
6.5.1.	Geometry of the problem with relative mutual displacement of transducers	318
6.5.2.	Formation of the amplitude–frequency characteristic	319
6.5.3.	Formation of the phase-frequency response	321
6.5.4.	The influence of the length of the transducers on the structure of	

	the frequency response	322
6.6.	Deformation of the wave front of surface magnetostatic waves in ferrite films magnetized by linearly inhomogeneous field	325
6.6.1.	General geometry of the problem	325
6.6.2.	Various cases of orientation of the emitting transducer	328
6.6.2.1.	Orientation corresponding to $\varphi = 30°$	328
6.6.2.2.	Other orientations	331
6.6.	General character of transformation of the area of distribution of SMSW when various parameters of the structure change	334
6.6.1.	Changing the orientation of the emitting transducer	334
6.6.2.	Frequency change	335
6.6.3.	Changing the gradient of the field	336
6.7.	Recommendations for optimizing the parameters of the transmission line of the SMSW	337

7. Use of magnetostatic waves in inhomogeneously magnetic ferrite films for information processing devices and other technical applications   341

7.1.	Brief overview of possible technical applications	341
7.2.	Wave guiding structures for SMSW on ferrite films magnetized by a shaft-type field	343
7.3.	Optimization of the shape of SMSW converters for devices on inhomogeneous magnetized ferrite films	346
7.4.	Multi-channel filter on ferrite film magnetized by a valley-type field	349
7.5.	Multi-channel filter on packed ferrite structures	351
7.6.	Microwave signal delay line on a ferrite film magnetized by a shaft-type field	355
7.7.	Measurements of parameters of yttrium iron garnet films with a complex anisotropy character	357
7.8.	Study of the spatial distribution of the magnetic field with the help of the sensor on the SMSW	359
7.9.	Use of the transmission line to SMSW to determine the orientation of the magnetic field	360

Bibliography   **365**
Index   **400**

Introduction

Magnetostatic waves (MSWs) in magnetodielectric media are fundamental for the creation of a number of highly efficient devices for analog information processing in the microwave range. These devices include various filters, delay lines, phase shifters, frequency converters, nonreciprocal and nonlinear devices, and others. The main medium where magnetostatic waves can propagate are films, plates, and bulk ferrite samples, first of all, yttrium-iron garnet (YIG), which has record low magnetic losses. At the same time, the propagation of MSWs in such media is accompanied by numerous and very diverse physical effects that sharply distinguish them from ordinary electromagnetic waves in isotropic media. First of all, the complex character of the dispersion properties of MSWs, the noncolinearity of the phase and group velocity vectors, and also the nonreciprocity of propagation should be noted. Such circumstances stimulate the use of a detailed and comprehensive study of the physical properties of MSWs under a wide variety of conditions. This monograph is devoted to a description of some aspects of such research. The information given in it in no way pretends to be exhaustive, so that it does not more than reflect the range of scientific interests of the authors. The main volume of the monograph is a summary and generalization of the main scientific and applied results obtained by the authors in the period from 1990 to 2015.

The monograph is the first in a series devoted to magnetostatic waves and related phenomena. In accordance with the name, this part contains the main material on the propagation of MSWs in inhomogeneously magnetized films and structures based on them. The next monograph is expected to be devoted to the more complex effects accompanying such propagation, including the interaction of MSWs with periodic structures, domains, and inhomogeneous fields of complex configurations. Separately, it is proposed to consider the question of converting magnetostatic waves into electromagnetic waves on field inhomogeneities, accompanied by the emission of electromagnetic waves into free space.

The monograph is mainly theoretical in nature. The experimental results are presented to the extent that they are the basis for the described physical phenomena.

In order to present the general picture of the science of MSW, the first chapter provides a brief review of the literature over the past three to four decades, that is, since the 60s of the 20th century, when MSW became available in experiments due to the appearance of yttrium iron garnet.

Certain experience of the authors' work with students and novice researchers shows that at the initial acquaintance with the subject, noticeable difficulties are caused by mastering the mathematical apparatus, a significant part of which is scattered in articles that do not always easily fit with each other. In order to overcome this situation, the monograph includes a second chapter, which is entirely devoted to a detailed presentation of the basic mathematical techniques for working with MCWs. Other more specific calculation methods are also discussed in the relevant chapters in sufficient detail.

Numerous co-authors took part in the works on the basis of the monograph. So in the ideological basis and interpretation of most of the works lie the ideas of A.V. Vashkovsky, in theoretical works an important role belongs to V.I. Zubkov, most of the experimental work was carried out jointly with E.G. Locke. A.Yu. Annenkov took part in the design and implementation of working models of a number of devices.

The participation of the co-authors of the works on which the monograph is based is reflected in more detail in the list of references.

The work was greatly facilitated by the stimulating attention and repeated useful remarks of S.V. Yakovlev.

Numerous discussions of the works were attended by: S.V. Gerus, Yu.I. Bespyatykh, I.E. Dikshtein, V.V. Tarasenko, V.D. Kharitonov, F.V. Lisovskiy, E.G. Mansvetova, G. V. Arzamastseva, A. V. Voronenko, D. G. Shakhnazaryan, V. V. Kildishev, L. A. Krasnozhen.

Numerous discussions with fruitful discussion took place with the participation of P.E. Zilberman.

A.F.Kabychenkov and V.V.Koledov took part in numerous useful discussions concerning the physics of phase transitions.

The most important role in creating favorable conditions for work, constant attention to it and repeated provision of administrative, economic and scientific assistance belongs to Yu.V. Gulyaev, Academician of the Russian Academy of Sciences (RAS).

The immediate execution of the work, including the writing of this monograph, was made possible thanks to the help of S.A. Nikitov, Corresponding Member of the RAS.

Almost all the work was carried out with the active participation of numerous technical personnel, without whose help it would have been absolutely impossible to carry out the above work.

The authors express their deep gratitude to all the listed participants and co-authors of the works, give a deep bow and bring the deepest gratitude. The greatest thanks to you, dear colleagues, friends and assistants!

List of frequently used abbreviations

FMR	–	ferromagnetic resonance.
MSW	–	magnetostatic wave.
SMSW	–	surface magnetostatic wave.
BMSW	–	bulk magnetostatic wave.
AFC	–	amplitude-frequency characteristic.
FF	–	ferrite film.
FM	–	ferrite–metal (structure).
FDM	–	ferrite–dielectric-metal (structure).
FF	–	ferrite–ferrite (structure).
YIG	–	yttrium–iron garnet.

1
Magnetostatic waves and domain structures in ferrite-garnet films (literature review)

This chapter is a brief overview of the main results of studying the propagation of magnetostatic waves (MSWs) and processes of domain dynamics in uniformly and inhomogeneously magnetized iron garnet films and structures based on them. Oscillations and waves in magnetically ordered media are considered in the magnetostatic approximation, the conditions of existence, dispersion and propagation of MSWs in magnetic films and structures, the propagation of MSWs in arbitrary directions along the film plane under conditions of homogeneous and inhomogeneous magnetization, as well as in the presence of a domain structure. The main possible applications of FMR (ferromagnetic resonance), MSW, and domains in ferrite films for information processing devices, as well as for measuring the parameters of films and other physical objects, are discussed.

The results of the work of the authors of this monograph, which constitute the main part of this presentation and are presented in the subsequent chapters, are excluded from the consideration carried out here.

1.1. Oscillations and waves in magnetically ordered media in the approximation of magnetostatics

The study of dynamic phenomena in magnetic media began more than half a century ago. The first works devoted to the dynamics of domain walls [1] were followed by numerous studies of various to the high-frequency properties of bulk single- and polycrystals of ferrites [6–10, 23–32], as well as artificial materials and composites based on ferrite particles [33–35], including for the problem of absorption of microwave radiation [36, 37].

The theoretical interpretation is usually based on the natural FMR model and domain wall resonance [24–32]. To calculate the properties of the ensemble of particles making up the composite, along with the traditional averaging methods [24–32, 33, 39], the Monte Carlo method is used [34].

The FMR phenomenon has been studied in detail in films of yttrium iron [39–44], yttrium–gallium, and mixed ferrite–garnets [4, 14, 45–49], including films with a large uniaxial anisotropy [47–50], low magnetization [51], large Faraday rotation [52, 53], and multilayer [54, 55]. FMR was studied in spinel ferrites [23, 56, 57], metal [11, 58–61], and amorphous films [62, 63]. In recent years, especially in connection with the discovery of the giant magnetoresistance effect, FMR has been widely studied in multilayer nanofilms [64-74], superlattices [75, 76], and granular structures [32, 44, 35, 77–81]. The nature of the bond between the layers and the state of the surface of such materials are manifested in the properties of spin-wave resonance [63, 69, 73, 74, 82–85]. A significant number of works are devoted to the use of FMR as a tool for measuring the parameters of films, primarily anisotropy and magnetization [13, 14, 45, 47–50, 59, 65, 86–90], as well as studies of magnetic relaxation processes [20, 41, 45, 91–95]. FMR is a powerful tool for studying various types of inhomogeneities in films [11, 14, 44, 83, 85, 88–90, 95–103]. Along with the above, FMR is effectively used to study magnetoelastic [104–107] and exchange interactions [66, 72, 74], domain structure [108], and parameters of media for magnetic recording [109–112]. Magneto-optical effects associated with FMR are of considerable interest [52, 53, 88, 113–116].

A more complex form of motion of the magnetization vector–inhomogeneous precession manifests itself in the form of standing oscillations and traveling waves. For sufficiently long waves with wavenumbers less than 10^4–10^5 cm^{-1}, magnetic dipole interaction is predominant, while the role of exchange interaction is negligible. These are magnetostatic exchangeless waves. For shorter waves with wave numbers exceeding 10^5–10^6 cm^{-1}, the exchange interaction becomes dominant. These are exchange spin waves.

types of oscillations and waves in a wide variety of magnets. The classical description of dynamic phenomena in magnets is based on the use of the Landau–Lifshitz equation [2–4]:

...

$$\frac{\partial \vec{M}}{\partial t} = -\gamma \left[\vec{M} \times \vec{H}_{eff} \right] + \frac{\alpha}{M_0} \left[\vec{M} \times \frac{\partial \vec{M}}{\partial t} \right] \quad (1.1)$$

Here $\vec{M}$ is the magnetization vector, M^0 is its absolute value (saturation magnetization), α is the damping parameter, γ is the gyromagnetic ratio, $\vec{H}_{eff}$ is the effective field equal to:

,

$$\vec{H}_{eff} = -\frac{\delta U}{\delta \vec{M}} = -\frac{\partial U}{\partial \vec{M}} + \sum_{p=1}^{3} \frac{\partial}{\partial x_p} \left[\frac{\partial U}{\partial \left(\partial \vec{M} / \partial x_p \right)} \right] \quad (1.2)$$

where U is the energy density of the magnet, δ is the sign of functional differentiation.

For a ferromagnetic medium, the energy density has the form:

$$U = U_{ex} + U_H + U_M + U_a + U_e + U_{me} + \ldots \quad (1.3)$$

In this expression: U_{ex} is the energy density of the inhomogeneous exchange interaction:

$$U_{ex} = \frac{A_0}{2aM_0^2} \sum_{i,k} \left(\frac{\partial M_i}{\partial x_k} \right)^2 = \frac{1}{2} q_0 \sum_{i,k} \left(\frac{\partial M_i}{\partial x_k} \right)^2 = \frac{A}{M_0^2} \sum_{i,k} \left(\frac{\partial M_i}{\partial x_k} \right)^2 \quad (1.4)$$

where A_0, q_0, A, is the exchange interaction constant in various forms of notation, a is the lattice constant of the magnetic crystal, U_H is the energy density of interaction of magnetization $\vec{M}$ with an external field $\vec{H}_e$
:

$$U_H = -\vec{M} \cdot \vec{H}_e \quad (1.5)$$

the density of the internal (dipole-dipole) interaction energy of the magnetic moments of the sample:

$$U_M = -\frac{1}{2} \vec{M} \cdot \vec{H}_m \quad (1.6)$$

where the field $\vec{H}_m$ is determined by the formula [5]:

,

$$\vec{H}_m = -\,\text{grad}\left(\int_V \frac{\text{div}\,\vec{M}}{R}dV - \oint_S \frac{\text{div}\,\vec{M}}{R}dS\right) \quad (1.7)$$

in which R is the distance from the inner point of the sample to the observation point, V and S are the volume and surface of the sampleI_s the energy density of magnetocrystalline anisotropy, for a uniaxial ferromagnet in a coordinate system with an axis along the easy magnetization axis (EMA), having the form:

$$U_a^{(o)} = \frac{K}{M_0^2}\left(M_y^2 + M_z^2\right) \quad (1.8)$$

where K is the constant of uniaxial anisotropy. For a cubic ferromagnet in a coordinate system whose axes coincide with the edges of a cubic crystallographic cell:

$$U_a^{(k)} = \frac{K_1}{M_0^4}\left(M_x^2 M_y^2 + M_y^2 M_z^2 + M_x^2 M_z^2\right) \quad (1.9)$$

where K_1 is the constant of cubic anisotropy (first), U_e the energy density of an elastically deformed body, for a cubic crystal in the same coordinate system has the form [2–4]:

$$U_e = \frac{1}{2}c_{11}\left(e_{xx}^2 + e_{yy}^2 + e_{zz}^2\right) + c_{12}\left(e_{xx}e_{yy} + e_{yy}e_{zz} + e_{xx}e_{zz}\right) +$$
$$+ 2c_{44}\left(e_{xy}^2 + e_{yz}^2 + e_{xz}^2\right) \quad (1.10)$$

where c_{ik} are the components of the elastic modulus tensor, e_{ik} are the components of the deformation tensor and U_{me} is the energy density of magnetoelastic interaction:

$$U_{me} = \frac{B_1}{M_0^2}\sum_p M_p^2 e_{pp} + \frac{B_2}{M_0^2}\sum_p \sum_{q\neq p} M_p M_q e_{pq} \quad (1.11)$$

where B_1, B_2 are the constants of magnetoelastic interaction.

The simplest form of motion of the magnetization vector is homogeneous precession – ferromagnetic resonance (FMR). This type of precession of magnetization was historically the first to be studied and investigated in greater detail. The results of FMR studies in various magnetic materials are described in detail in numerous reviews and monographs [2–4, 6–22]. Considerable attention is paid

1.2. Conditions of existence and dispersion of MSWs in magnetic films and structures on their basis

Let us consider the traditional scheme for describing wave processes for the case of exchangeless magnetostatic waves. In this case, in expression (1.2) for the effective field, the second term can be neglected. The field acting on a magnetic crystal

$$\vec{H} = H_i \vec{e}_z + \vec{n} e^{i\omega t} \qquad (1.12)$$

corresponds to magnetization

$$\vec{M} = M_0 \vec{e}_z + \vec{m} e^{i(\omega t - \vec{k}\vec{r})} \qquad (1.13)$$

where H_i is the static internal field in the sample, $\vec{e}_z$ is the unit vector in the direction of the axis, h and ω are the amplitude and frequency of the alternating field, m is the amplitude of the variable magnetization, $\vec{k}$ is the wave vector of the magnetization wave, $\vec{r}$ is the vector from the origin to the point of observation of the wave. The equations describing magnetostatic oscillations and waves are as follows:

$$i\omega \vec{m} = -\gamma \left[\vec{e}_z \times \left(M_0 \vec{h} - H_i \vec{m} \right) \right] \qquad (1.14)$$

$$\operatorname{rot} \vec{h} = 0 \qquad (1.15)$$

$$\operatorname{div}\left(\vec{h} + 4\pi \vec{M} \right) = 0 \qquad (1.16)$$

Let us choose the coordinate system in such a way that the plane coincides with the film plane, and the axis Ox is perpendicular to it, and the field is directed along the axis Oz. Equation (1.15) allows one to introduce the magnetostatic potential using the relation. In this case, from (1.16) we find the equation for the potential inside the sample (Walker's equation) [117]:

$$(1+\eta)\left(\frac{\partial^2}{\partial x^2} + \frac{\partial^2}{\partial y^2} \right)\psi_i + \frac{\partial^2 \psi_i}{\partial z^2} = 0 \qquad (1.17)$$

and outside the sample (Laplace equation):

$$\left(\frac{\partial^2}{\partial x^2} + \frac{\partial^2}{\partial y^2} + \frac{\partial^2}{\partial z^2} \right)\psi_e = 0 \qquad (1.18)$$

These formulas contain the following notation:

$$\eta = \frac{\Omega_H}{\Omega_H^2 - \Omega^2} \quad (1.19)$$

$$\nu = \frac{\Omega}{\Omega_H^2 - \Omega^2} \quad (1.20)$$

$$\Omega_H = \frac{H_i}{4\pi M_0} \quad (1.21)$$

$$\Omega = \frac{\omega}{4\pi\gamma M_0}. \quad (1.22)$$

From the standard boundary conditions at the interface of two media, coinciding with the plane, we obtain the boundary conditions for the potentials:

$$\left[(1+\eta)\frac{\partial\psi_i}{\partial x} - i\nu\frac{\partial\psi_i}{\partial y}\right]\bigg|_{x=0} = \frac{\partial\psi_e}{\partial x}\bigg|_{x=0} \quad (1.23)$$

$$\psi_i\big|_{x=0} = \psi_e\big|_{x=0} + \text{const.} \quad (1.24)$$

By separating variables, we obtain a solution in the form [118]:

$$\psi_{i,e} = X_{i,e}\exp(ik_y y)\cdot\exp(ik_z z) \quad (1.25)$$

and outside the ferrite environment:

$$X_e^{(\pm)} = C\cdot\exp(\pm ik_x^e x) \quad (1.26)$$

and inside:

$$X_i = A\cdot\exp(\alpha k_x^i x) + B\cdot\exp(-\alpha k_x^i x) \quad (1.27)$$

where: A, B are arbitrary constants k_y, k_z are the components of the wave vector along the axes Oy and Oz, k_x^i, k_x^e are the components of the wave vector along the axis Ox inside and outside the ferrite medium.

Introducing the angle counted from the axis Oy in the direction to the axis, we obtain the dispersion relation in the form:

$$\beta - 1 - 2\mu\alpha\,\text{cth}(kd\alpha) = 0 \quad (1.28)$$

where:

$$\alpha = \sqrt{\frac{\sin^2\phi}{\mu} + \cos^2\phi} \quad (1.29)$$

$$\beta = \left(v^2 - \mu^2 + \mu\right)\cos^2\phi - \mu \qquad (1.30)$$

$$\mu = 1 + \eta \qquad (1.31)$$

η and v are determined by the formulas (1.19) are (1.20), k is the length of the wave vector of the MSW, d is the thickness of the ferrite film.

Dispersion relation (1.28) in the case $(1 - \eta) < 0$ describes bulk vibrations and waves, and in the case $(1 + \eta) > 0$ the surface vibrations. The specific form of the solution and the dispersion relation is determined by the sample geometry. If all sample sizes are finite, a discrete spectrum of magnetostatic precession types is observed. A solution of this type for $(1 + \eta) < 0$ (volumetric vibrations) for various types of ellipsoids was obtained in [1, 117, 119–123]. If one (or two) of the sample sizes are infinite, a continuous spectrum is observed, consisting of a countable set of branches [124–128]. In the case $(1 + \eta) > 0$ (surface waves), solutions were obtained for a semi-infinite medium, as well as for bounded samples of various shapes [118, 127, 129] with different surface states [130, 131]. Surface magnetostatic waves, unlike bulk ones, are nonreciprocal, that is, the direction of their propagation is uniquely determined by the direction of the magnetic field.

The properties of both exchangeless and converted magnetostatic waves (MSWs) are described in numerous reviews and monographs [2-4, 6, 9, 11, 12, 18, 132–143, etc.]. Historically, the first to be discovered were standing [117, 144], and shortly after that – running bulk and surface exchangeless MSWs [118, 145, 146] in rods and plates. It turned out that for the effective excitation of inhomogeneous types of precession, the losses in the material of the medium must be sufficiently small. This condition was satisfied by pure yttrium iron garnet (YIG) [119] or with additives [120]. The first works were followed by extensive studies of the dispersion properties of magnetostatic waves (MSW) under various conditions [18, 42, 118, 143, 147–162], including in films of spinel ferrites [163], in limited samples [149], in anisotropic films [50, 91, 164–171], with an arbitrary orientation of the external field [156, 160, 164, 165, 167], various temperature conditions [161, 172], in the millimeter range [141], a nonstationary medium [138] and also when exposed to a light field [113, 173].

In YIG films with strong surface anisotropy (spin pinning), exchange [174, 175] studied were mixed dipole-exchange waves [15,

16, 176–185], including those in submicron [186] and inhomogeneous by thickness of films [83, 175, 187–197]. For the theoretical description of the observed phenomena, the method of spin-wave Green's functions turned out to be very effective [15, 16, 183].

Along with exchange waves, magnetoelastic waves were also studied in detail in bulk samples [2–4, 132, 133, 136, 198–205] and films [136, 205–212], including those with exchange [206], as well as the generation of phonons magnetic films [213].

In addition to single-layer ferrite films, the dispersion of MSWs has been widely studied in various multilayer structures [2–4, 9, 134, 135]. In this case, the initial task was to obtain a linear or other predetermined dispersion law of the MSW [214–226]. In the process of research, the regions of existence in frequency, field and wave vector were determined for forward and backward, bulk and surface waves. The features of dispersion and nonreciprocal properties of internal and external waves are considered [217]. The mutual transformation of forward and backward, as well as internal and external waves has been studied [218–221]. As a result, it was shown that varying the thickness and magnetization of individual layers makes it possible to obtain practically any desired dispersion law for MSWs [219, 220]. The dispersion of MSWs in FM and FDM structures [134, 135, 162, 222], including finite sizes [223], for an arbitrary field direction [224, 225], in a periodically inhomogeneous ferrite plate [226] has been extensively studied. Slot MSWs propagating in the gap between two ferrite films [227], MSWs in multilayer films with anisotropy [228, 229] and exchange-coupled films [230], as well as dipole-exchange films [17, 74, 135, 231] and magnetoelastic waves [232].

The possibility of amplifying traveling MSWs by a transport current has stimulated numerous studies on the ferrite-semiconductor structure [233–244]. A similar possibility was also considered in magnetic semiconductors [225–250]. The spectrum of MSWs was also studied in layered ferrite – ferroelectric structures [226, 251, 252]. The discovery of high-temperature superconductivity (HTSC) [253–255] gave rise to many works on the study of the dispersion properties of MSWs and the interaction of MSWs with Abrikosov vortices, including in the presence of a transport current in the ferrite-HTSC structure [256–266].

A considerable number of works have been devoted to the excitation of MSWs by metal microstrip converters. Without claiming to be complete, the following can be noted [267, 268, 269, 270, 271, 272, 273].

1.3. Spreading of SMSW (surface magnetostic waves) in an arbitrary direction along the film plane

Most of the works cited above consider the propagation of MSWs along or perpendicular to the field. At the same time, with the appearance of high-quality films of yttrium–iron garnet with an area of tens of square [120] centimeters, it became possible to realize the planar propagation of MSWs in any directions along the film plane. When an isotropic film is magnetized by a field perpendicular to its plane, the MSW propagation is also isotropic due to the equivalence of any direction in the film plane relative to the field. The deviation of the field from the normal and, moreover, the magnetization of the film along its plane violates the indicated equivalence, as a result of which the propagation of MSW in the plane of the film becomes anisotropic. The above solution (1.25)–(1.27) for a tangentially magnetized film already describes the propagation of MSWs in different directions in the plane. In this case, due to the gyrotropic properties of the ferrite medium, the directions of the vectors of the phase and group velocities of the MSW are different. This difference becomes especially significant for a finite-width wave beam. In this case, the group velocity vector determines the direction of the wave energy transfer, that is, the direction of propagation of the beam itself, and the phase velocity vector determines the direction of the wave front inside this beam. The direction of the phase velocity vector coincides with the direction of the wave vector of the wave, the direction of the group velocity vector is determined by the formula in which the dependence is determined by the dispersion relation (1.28).

Earlier studies, which indicated the discrepancy between the vectors of the phase and group velocities of the MSW, were carried out on relatively small YIG plates [274, 275].

Large YIG films made it possible to study MSW wave beams under conditions when the propagation medium length significantly exceeds the transverse dimension of the beam, which made it possible to use the methods of geometric optics very effectively [276–289]. It was shown both theoretically and experimentally that the admissible directions of propagation of surface magnetostatic waves (SMSW) lie inside a sector symmetric with respect to the perpendicular to the direction of the field, and the sector boundaries are determined by the cutoff angles, the smaller, the higher the SMSW (surface magnetostatic wave) frequency. At a given frequency, the longest waves propagate perpendicular to the direction of the field, the shortest – near the cutoff angles.

A convenient tool for a visual qualitative analysis of the anisotropic properties of MSWs is the method of isofrequency curves [274, 290] (also called equifrequency lines of dispersion surfaces of MSWs [134]) plotted on the plane of wave vectors. For a fixed frequency, the isofrequency curve is the locus of the points of the end of the wave vector of the MSW, and the direction of propagation of the wave beam coinciding with the direction of the group velocity vector is determined by the perpendicular to the tangent to the isofrequency curve constructed through the point corresponding to the end of the wave vector [274].

The most consistent application of the isofrequency curve method for analyzing the relationship between the directions of the phase and group velocities for SMSW in two-dimensional geometry is described in [291, 292]. Based on the analysis of the isofrequency dependences in two-dimensional geometry, the properties of propagation, reflection and refraction of waves in anisotropic media, a particular case of which are magnetic films magnetized in the plane, are considered. The interpretation of the isofrequency curve as the locus of the end of the wave vector of a wave emanating from the origin is presented, and the group velocity vector at each point of the isofrequency curve is determined by the normal to this curve at the mentioned point. The basic properties of isofrequency curves are given, such as asymptotes, inflection points, central and axial symmetry, extremum points in the polar coordinate system. The sectorial limitation of the existence of waves, nonreciprocity and unidirectionality of propagation, and the possible negative nature of reflection and refraction are revealed. Examples of the properties of isofrequency curves for a ferrite film with free surfaces and for a ferrite-dielectric-metal structure at some characteristic values of the dielectric layer thickness are considered.

Some issues of reflection and refraction of SMSWs for three-dimensional geometry were noted in [293].

The general anisotropic properties of SMSW wave beams were studied in [276, 277, 280, 282, 284, 287], including for radio pulses in [286]. In [279, 283], the spatial-frequency properties of the energy flux of SMSWs were analyzed, and in [278] the same was done for inverse bulk MSWs. The anisotropic properties of wave beams have been generalized to the case of an arbitrary field direction [285, 288, 289], including in relation to the problems of controlling the spectrum and group velocity of MSWs [288], as well as increasing the thermal stability of MSW devices [289]. Similar phenomena in two-layer ferrite films were considered in Refs. [294, 295].

The discrepancy between the directions of the phase and group velocity vectors significantly changes the conditions of reflection and refraction of the MSW at the interfaces between two media with different parameters. When crossing the interface or reflecting from it, the value of the projection of the wave vector onto this interface is retained, and its perpendicular component on different sides of the interface is determined by the frequency and dispersion laws of the MSW in the corresponding media. The method of isofrequency curves is again convenient for determining the angles of reflection and refraction. Reflection occurs in the same medium from which the wave comes, therefore the dispersion laws for the incident and reflected waves are the same and the properties of the reflecting medium, without changing the orientation dependences, determine only the reflection coefficient. Refraction, on the contrary, occurs in a different medium compared to the medium of incidence, therefore, both the angle and the refractive index depend on the properties of the refracting medium.

The dependences of the angles of reflection and refraction of the MSW on the angle of incidence for different orientations of the straight edge of the film, the step of its thickness, or a metal strip superimposed on the film were obtained in Refs. [139, 291]. The reflection and refraction properties of MSWs in anisotropic magnets are considered in Refs. [292, 296, 298, 378, 379].

In Refs. [299, 300], the reflection coefficients of surface and direct bulk MSWs from the edge of a metal strip superimposed on a film were obtained for the case of normal wave incidence. In papers [301, 302, 303, 304] these results were generalized to the case of a metal half-plane and a step of the film thickness of an arbitrary orientation. In papers [301, 302, 303, 304] these results were generalized to the case of a metal half-plane and a step of the film thickness of an arbitrary orientation.

Study [305] carried out a detailed experimental examination of the refraction of surface magnetostatic waves passing under a metal strip of finite width. Some theoretical considerations based on the model of isofrequency surfaces for a ferrite film with free surfaces and a ferrite–dielectric–metal (FDM) structure are presented.

For some special cases, a good agreement of theoretical positions with experiment was demonstrated. The definition of forward and backward waves based on the sign of the scalar product of the wave vector and the group velocity vector is noted: if the sign is positive, the wave is forward, and if it is negative, it is backward.

At the same time, despite the possibility of forming rather narrow wave beams, in real experiments the length of the exciting converter is comparable to or exceeds the length of the MSW by no more than several tens of times, which leads to noticeable 'diffraction' phenomena, which were first noted for backward bulk waves [313] and in more detail for bulk MSW [314–316], for the structure of FDM [317] and SMSW [318–320].

Comment. Generally speaking, the 'diffraction' properties of SMSWs given in the cited works are not such in the classical definition, since the splitting of the wave beam is caused not by the interference interaction of Huygens point sources at the transducer aperture, but by the noncolinear dependence between the vectors of the phase and group velocities. Thus, near the cutoff angle, there is a significant concentration of the SMSW flux density, which manifests itself as a splitting of the wave beam into two components located symmetrically relative to the normal to the field direction. In this regard, it is more logical to call the phenomenon of wave beam splitting due to thickening of the SMSW flux density not 'diffraction divergence', but 'dispersive divergence' or 'dispersive splitting', since it is solely due to the dispersion properties of SMSWs. However, in the cited works, when describing the splitting, the not entirely correct term 'diffraction divergence' is traditionally used, which should be paid special attention to.

Thus, for surface waves with a transducer length comparable to the SMSW length, instead of the classical diffraction pattern formed by one central beam and several weaker lateral ones, the main beam is split into two, propagating at angles close to the cutoff angles [139, 282, 314–316]. The dispersion divergence behaves similarly in the FDM structure [317, 320].

In the FM structure, the channelization of the SMSW energy is possible, due to the total internal reflection of the wave from the channel edges [321]. In a ferrite waveguide, which is a narrow ferrite strip, propagation of waveguide modes is possible, formed by the reflection of split beams from the side walls of the waveguide [322]. The propagation of limited packets of MSWs in an anisotropic medium is considered in papers [286, 287]. It was shown in [281,

323, 324] that an inhomogeneous field of a special shape can significantly change the propagation of split SMSW beams, in particular, lead to their self-intersection at a certain point in the film. The propagation of split beams in a nonlinear regime was studied in Ref. [325].

The classical diffraction of SMSWs, analogous to the Fraunhofer diffraction known in optics, was studied in Refs [326, 327, 328]. In the magnetostatic approximation, the diffraction of a surface magnetostatic wave propagating in a ferrite film is considered when it is incident on a slit in an opaque screen. The calculation of the distribution of the amplitude of the wave transmitted through the slit at an arbitrary point of the film is based on the integration of the potential of point sources obeying the Huygens principle and uniformly distributed over the width of the slit. The multi-lobed nature of the diffraction pattern with a strong predominance of the central lobe corresponding to the wave propagation in the direction determined by the laws of geometric optics is revealed. An additional lobe associated with the reflection of the initial wave from the slit is noted. In papers [329, 330], a similar analysis was carried out for backward waves. In [331], theoretical and experimental (probe method) studies were performed.

1.4. Distribution of SMSW in ferrite films and structures under the conditions of inhomogeneous magnetization

Even the first experimental observations of traveling MSWs [145, 146] revealed the smallness of their length and velocity compared to EMA of the same frequency. This circumstance, which is very attractive for the creation of small-size microwave delay lines [332], has stimulated many studies on the properties of these waves. An effective method of excitation and reception of MSWs was the transformation of their length during propagation in an inhomogeneous field. To implement this method, the transducer was located at the place of the sample where the length of the MSW approached the length of the EMA, which ensured high efficiency of its operation. Long MSWs excited in this way, propagating in an inhomogeneously magnetized sample, became short and their velocity decreased by one or two orders of magnitude, which provided a long (in comparison with EMA) delay time. The reverse transformation of the delayed MSWs into EMAs was carried out in a similar way on the field inhomogeneity in another place of the sample. Early work was carried out on axially magnetized rods or normally magnetized disks, in which reverse or forward bulk MSW propagated along the axis of the rod or disk radii [145, 146, 332–338]. The theoretical model developed on the basis of the WKB method for backward body waves in an axially magnetized cylinder made it possible to calculate the conversion efficiency expressed in terms of the 'bond length' – a parameter determined by the smoothness of the field change along the axis of the rod [333–336]. In addition to purely dipole MSWs, magnetoelastic waves have also been extensively studied [145, 146, 203, 335, 339].

In all the works listed above, the wave vector of the MSW always remained parallel (for a cylinder) or perpendicular (for a disk) to the field direction. In [274], an apparatus suitable for analyzing the propagation of MSWs in arbitrary directions with respect to an inhomogeneous field was developed and tested for the first time. The method is based on an analogy between an MSW and a mechanical particle moving in an inhomogeneous potential. Equations are derived, similar to Hamilton's equations in mechanics, which make it possible to calculate the trajectory of a wave beam provided that the dispersion law of the MSW is known. This method was used in Ref. [275] to consider the propagation of SMWs along the surface of a tangentially magnetized plane sample under the condition of a small change in the field at the wavelength. In the Cartesian coordinate system, the plane of which coincides with the plane of the sample, when using the polar system for the components of the wave vector, the system of the above equations has the form:

$$\frac{dk}{dy} = k\left(\frac{\partial k}{\partial y}\cos\varphi + \frac{\partial k}{\partial z}\sin\varphi\right)\left(k\cos\varphi + \frac{\partial k}{\partial \varphi}\sin\varphi\right)^{-1} ; \quad (1.32)$$

$$\frac{d\varphi}{dy} = -\left(\frac{\partial k}{\partial y}\sin\varphi - \frac{\partial k}{\partial z}\cos\varphi\right)\left(k\cos\varphi + \frac{\partial k}{\partial \varphi}\sin\varphi\right)^{-1} ; \quad (1.33)$$

$$\frac{dz}{dy} = \left(k\sin\varphi - \frac{\partial k}{\partial \varphi}\cos\varphi\right)\left(k\cos\varphi + \frac{\partial k}{\partial \varphi}\sin\varphi\right)^{-1} ; \quad (1.34)$$

where: k is the length of the SMSW wave vector, φ is the angle between the SMSW wave vector and the axis Oy, and the partial derivatives included in the right-hand side of the above equations are determined from the dispersion relation $f = (\omega, k, H(y,z)) = 0$, for example, of the form (28). Under general initial conditions, the system of equations (1.32)–(1.34) is integrated only numerically (for example, by the Euler or Runge–Kutta method [340]), as a result of which the first equation gives the length of the MSW wave vector, the second gives the direction of the MSW phase front through the angle and the third – the trajectory of the MSW, on the plane in the form. Knowledge of the set of the given dependences allows one to get a complete picture of the propagation of SMSW over the plane of the sample.

The use of the Hamilton method made it possible in [274] to calculate the trajectories of magnetoelastic bulk MSWs in an axially magnetized rod over the entire area of its section, and in [275] the trajectories of surface MSWs in a plate in the form of a rectangular parallelepiped also over the entire area of the plate. In both cases, the external field was uniform, and the inhomogeneity of the internal field was due to demagnetizing fields.

The creation of a field inhomogeneity due to the shape of the sample was characteristic of almost all early studies performed on bulk single-crystal samples [145, 146, 203, 332–339]. For thin epitaxial YIG films, this method is unacceptable; therefore, an inhomogeneous field in them is formed with the help of external magnets [341–348]. Excitation of direct bulk MSWs under such conditions is considered in [349, 350]. The trajectories of bulk MSWs in a linearly inhomogeneous field were investigated in [346-348]. Similar problems for a radially symmetric field were solved in [351, 354]. In Ref. [355], the transformation of an MSW upon reflection from a field inhomogeneity was considered. In [349], the delay time was investigated for the propagation of direct bulk MSWs in an inhomogeneous field.

A large number of works on inhomogeneous fields are devoted to the study of the propagation of surface MSWs. In papers [275, 349], an inhomogeneous field created by the shape of a sample in the form of a rectangular plate is considered. The trajectories of the SMSW near the ends and in the centre of the plate are calculated, the regions of existence and localization of the SMSW are found. The inhomogeneous field created by external magnets provides much more opportunities for controlling the propagation of SMSWs [343, 297]. The radially symmetric field makes it possible to control the localization of propagation [351–354], to rotate the trajectories by significant angles, and to change the depth of penetration of the SMSW into the film thickness [341, 342]. In [343], the trajectories of the SMSW are closed in the form of a ring, which made it possible to create a high-Q resonator based on the SMSW. In [356], SMSWs were considered with allowance for the exchange interaction.

The calculation of the SMSW propagation is carried out using integral equations solved numerically [357]. Papers [358–360] are devoted to the study of the propagation of SMSW in rectilinear channels generated by an external field. The relationship between adjacent channels is also considered in [358, 360]. In works [351–354] channels were investigated, including annular ones, formed by a radially symmetric field. The propagation of wave beams of limited width is considered in papers [323–325]. The possibility of creating an inhomogeneous field of such a shape, which can lead to self-intersection of the split-off branches of the split beam, is shown.

The presence of magnetoelastic coupling in ferrites makes it possible to use elastic deformations of the sample to create inhomogeneous internal fields. In [85, 106, 361, 362], the FMR, spin-wave resonance, and dispersion of MSWs in films deformed by static inhomogeneous external stresses were studied.

1.5. Distribution of SMSW in ferrite films and structures with periodic inhomogneities

Along with a monotonic change in the field, the scattering of SMSWs by static magnetic lattices with a periodic structure was also considered. A detailed experimental study of the propagation of SMSWs in such structures was carried out in [140, 363–365]. In this case, the magnetic grating was made in the form of a highly coercive tape recorder superimposed on the YIG film with a periodic signal recording. The nontransmission bands corresponding to the coincidence of the period of the grating with the length of the propagating SMSW were found, and the reflection and refraction of the SMSW by such gratings was considered for an arbitrary orientation of the grating in the plane of the magnetic film.

The theoretical analysis of such problems was carried out by numerical methods based on isofrequency curves. The results obtained were interpreted using a magnonic crystal model [365–367]. One of the results of this research was the method proposed in [368] for measuring the parameters of magnetic films, such as the saturation magnetization and the uniaxial and cubic anisotropy constants.

Papers [369, 358, 360, 370–374, 366, 375] are devoted to the study of the propagation of SMSWs in channels formed by an inhomogeneous field created by a highly coercive magnetic tape with a recording. In [369], SMSW modes in one channel were considered, and in [358], in two parallel coupled channels. Periodic transfer of wave energy during the transition from one channel to another and back is noted. In [360], numerical modeling of the propagation of SMSWs in a two-channel structure, performed by the finite difference method, is described. In [370, 371], the transformation of SMSWs channeled by a stepped bias field is considered. In [372], the interaction of bulk and surface MSWs in a channel created by an inhomogeneous field was considered. The work [373] is devoted to the description of the experimental technique of MSW propagation in channels, based on scanning the surface of a ferrite film with a magnetic probe. Papers [374, 366, 375] are devoted to the study of dispersion properties and the formation of SMSW beams, including in a magnonic crystal with nonreciprocity. The most detailed studies listed above are discussed in [367].

A significant number of works are devoted to the propagation of SMSWs in periodic structures formed by mechanical inhomogeneities of a ferrite film. In an early work [376], non-transmission bands were considered during the passage of an MSW over the surface of a film covered with transverse grooves. Such studies were further developed, including taking into account the converted nature of waves, in [377–379]. [380] considered the tunneling of MSWs through a local inhomogeneity created by a magnetic field. [381] considered magnetostatic spin waves in two-dimensional periodic structures – magnetophotonic crystals.

An important place among the listed works is occupied by the study of the propagation of SMSWs in the presence of periodic structures made in the form of metal gratings superimposed on the film surface. A distinctive feature of a number of works was the lack of commensurability between the SMSW length and the lattice period. Thus, in the case when the lattice period was significantly less than the length of the SMSW, instead of the nontransmission bands, a change was observed in the smooth dispersion curves of the SMSW without the appearance of any discontinuities. Thus, in experimental works [382, 383], where the period of the grating made by the photolithography method was tens of microns, and the length of the SMSW was up to several millimeters, it was shown that the dispersion curve for the ferrite-lattice structure lies above the curve for a free film and has the opposite character, that is, it decreases with increasing wavenumber, and the decline is steeper than for the ferrite-metal structure. The ideological interpretation of the phenomena observed in [382, 383] was based on the consideration of a metal lattice as a homogeneous medium with parameters averaged over the volume and having a negative dielectric constant.

In [384], based on the representation of the dielectric constant for a lattice of thin straight conductors in a form similar to the dielectric constant for an electron plasma in a metal [385], a dispersion relation was obtained for a ferrite–dielectric structure surrounded by half-spaces with negative dielectric constant. It is shown that in such a structure, surface waves can propagate along the surface of a ferrite plate, the more pronounced, the greater the ferrite thickness compared to the dielectric thickness [377–380].

In [386], a new method for determining the length of the magnetostatic wave was proposed for the ferrite – lattice composite structure of metal strips, based on the movement of the metal lattice in a direction perpendicular to the film plane. A significant difference was revealed between the action of a grating on a propagating wave compared to the action of a metal plane and a medium with a negative dielectric constant.

In contrast to Refs. [382, 383], the location of the dispersion branch of surface waves with a grating was noted at the same level as without a grating, and a hypothetical assumption about its dissipative nature was made regarding the falling high-frequency branch observed in Refs. [382, 383].

The ideology of the observed phenomena was further developed in [387], where a model was proposed according to which a wave propagating in a ferrite film under the grating, passing through the regions corresponding to the grating lines and the gaps between them, is alternately localized along the upper or lower surface of the ferrite film, in as a result, it acquires an additional phase incursion, leading to a displacement of the dispersion curve towards large wave numbers. The most detailed issues mentioned were considered in [292].

A number of works are devoted to dynamic inhomogeneities created by an elastic wave, which is an analogue of a traveling periodic lattice. In [388], the first observation of the scattering of backward bulk MSW by elastic bulk waves was reported. In [389, 390], a similar effect was considered for SMSW and surface acoustic waves (SAW), and in [391] – for exchange bulk MSW and bulk elastic waves under conditions of collinear propagation of both types of waves.

Papers [209, 210, 392–395] are devoted to the study of the scattering of SMSW by SAW for arbitrary directions of propagation, and the method of isofrequency curves is effectively used to interpret the experimental results. The reflection and refraction of the SMSW at various angles with respect to the mutual direction of propagation of the initial SMSW and the elastic wave are investigated. Changes in the frequency of propagating SMSWs due to interaction with an elastic wave, transformation of SMSWs into reverse bulk, and anisotropic-dipole MSWs are noted. The effect of a conducting plane near the surface of a magnetic film is considered, which is similar to the formation of a ferrite-dielectric-metal structure. On the basis of the results obtained, methods for measuring the parameters of magnetic films and dispersion characteristics of SMSWs are proposed, and the possibility of using the observed effects in microwave devices is discussed.

An important place among the listed works is occupied by the study of the propagation of SMSWs in the presence of periodic structures made in the form of metal gratings superimposed on the film surface. A distinctive feature of a number of works was the lack of commensurability between the SMSW length and the lattice period. Thus, in the case when the lattice period was significantly less than the length of the SMSW, instead of the nontransmission bands, a change was observed in the smooth dispersion curves of the SMSW without the appearance of any discontinuities. Thus, in experimental works [382, 383], where the period of the grating made by the photolithography method was tens of microns, and the length of the SMSW was up to several millimeters, it was shown that the dispersion curve for the ferrite-lattice structure lies above the curve for a free film and has the opposite character, that is, it decreases with increasing wavenumber, and the decline is steeper than for the ferrite–metal structure. The ideological interpretation of the phenomena observed in [382, 383] was based on the consideration of a metal lattice as a homogeneous medium with parameters averaged over the volume and having a negative dielectric constant.

In [384], based on the representation of the dielectric constant for a lattice of thin straight conductors in a form similar to the dielectric constant for an electron plasma in a metal [385], a dispersion relation was obtained for a ferrite–dielectric structure surrounded by half-spaces with a negative dielectric constant. It is shown that in such a structure, surface waves can propagate along the surface of a ferrite plate, the more pronounced, the greater the ferrite thickness compared to the dielectric thickness. [377–379]. [380]

In [386], a new method for determining the length of the magnetostatic wave was proposed for the ferrite–grating composite structure of metal strips, based on the movement of the metal lattice in a direction perpendicular to the film plane. A significant difference was revealed between the action of a grating on a propagating wave compared to the action of a metal plane and a medium with a negative dielectric constant.

In contrast to Refs [382, 383], the location of the dispersion branch of surface waves with a grating was noted at the same level as without a grating, and a hypothetical assumption about its dissipative nature was made regarding the falling high-frequency branch observed in Refs [382, 383].

The ideology of the observed phenomena was further developed in [387], where a model was proposed according to which a wave propagating in a ferrite film under the grating, passing through the regions corresponding to the grating lines and the gaps between them, is alternately localized along the upper or lower surface of the ferrite film, in as a result, it acquires an additional phase incursion, leading to a displacement of the dispersion curve towards large wave numbers. The most detailed issues mentioned were considered in [292].

A number of works are devoted to dynamic inhomogeneities created by an elastic wave, which is an analogue of a traveling periodic lattice. In [388], the first observation of the scattering of backward bulk MSW by elastic bulk waves was reported. In [389, 390], a similar effect was considered for SMSW and surface acoustic waves (SAW), and in [391] – for exchange bulk MSW and bulk elastic waves under conditions of collinear propagation of both types of waves.

Papers [209, 210, 392–395] are devoted to the study of the scattering of SMSW by SAW for arbitrary directions of propagation, and the method of isofrequency curves is effectively used to interpret the experimental results. The reflection and refraction of the SMSW at various angles with respect to the mutual direction of propagation of the initial SMSW and the elastic wave are investigated. Changes in the frequency of propagating SMSWs due to interaction with an elastic wave, transformation of SMSWs into reverse bulk, and anisotropic-dipole MSWs are noted. The effect of a conducting plane near the surface of a magnetic film is considered, which is similar to the formation of a ferrite–dielectric–metal (FDM) structure. On the basis of the results obtained, methods for measuring the parameters of magnetic films and dispersion characteristics of SMSWs are proposed, and the possibility of using the observed effects in microwave devices is discussed.

1.6. Conversion of a magnetostatic wave into electromagnetic on the field inhomogeneity

An important phenomenon accompanying the propagation of magnetostatic waves in an inhomogeneously magnetized medium is the transformation of a magnetostatic wave into an electromagnetic one, which manifests itself as radiation of electromagnetic waves from ferrite.

The possibility of such radiation for bulk MSWs propagating in an inhomogeneous field formed due to demagnetizing factors of the sample was pointed out in papers [333, 334, 335], which served as the basis for the creation of ferrite delay lines [332]. As a mechanism providing radiation, it was indicated that the MSW length approaches the length of an electromagnetic wave in free space, where the MSW length in ferrite changed during its propagation in an inhomogeneous field. The energetic reciprocity of the effect was noted, that is, on the field inhomogeneity, both the transformation of a magnetostatic wave into an electromagnetic one and an electromagnetic wave into a magnetostatic one could be carried out, which was achieved by changing the direction of wave propagation to the opposite.

For SMSW, the emission of electromagnetic waves from a ferrite film magnetized by a linearly increasing field was observed, apparently for the first time, in Refs. [396, 397]. In papers [398-400] it was shown that in the plane perpendicular to the plane of the film, the radiation pattern of such an antenna has two lobes, inclined forward along the SMSW propagation.

In [401], based on an analogy with the radiation arising from an inhomogeneous dielectric waveguide during the transition of an electromagnetic wave from a homogeneous section of such a waveguide to an inhomogeneous one [402], a possible mechanism for converting a surface magnetostatic wave into an electromagnetic one was proposed. Surface waves in a magnetic plate are considered on the basis of Maxwell's equations, the inhomogeneity of the magnetizing field and the associated transformation of the wavelength with the approach of its length to a similar length in free space are taken into account. On the basis of the mentioned analogy, it is shown that radiation originates from a section of a magnetic film, for which the condition of equality of the bias field to the resonance value is satisfied, where the junction of the homogeneous and inhomogeneous sections of the magnetic waveguide occurs. The decisive role of the magnetic film substrate in the formation of the radiation pattern is noted: a decrease in the substrate thickness leads to a narrowing of the lobe of the pattern and its inclination forward along the wave. The above mechanism and the consequences following from it are described in most detail in [292].

1.7. Domain structures in ferrite films, FMR and MSW under the conditions of the existence of domain structures

In all the studies considered above, ferrite was assumed to be magnetized either uniformly or weakly inhomogeneous, so that the field changes little along the MSW length. In this case, the distribution of magnetization over the sample is also close to uniform. At the same time, along with the above, since ancient times (see, for example, [1]), works devoted to the study of the precession of magnetization under conditions of a sharply inhomogeneous distribution of magnetization, namely, in the presence of a domain structure, when the magnetization, while maintaining relative homogeneity within domains, at their boundaries, changes very much. The static state of the domain structure is due to the competition between the energies of the demagnetizing fields and domain walls, and the thickness of the domain wall is due to the competition between the fields of anisotropy and inhomogeneous exchange interaction. Typical domain widths are units and tens of microns, and the thickness of domain walls is tenths and hundredths of a micron, that is, the width of domains, as a rule, is one to two orders of magnitude less than the length of the MSW, and the thickness of domain walls is another two to three orders of magnitude smaller.

The static state of the domain structure has been considered in numerous reviews and monographs [5, 12, 403]. The study of the physics of domains in garnet–ferrite films has advanced greatly in connection with the study of cylindrical magnetic domains (CMDs) [403]. A necessary condition for the existence of a CMD is the presence of a large uniaxial anisotropy in the film, the field of which significantly exceeds the saturation magnetization of the film material. At the same time, in such films, the decay of the precession of the magnetization is very large, which limits the possibilities of MSW propagation. In films with low damping, primarily in YIG, the uniaxial anisotropy field is much lower than the saturation magnetization, and in addition, there is a noticeable cubic anisotropy; therefore, the domain structure is very complex and has not yet been sufficiently studied.

Considerable attention has also been paid to the dynamic behavior of magnetization in materials with domains [2–5, 403]. In materials with CMD, mainly the translational motion of domain boundaries [403] was studied, while oscillatory processes [2–4] were considered less. In films with domains, there are two main vibrational processes - translational vibrations (resonance) of domain walls (RDW) and precession of magnetization (FMR) within domains [2–4]. Surface MSWs propagating along stationary and moving domain walls have also been predicted theoretically [404, 405].

Historically, the first dynamic phenomenon in materials with domains was the resonance of domain walls [1], to which numerous studies are devoted, summarized, for example, in [2–4, 7, 8]. The resonance of domain walls in the microwave range was considered in [406]. In [407], bending vibrations of domain walls in YIG were studied, and in [408], general questions of RDW in anisotropic magnetics were studied. Work [409] is devoted to the excitation of oscillations of domain walls using acoustic pumping. In [410], collective excitations of domain walls similar to MSWs were considered.

Large-amplitude oscillations of domain boundaries, accompanied by self-organization phenomena ('anger' state), are considered in [411–418]. [419] described the unidirectional motion of domains excited by an alternating field in a mechanically stressed YIG crystal.

Along with the resonance of domain walls, immediately after the discovery of FMR in saturated samples, its study began in samples with domains [420]. The main results of these studies are also summarized in [2–4, 7, 8]. FMR in multidomain plates and films of ferrites, including those with CMD, as well as cubic and uniaxial anisotropy was investigated in [38, 108, 421–423]. MSWs in YIG films and plates with domains were studied in Refs [424–427].

A rather detailed study of the regions of existence and dispersion properties of MSWs for an arbitrary direction of the field and the presence of cubic and uniaxial anisotropy in the film was performed in Refs. [428–430]. An unsaturated single-domain state is considered:

the influence of domains is not taken into account, and domain blocks are assumed to be so large that the wave propagates only within one block. Waves are considered in the magnetostatic approximation for an arbitrary direction of the field, and the cubic and uniaxial anisotropies are small in comparison with the saturation magnetization of the film. The conditions for the existence of surface and bulk waves are investigated, and the characteristic cutoff frequencies of the spectrum and their dependence on the field are found. It is shown that, upon tangential magnetization of the film near the saturation field, the MSW frequency decreases significantly, after which it increases again (which is also typical for FMR [2-4, 431]). [432] considered MSWs in Ga- and Sc-substituted YIG films with domains, and [429] considered MSWs in YIG films of submicron thickness. Work [428] is devoted to the study of magnetoelastic waves in the presence of a domain structure, and [433] predicted the possibility of the existence of MSWs in orthoferrite plates with domains.

The nonlinear effects accompanying the propagation of MSWs in a YIG film with domains were considered in Ref. [434].

1.8. Features of magnetostatic waves in the long-wave limit

The previous sections considered the properties of magnetostatic waves, which are electromagnetic waves in the magnetostatic approximation. According to the classical estimate [4], the use of the magnetostatic approximation is legitimate with a fairly high degree of accuracy (on the order of fractions of a percent) at wave numbers in the range from 10^2 cm^{-1} 10^4 to cm^{-1}. At lower wavenumbers, the electric field of the wave comes into play, at large, exchange interaction. Aside from the influence of exchange, as applied to real experiments, let us consider in which case it becomes necessary to take into account the electric field of the wave. Thus, in a typical experiment, the SMSW is excited by wire antennas with a length of about 0.5 cm and a diameter of about 0.001 cm superimposed on the plane of the YIG film [292, 297, etc.]. In this case, the length of the SMSW excited by such converters is from 10^{-1} cm to $7 \cdot 10^{-3}$ cm [ibid.]. Such wavelengths correspond to wavenumbers from 60 cm^{-1} to 900 cm^{-1}. With a certain stretch, one can observe SMSWs having a length of the order of 0.5 cm, which corresponds to a wavenumber of 12 cm^{-1}; however, the excitation efficiency of such waves is relatively low.

The typical frequency of the SMSWs excited in the experiment is from 1 to 10 GHz. Below this frequency, the field required for the existence of the SMSW should be less than 80 Oe, which becomes insufficient to saturate the YIG film (that is, the absence of domains in it). For a frequency higher, a field of the order of 3000 Oe is required, so that the magnetic system becomes rather bulky and inconvenient for use in real devices. In most experiments, the frequency is chosen near $2 \div 4$ GHz, so that the length of the electromagnetic wave in this range is from 7.5 to 15 cm. That is, the wave number of such waves is $0.4 \div 0.8$ cm^{-1}.

So, one can see that the length of the SMSWs actually observed in experiments is approximately two to three orders of magnitude less than the length of electromagnetic waves in the same frequency range.

Thus, comparing the range of typical wave numbers of SMSWs ($60 \div 900$ cm^{-1}) given here with the allowable one in the magnetostatic approximation ($10^2 \div 10^4$ cm^{-1}), we can assume that in real experiments the magnetostatic approximation is fulfilled with sufficient accuracy and the contribution of the electric field, if then very insignificant.

Nevertheless, it should be noted that there are a number of phenomena where taking the electric field into account can be essential. These phenomena include, first of all, the dispersion properties of rather long SMSWs (of the order of 1 cm or more), then the energy fluxes of propagating waves, as well as the efficiency of conversion of SMSWs into electromagnetic waves due to the inhomogeneity of the constant field. So, on the YIG films with a diameter of up to 7 cm that are commercially available today, with an appropriate design of the transducers, it is possible to excite SMSWs up to several centimeters long, which provides wavenumbers up to 1 cm^{-1}. In the composition of a sufficiently long SMSW, the electric field can be quite noticeable, comparable to the magnetic one, which requires consistent accounting of the wave energy flux based on the total Poynting vector. It can be expected that the conversion of the SMSW into an electromagnetic wave is the more efficient, the closer the SMSW wavelength approaches the length of the electromagnetic wave in free space.

Thus, consideration of surface magnetic waves on YIG films in the long-wavelength limit may require taking into account their electromagnetic character on the basis of the full Maxwell equations. Without going into the details of the results obtained, we briefly list the currently available most notable works in this direction.

So in works [435–437] on the basis of full Maxwell's equations the dispersion properties of forward and backward electromagnetic waves in ferrite-dielectric structures are considered.

In papers [438, 439], a magnetic wall was added to the consideration, that is, such a hypothetical surface, the magnetic field on which is equal to zero. In [384], surface magnetostatic waves in a ferrite–dielectric (FD) structure surrounded by half-spaces with negative permittivity are considered. In [440–442], the structure of the fields of electromagnetic waves in the ferrite-dielectric and ferrite–metal structures was determined, including taking into account the magnetic wall. The isofrequency dependences of electromagnetic waves in an unbounded ferrite medium were constructed in [443]. Papers [444–446] are devoted to the calculation of the energy fluxes of an electromagnetic wave in a ferrite–dielectric structure. Thus, in early works [444, 445], the calculation was performed on the basis of a 'modified' magnetostatic approximation, in which the electric field required to obtain the Poynting vector was found from Maxwell's equation, which determines such an electric field through a magnetic field. In [446], on the basis of the full Maxwell equations, a more correct sequential account of the relationship between the energy and dispersion characteristics of magnetostatic waves in the ferrite–dielectric–metal structure was performed under conditions when the dispersion curve has one or two extremum points. The Poynting vector and partial wave power fluxes in ferrite, dielectric and semi-infinite free space are found. It is shown that in the presence of extrema of the dispersion curve, a mutually opposite direction of partial flows is possible. Insufficient correctness of the earlier interpretations of the wave power flux, based on the equations of magnetostatics, is noted, which is especially pronounced at small wave numbers. The necessity of turning in these cases to a solution based on the full Maxwell equations is indicated.

In [401, 445, 447], based on an analogy with radiation arising from an inhomogeneous dielectric waveguide, an analysis of such radiation from ferrite during the propagation of an electromagnetic wave in an inhomogeneous field was carried out. The obtained values of the wave numbers are of the order of $0.1 \div 3.0$ cm^{-1}, which is beyond the scope of magnetostatics and also testifies in favour of the use of the full Maxwell equations.

1.9. Use of FMR, MSW and domains in ferritle films for information processing devices

Magnetostatic waves (MSW) are the basis for creating numerous devices for analog information processing in the microwave range, the main stages of development of which are described in numerous reviews and monographs [18, 20, 56, 58, 214, 215, 448–458]. Reviews [19, 459, 460] are devoted to the growth and properties of films for such devices. Some general principles for calculating devices are described in [142, 461–463], issues of thermal stability are touched upon in [464]. The use of inhomogeneous fields and methods of geometric optics to create devices is described in papers [274, 309–311, 449].

One of the most traditional areas of application for garnet–ferrite films is transmission lines, which are strip lines on ferrite substrates [465, 466] or waveguides for MSWs made in the form of a ferrite strip [323, 467–470]. An advantage of such lines is the ability to control propagation characteristics not only by the field and current elements [268–273, 470], but also by light [471, 472]. Work [216] describes circulators for integrated circuits on ferrite films.

Considerable attention in the literature is paid to the issues of effective excitation of MSWs [137], and, along with general issues of excitation by a metal strip [137, 474, 475], the radiation resistance and current density in a microstrip line on a ferrite substrate, as well as equivalent circuits of emitters of MSWs [269–273]. Various designs of emitters based on strip lines are considered in works [137–273]. Excitation of exchange MSWs was considered in [137, 476]. For some emitters of a special shape, both the efficiency of excitation of the MSW [477–480] and the calculation of frequency characteristics [481] are considered. In papers [306–308], the possibility of creating converters providing efficient spatial focusing of MSW beams is shown.

Historically, the first application of MSWs was microwave signal delay lines based on bulk YIG single crystals [145, 146, 332–338]. The diminutiveness and the ability to control the delay time by changing the field were among the most important advantages of such lines [332].

The appearance of YIG films has advanced this work to a qualitatively new level [435]. The ability to create complex films based on films, including multilayer structures, made it possible to achieve any desired properties of dispersion, frequency band, input impedance, matching and deformation of microwave pulses, combined with low insertion loss [482–490]. Generators [491–494] and modulators [495, 496] of the microwave signal are important applications of the delay lines.

Another traditional field of application of MSWs is microwave filters [43, 497–503]. In papers [500, 502, 504, 505] methods of forming the amplitude–frequency and phase characteristics of such filters are considered. In works [306–308] methods of filtering a microwave signal based on focusing converters MSW are described.

MSWs are widely used to create resonators [506–509]. [343] described ring resonators using waveguide propagation of SMSWs in a radially inhomogeneous field. Resonators with periodic reflecting metal gratings are described in Ref. [510]. Other properties of MSWs in periodic structures are considered in works [511–513]. An important feature of a number of devices based on YIG films are MSSV absorbers [514].

Papers [515, 516] describe the use of MSWs in microwave antennas. An antenna device containing a YIG plate with domains is described in [517]. Some nonlinear devices on YIG films are described in papers [518, 519]. The processing of optical signals using an MSW is described in [520, 521].

New broad possibilities for the design of devices based on MSWs are opened by inhomogeneous bias of films [100, 274, 341–344, 516]. Methods for the synthesis of magnetic systems for some of these devices are described in [342, 344].

The study of FMR and MSW in magnetic films is a powerful tool for measuring the parameters of films, both ferrite garnets [13, 14, 41, 45-49, 53, 89, 90, 93, 522, 523], and other magnetic materials [115 , 522, 523], including media for magnetic recording [109–112] and HTSC films [97, 258]. The FMR method makes it possible to measure the local parameters of a material [14, 89, 90, 99, 100]; additional possibilities are provided by photothermal modulation of film parameters [88, 116].

1.10. Basic issues for further explanation

At the end of the review, it should be noted the main issues are not covered by the cited works, that is, those that are supposed to be developed in more detail in the following chapters of the monograph. Here we give only a brief formulation of the main problems, the illumination of which will be the main subject of further presentation.

It is proposed to present a generalizing dispersion relation for metal–dielectric–ferrite–dielectric-metal (MDFDM) and metal-dielectric–ferrite–ferrite–dielectric-metal (MDFFDM) structures, a particular case of which would be all other simpler dispersion relations.

A more complete study of the ferrite–dielectric–metal (FDM) structure is planned, including under the conditions of the existence of two extrema on the dispersion curve, between which the waves are inverse.

It is supposed to carry out a study of anisotropic structures in the case of an arbitrary direction of the magnetization vector relative to the field during an orientational phase transition.

A convenient and intuitive method for a qualitative study of the trajectories of wave beams in inhomogeneously magnetized films will be proposed.

Based on the numerical solution of the system of differential equations of wave propagation, the trajectories of wave beams in inhomogeneous fields of various configurations will be investigated. Special attention will be paid to the ferrite–dielectric–metal (FDM) structure, including the possible conditions for the existence and propagation of backward waves. The wave phase, orientations of the phase and group velocity vectors, and the delay time for an arbitrary direction of propagation of the MSW will be studied.

It is proposed to study the propagation of SMSWs under conditions of complex configurations of the magnetizing field, including periodic and doubly periodic in two coordinates.

A separate issue will be the interpretation of the effect of converting magnetostatic waves into electromagnetic waves due to field inhomogeneity, including the analysis of the radiation directivity diagram for various structures and designs of converters.

The propagation of MSWs in YIG films with a domain structure will be studied in more detail. In order to interpret the phenomena observed in the experiment, the dispersion properties of magnetostatic waves in an anisotropic medium under the conditions of an orientational transition will be considered.

Methods for the synthesis of magnetic systems and the creation of devices operating in inhomogeneous fields will be presented, including methods for the active use of inhomogeneous fields to form the characteristics of devices and minimize the size of magnetic systems. Methods for measuring the parameters of YIG films that are important for devices with domains, as well as methods for measuring inhomogeneous fields used in MSW devices, will be discussed.

1.11. Some new directions of research of MSW

In conclusion of this chapter, we mention some areas of research on magnetostatic and spin waves that are gaining momentum in recent years. Note that the selection of the sources cited below does not in the least pretend to be complete, but only reflects some of the current interests of the authors.

First of all, we note that until recently, the vast majority of studies of MSWs were carried out under conditions where wave attenuation during propagation was not considered. Sufficiently sensitive equipment was used in the experiments, and the theory turned to nondissipative media. At the same time, it was already shown in [524] that in a medium with dissipation, when losses are taken into account by introducing an imaginary addition to the frequency, a branch of 'dissipative' SMSWs appears, which has an inverse character. A dissipation-induced limitation of the spectra of forward and reverse SMSWs in terms of the wave number was revealed. These questions were considered in more detail in [525].

In comparatively recent papers [526, 527], indicatrices (isofrequency curves) of straight SMSWs in a dissipative medium were constructed and the direction of the group velocity of a wave was investigated when the orientation of its wave vector was changed.

It was shown in [528, 529] that successive allowance for the SMSW damping based on the magnetic susceptibility leads to a complex character of not the frequency, but the wave number. In [530], dispersion relations were obtained for the real and imaginary parts of the wave number, and in [531, 532] some of their properties were considered.

Note that in Ref. [386], a cautious assumption was made that the branch of backward waves observed in Refs [382, 383] is dissipative, but the character of the wave attenuation for such a branch has not been investigated.

Another important area of research in MSWs is the widespread use of optical methods to influence the characteristics of MSWs. As a typical scheme, a sufficiently powerful pump laser is used, which provides an active effect on the magnetic material, in combination with another low-power laser acting as a magneto-optical probe, registering the precession of magnetization according to the Kerr effect [533]. For example, the action of a pulsed laser, usually carried out in a pulsed mode, can change the magnetization of the medium through heating [534, 535] or the inverse Faraday effect [536], which was proposed as early as [173].

Another type of optical impact is the excitation of acoustic pulses using a femtosecond laser, which heats a thin metal film, the thermal expansion of which creates a mechanical pulse transmitted to the magnetic medium, where such a pulse due to magnetoelastic interaction excites precession of magnetization [537–540]. In [541], the possibility of direct excitation of magnetostatic waves by light from a femtosecond laser due to the inverse Faraday effect was demonstrated, and the MSWs excited by a probe laser are used to locally measure the parameters of magnetic films, such as the uniaxial and cubic anisotropy constants.

It was shown in [528, 529] that successive allowance for the MSSW damping based on the magnetic susceptibility leads to a complex character of not the frequency, but the wave number. In [530], dispersion relations were obtained for the real and imaginary parts of the wave number, and in [531, 532] some of their properties were considered.

Note that in Ref. [386], a cautious assumption was made that the branch of backward waves observed in Refs [382, 383] is dissipative, but the character of the wave attenuation for such a branch has not been investigated.

Another important area of research in MSWs is the widespread use of optical methods to influence the characteristics of MSWs. As a typical scheme, a sufficiently powerful pump laser is used, which provides an active effect on the magnetic material, in combination with another low-power laser acting as a magneto-optical probe, registering the precession of magnetization according to the Kerr effect [533]. For example, the action of a pulsed laser, usually carried out in a pulsed mode, can change the magnetization of the medium through heating [534, 535] or the inverse Faraday effect [536], which was proposed as early as [173].

Another type of optical impact is the excitation of acoustic pulses using a femtosecond laser, which heats a thin metal film, the thermal expansion of which creates a mechanical pulse transmitted to the magnetic medium, where such a pulse due to magnetoelastic interaction excites precession of magnetization [537–540]. In [541], the possibility of direct excitation of magnetostatic waves by light from a femtosecond laser due to the inverse Faraday effect was demonstrated, and the MSWs excited by a probe laser are used to locally measure the parameters of magnetic films, such as the uniaxial and cubic anisotropy constants.

2

Mathematical apparatus used in calculating the properties of magnetostatic waves

This chapter is devoted to a review of the basic mathematical apparatus used in solving problems related to the propagation of magnetostatic waves in inhomogeneously magnetized iron garnet films. The derivation of the basic equations of the motion of magnetization is given, the dispersion properties of various structures containing ferrite films are obtained, and the Hamilton–Auld method is considered in detail, which is used to calculate the trajectories of wave beams under conditions of an inhomogeneous field.

When considering the above issues, we will mainly follow the works [2-4, 134, 118, 274, 275, 542, 543], the rest of the references are indicated in the text.

2.1. Landau–Lifshitz equation

The phenomenological consideration of the dynamics of magnetic phenomena is based on the equation of motion of the magnetization vector under the action of its own gyroscopic forces and external fields. This is the Landau–Lifshitz equation proposed by the authors in 1935 to interpret the phenomena associated with the resonance of domain walls [1]. Let us briefly consider the derivation of this equation.

The Landau–Lifshitz equation is obtained from the equation of motion of a rigid body rotating around an axis [542]:

$$\frac{d\vec{N}}{dt} = \vec{P} \qquad (2.1)$$

where $\vec{N}$ is the mechanical moment of momentum of the body, $\vec{P}$ is the mechanical moment of the force acting on the body (instead of the traditional designation it is used so as not to be confused with the magnetic moment or magnetization). In this case, the moment of force by definition is equal to [542]:

$$\vec{P} = \left[\vec{R} \times \vec{F}\right] \qquad (2.2)$$

where $\vec{R}$ is the radius vector of the point of application of the force, $\vec{F}$ is the acting force itself.

Substituting (2.2) into (2.1), we obtain:

$$\frac{d\vec{N}}{dt} = \left[\vec{R} \times \vec{F}\right] \qquad (2.3)$$

This is the basic equation from which the Landau–Lifshitz equation will be derived below. To do this, it is necessary to express all the quantities included in it in terms of magnetic parameters: the magnetic field and the magnetic moment (or magnetization as the magnetic moment of a unit volume).

We will take an electron moving in a circular orbit as the basis for consideration. According to the classical definition [543], the magnetic moment created by the current flowing in a flat closed loop S is:

$$\vec{M} = \frac{1}{c} \cdot I \cdot \vec{S} \qquad (2.4)$$

where the vector $\vec{S}$ is determined by the ratio:

$$\vec{S} = S \cdot \vec{n}_s \qquad (2.5)$$

in which $\vec{n}_s$ is a unit vector normal to the contour plane. As such a contour, we will further take a circular orbit of motion of an electron of radius, the area r of which is equal to:

$$S = \pi \cdot r^2. \qquad (2.6)$$

Suppose that the frequency of motion of an electron in its orbit is f. Then the current along the loop, defined as the amount of charge flowing through the cross section of the orbit, is equal to:

$$I = e \cdot f \qquad (2.7)$$

where e is the electron charge.

Substituting (2.5)–(2.7) into (2.4), we obtain the total magnetic moment in the form:

$$\vec{M} = \frac{1}{c} e f \pi r^2 \cdot \vec{n}_s \qquad (2.8)$$

Let us now consider the mechanical moment of momentum of an electron moving in the same orbit. According to the classical definition [542], the angular momentum of one particle is:

$$\vec{N} = m[\vec{r} \times \vec{v}]. \qquad (2.9)$$

where m is the mass of the particle, in this case, the electron, $\vec{v}$ is the speed of the electron in the orbit, directed tangentially to the line of the orbit along the unit vector $\vec{n}_t$.

We find the absolute value of the velocity as the ratio of the length of the orbit to the time of revolution of the electron in the orbit:

$$v = \frac{2\pi r}{1/f} = 2\pi r f. \qquad (2.10)$$

For a flat orbit, vectors $\vec{r}$ and $\vec{v}$ are perpendicular, so the vector is directed perpendicular to the orbital plane along the vector n_s. Substituting (2.10) into (2.9), we obtain:

$$\vec{N} = m \cdot 2\pi r^2 f \cdot \vec{n}_s. \qquad (2.11)$$

Thus, both the magnetic moment $\vec{M}$ (2.8) and the angular momentum $\vec{N}$ (2.11) are directed along the same vector $\vec{n}_t$. Expressing this vector from (2.8) and (2.11) separately and equating the obtained expressions, we find:

$$\frac{c}{e f \pi r^2} \cdot \vec{M} = \frac{1}{m \cdot 2\pi r^2 f} \cdot \vec{N}$$

(2.12)

whence we get:

$$\vec{M} = \frac{e}{2mc} \cdot \vec{N}$$

(2.13)

or:

$$\vec{N} = \frac{2mc}{e} \cdot \vec{M}$$

(2.14)

Thus, the mechanical moment of momentum of an electron moving in an orbit, included in the left-hand side of equation (2.3), is expressed in terms of the magnetic moment.

Consider now the quantities on the right-hand side of the same equation. There are two of them: the shoulder of the force $\vec{R}$ and the force itself $\vec{F}$ — acting on the magnetic moment, which will now be considered as the dipole moment $\vec{M}$ of a system of two magnetic charges of different signs. The diagram of the interaction of charges with the field is shown in the figure. Here are $\pm m$ are magnetic charges equal in magnitude and opposite in sign, $\vec{R}_{1,2}$ are the terms of the vector of the magnetic dipole moment, $\vec{H}$ is the field acting on the magnetic dipole, and $\vec{F}_{1,2}$ are the forces caused by this field. Due to the symmetry of the system $\vec{R}_1 = \vec{R}_2$, $\vec{F}_1 = -\vec{F}_2$. The O indicates the axis of rotation of the dipole.

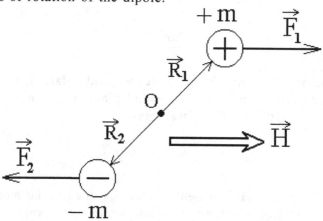

Fig. 2.1. Interaction of charges with the field.

According to the definition of the magnetic dipole moment:

$$\vec{M} = \vec{d}_m = m \cdot (\vec{R}_1 - \vec{R}_2) \quad (2.15)$$

whence, by virtue of the relation, we obtain:

$$\vec{M} = 2m \cdot \vec{R}_1 \quad (2.16)$$

In this case, the radius vector of the point of application of the force takes the form:

$$\vec{R}_1 = \frac{1}{2m} \cdot \vec{M} \quad (2.17)$$

According to the definition of the concept of field strength, the force of the action of a magnetic field on a magnetic charge is:

$$\vec{F}_1 = m \cdot \vec{H} \quad (2.18)$$

In this case, the moment of force is equal to:

$$\vec{P}_1 = \left[\vec{R}_1 \times \vec{F}_1\right] = \frac{1}{2} \cdot \left[\vec{M} \times \vec{H}\right] \quad (2.19)$$

Similarly, the force acting on another magnetic charge (of a different sign) has a moment equal to:

$$\vec{P}_2 = \left[\vec{R}_2 \times \vec{F}_2\right] = \frac{1}{2} \cdot \left[\vec{M} \times \vec{H}\right] \quad (2.20)$$

Since two forces are applied to the dipole, turning it in the same direction, their action is added and the resulting moment of force takes the form:

$$\vec{P} = \vec{P}_1 + \vec{P}_2 = \left[\vec{M} \times \vec{H}\right] \quad (2.21)$$

Hence, according to (2.2), we obtain:

$$[\vec{R} \times \vec{F}] = [\vec{M} \times \vec{H}]$$
(2.22)

Thus, the vector product included in the right-hand side of equation (2.3) is also expressed in terms of magnetic quantities - the magnetic moment and the magnetic field.

Now, substituting (2.14) and (2.22) into (2.3), we obtain the equation of motion of the magnetic moment vector of an electron moving in a circular orbit in the form:

$$\frac{d\vec{M}}{dt} = \frac{e}{2mc} [\vec{M} \times \vec{H}]$$
(2.23)

According to quantum mechanical concepts, the spin moment of an electron is half the orbital angular momentum, that is, relation (2.14) takes the form:

$$\vec{N} = \frac{mc}{e} \cdot \vec{M}$$
(2.24)

In this case, instead of (2.23), we obtain:

$$\frac{d\vec{M}}{dt} = \frac{e}{mc} [\vec{M} \times \vec{H}]$$
(2.25)

from which, by introducing the notation:

$$\gamma = -\frac{e}{mc}$$
(2.26)

we obtain the classical Landau-Lifshitz equation:

$$\frac{d\vec{M}}{dt} = -\gamma [\vec{M} \times \vec{H}]$$
(2.27)

Here, when calculating the value by formula (2.26), one must take into account that the electron charge is negative, as a result of which it turns out to be positive.

2.2. Dynamic sensitivity of a magnetic medium

When a magnetic field is applied to a medium with magnetic properties, the magnetization vector changes its position in accordance with the acting field. In a ferromagnetic medium, where the spins of individual atoms are linked by exchange interaction into a single continuous complex, it can be assumed that the total magnetization thus averaged is proportional to the applied magnetic field. The aspect ratio in this case is the magnetic susceptibility.

If a constant magnetic field is applied to an isotropic medium, then the magnetization vector inside the medium in a static state is oriented along this field. If now, in addition to the constant one, a small alternating field is applied, then the magnetization vector begins to precess around the constant field in accordance with the law given by the Landau–Lifshitz equation.

In the general case, this precession is nonlinear, so that the magnetization is proportional to the field not to the first, but to a higher degree. However, at a small amplitude of the alternating field, the precession can be considered linear, to describe which in the Landau–Lifshitz equation it is necessary to perform linearization. Let us show how the linearization is performed in this case.

Expanding the vector products in the Landau–Lifshitz equation, we obtain:

$$\dot{M}_x = -\gamma \left(M_y H_z - M_z H_y \right) + \frac{\alpha}{M_0} \cdot \left(M_y \dot{M}_z - M_z \dot{M}_y \right) \quad (2.28)$$

$$\dot{M} = -\gamma \left(M_z H_x - M_x H_z \right) + \frac{\alpha}{M_0} \left(M_z \dot{M}_x - M_x \dot{M}_z \right); \quad (2.29)$$

$$\dot{M}_z = -\gamma \left(M_x H_y - M_y H_x \right) + \frac{\alpha}{M_0} \left(M_x \dot{M}_y - M_y \dot{M}_x \right); \quad (2.30)$$

Now we assume:

$$\vec{H} = \{ h_x;\ h_y;\ H_0 + h_z \} \quad (2.31)$$

$$\vec{M} = \{ m_x;\ m_y;\ M_0 + m_z \} \quad (2.32)$$

where H_0 and are M_0 constants.

Substituting (4)–(5) in (1)–(3) gives:

$$\dot{m}_x = -\gamma\left[m_y(H_0+h_z)-(M_0+m_z)h_y\right] + \frac{\alpha}{M_0}\cdot\left[m_y\dot{m}_z-(M_0+m_z)\dot{m}_y\right] \quad (2.33)$$

$$\dot{m}_y = -\gamma\left[(M_0+m_z)h_x - m_x(H_0+h_z)\right] + \frac{\alpha}{M_0}\cdot\left[(M_0+m_z)\dot{m}_x - m_x\dot{m}_z\right] \quad (2.34)$$

$$\dot{m}_z = -\gamma\left(m_x h_y - m_y h_x\right) + \frac{\alpha}{M_0}\cdot\left(m_x\dot{m}_y - m_y\dot{m}_x\right) \quad (2.35)$$

These are the complete equations of motion for the magnetization vector with damping taken into account.

Now let's put:

$$h_x \sim h_y \sim h_z \sim m_x \sim m_y \sim m_z \sim \dot{m}_x \sim \dot{m}_y \sim \dot{m}_z < H_0 \sim M_0 \quad (2.36)$$

Linearization consists in neglecting all terms of order,, in expressions (2.33)–(2.35), as a result of which, instead of complete equations (2.33)–(2.35), we obtain linearized:

$$\dot{m}_x = -\gamma\left(m_y H_0 - M_0 h_y\right) - \alpha\dot{m}_y \quad (2.37)$$

$$\dot{m}_y = -\gamma\left(M_0 h_x - H_0 m_x\right) + \alpha\dot{m}_x \quad (2.38)$$

$$m_z = 0. \quad (2.39)$$

The third equation here speaks of the constancy of the z-component of the magnetization vector, that is, it can be ignored in dynamics. Let us now put the harmonic dependence on time for the field and magnetization in the form:

$$\vec{h} = \vec{h}_0 \cdot e^{i\omega t} \quad (2.40)$$

$$\vec{m} = \vec{m}_0 \cdot e^{i\omega t} \qquad (2.41)$$

Substituting (2.40)–(2.41) into (2.37)–(2.38), performing differentiation, opening brackets and transferring unknown quantities (magnetization components) to the left side, and known (field components) to the right, we obtain:

$$i\omega m_{0x} + (\gamma H_0 + i\omega\alpha)\cdot m_{0y} = \gamma M_0 h_{0y} \qquad (2.42)$$

$$i\omega m_{0y} - (\gamma H_0 + i\omega\alpha)\cdot m_{0x} = -\gamma M_0 h_{0x} \qquad (2.43)$$

Here, the subscripts "0" at variable values mean that the field and magnetization components are amplitudes at time factors. Keeping this in mind, they can be omitted to simplify the notation:

$$i\omega m_x + (\gamma H_0 + i\omega\alpha)\cdot m_y = \gamma M_0 h_y \qquad (2.44)$$

$$i\omega m_x + (\gamma H_0 + i\omega\alpha)\cdot m_y = \gamma M_0 h_y \qquad (2.45)$$

Expanding the brackets in (2.44)–(2.45), writing the variables in the order of the indices, and changing the sign of the second equation, we get:

$$; \quad = -i\omega\gamma M_0 h_y - \gamma M_0(\gamma H_0 + i\omega\alpha)h_x$$

$$i\omega m_x + (\gamma H_0 + i\omega\alpha)\cdot m_y = \gamma M_0 h_y \qquad (2.46)$$

$$(\gamma H_0 + i\omega\alpha)\cdot m_x - i\omega m_y = \gamma M_0 h_x \qquad (2.47)$$

This is a system of two linear equations with two unknowns. Let's solve it by the extended matrix method (Horner's scheme)

$$D_0 = \begin{vmatrix} i\omega & (\gamma H_0 + i\omega\alpha) \\ (\gamma H_0 + i\omega\alpha) & -i\omega \end{vmatrix} = \omega^2 - (\gamma H_0 + i\omega\alpha)^2 \qquad (2.48)$$

$$; \quad D_m = \begin{vmatrix} \gamma M_0 h_y & (\gamma H_0 + i\omega\alpha) \\ \gamma M_0 h_x & -i\omega \end{vmatrix} = \qquad (2.49)$$

$$= -i\omega\gamma M_0 h_y - \gamma M_0(\gamma H_0 + i\omega\alpha)h_x$$

$$D_{my} = \begin{vmatrix} i\omega & \gamma M_0 h_y \\ (\gamma H_0 + i\omega\alpha) & \gamma M_0 h_y \end{vmatrix} = i\omega\gamma M_0 h_x - \gamma M_0(\gamma H_0 + i\omega\alpha)h_y.$$

$$(2.50)$$

As a result, we get:

;

$$m_x = \frac{D_{mx}}{D_0} = -\frac{\gamma M_0(\gamma H_0 + i\omega\alpha)}{\omega^2 - (\gamma H_0 + i\omega\alpha)^2} \cdot h_x - \frac{i\omega\gamma M_0}{\omega^2 - (\gamma H_0 + i\omega\alpha)^2} \cdot h_y \quad (2.51)$$

$$m_y = \frac{D_{my}}{D_0} = \frac{i\omega\gamma M_0}{\omega^2 - (\gamma H_0 + i\omega\alpha)^2} \cdot h_x - \frac{\gamma M_0(\gamma H_0 + i\omega\alpha)}{\omega^2 - (\gamma H_0 + i\omega\alpha)^2} \cdot h_y \quad (2.52)$$

Here we can introduce the susceptibility tensor, which in the classical representation has the form [2–4, 134]:

,

$$\ddot{\chi} = \begin{pmatrix} \chi & i\chi_a & 0 \\ -i\chi_a & \chi & 0 \\ 0 & 0 & 0 \end{pmatrix} \quad (2.53)$$

where:

$$\chi = -\frac{\gamma M_0(\gamma H_0 + i\omega\alpha)}{\omega^2 - (\gamma H_0 + i\omega\alpha)^2}; \quad (2.54)$$

$$\chi_a = -\frac{i\omega\gamma M_0}{\omega^2 - (\gamma H_0 + i\omega\alpha)^2} \quad (2.55)$$

In this case, (2.51) and (2.52) is written in the form:

$$\begin{pmatrix} m_x \\ m_y \\ m_z \end{pmatrix} = \ddot{\chi} \cdot \begin{pmatrix} h_x \\ h_y \\ h_z \end{pmatrix} \quad (2.56)$$

or, taking into account the zero (as a result of linearization) values of the components of the field vectors and magnetization, in a truncated form:

$$\begin{pmatrix} m_x \\ m_y \end{pmatrix} = \begin{pmatrix} \chi & i\chi_a \\ -i\chi_a & \chi \end{pmatrix} \cdot \begin{pmatrix} h_x \\ h_y \end{pmatrix}, \quad (2.57)$$

i.e:

$$m_x = \chi \cdot h_x + i\chi_a \cdot h_y; \quad (2.58)$$

$$m_y = -i\chi_a \cdot h_x + \chi \cdot h_y \quad (2.59)$$

It is seen from the expressions presented here that the susceptibility due to the gyromagnetic properties of the medium has a tensor character. Moreover, the permeability, defined as

$$\ddot{\mu} = 1 + 4\pi\ddot{\chi},\tag{2.60}$$

is also a tensor. This circumstance is taken into account further when deriving the Walker equation and solving the Damon–Eshbach problem (as well as similar problems for various types of structures). Here are some possible improvements. In the record of the system (2.46)–(2.47), the traditional notation can be used:

$$\omega_H = \gamma H_0,\tag{2.61}$$

as a result, we get the same system in the form:

$$i\omega m_x + (\omega_H + i\omega\alpha)\cdot m_y = \gamma M_0 h_y;\tag{2.62}$$

$$(\omega_H + i\omega\alpha)\cdot m_x - i\omega m_y = \gamma M_0 h_x.\tag{2.63}$$

In this case, expressions (2.51)–(2.52) after opening the brackets in the denominators, take the form:

$$m_x = \frac{\gamma M_0(\omega_H + i\omega\alpha)}{\omega_H^2 - (1+\alpha^2)\omega^2 + 2i\alpha\omega\omega_H}\cdot h_x + \frac{i\omega\gamma M_0}{\omega_H^2 - (1+\alpha^2)\omega^2 + 2i\alpha\omega\omega_H}\cdot h_y \tag{2.64}$$

$$m_y = -\frac{i\omega\gamma M_0}{\omega_H^2 - (1+\alpha^2)\omega^2 + 2i\alpha\omega\omega_H}\cdot h_x + \frac{\gamma M_0(\omega_H + i\omega\alpha)}{\omega_H^2 - (1+\alpha^2)\omega^2 + 2i\alpha\omega\omega_H}\cdot h_y \tag{2.65}$$

Susceptibility tensor components take the form:

$$\chi = \frac{\gamma M_0(\omega_H + i\omega\alpha)}{\omega_H^2 - (1+\alpha^2)\omega^2 + 2i\alpha\omega\omega_H};\tag{2.66}$$

$$\chi_a = \frac{\omega\gamma M_0}{\omega_H^2 - (1+\alpha^2)\omega^2 + 2i\alpha\omega\omega_H}\tag{2.67}$$

These are classic expressions that coincide with those given in [2–4]. With a view to further consideration, it is convenient to introduce normalized notation:

$$\Omega = \frac{\omega}{4\pi\gamma M_0};\qquad(2.68)$$

$$\Omega_H = \frac{H_0}{4\pi M_0}\qquad(2.69)$$

In this case, the system (2.46) and (2.47) after dividing by takes the form:

$$i\Omega m_x + (\Omega_H + i\Omega\alpha)\cdot m_y = \frac{1}{4\pi}h_y;\qquad(2.70)$$

$$(\Omega_H + i\Omega\alpha)\cdot m_x - i\Omega\omega m_y = \frac{1}{4\pi}h_x\qquad(2.71)$$

Magnetization components (2.51) and (2.52) take the form:

$$m_x = \frac{1}{4\pi}\cdot\frac{(\Omega_H + i\alpha\Omega)}{\Omega_H^2 - (1+\alpha^2)\Omega^2 + 2i\alpha\Omega\Omega_H}\cdot h_x +\qquad(2.72)$$

$$+\frac{1}{4\pi}\cdot\frac{i\Omega}{\Omega_H^2 - (1+\alpha^2)\Omega^2 + 2i\alpha\Omega\Omega_H}\cdot h_y\qquad(2.73)$$

In this case, the components of the susceptibility tensor are equal:

$$\chi = \frac{1}{4\pi}\cdot\frac{(\Omega_H + i\alpha\Omega)}{\Omega_H^2 - (1+\alpha^2)\Omega^2 + 2i\alpha\Omega\Omega_H};\qquad(2.74)$$

$$\chi_a = \frac{1}{4\pi}\cdot\frac{\Omega}{\Omega_H^2 - (1+\alpha^2)\Omega^2 + 2i\alpha\Omega\Omega_H}\qquad(2.75)$$

In the absence of damping, that is, at, these expressions turn into the following:

$$\chi = \frac{1}{4\pi}\cdot\frac{\Omega_H}{\Omega_H^2 - \Omega^2}\qquad(2.76)$$

$$\chi_a = \frac{1}{4\pi}\cdot\frac{\Omega}{\Omega_H^2 - \Omega^2}\qquad(2.77)$$

These notation expressions are used further in the solution of the Damon–Eshbach problem.

2.3. Walker's equation

Electromagnetic phenomena in a medium with magnetization occur in accordance with the magnetic permeability of this medium, which in the general case has a tensor character.

In an isotropic medium, the use of classical equations of electrodynamics (Maxwell's equations), after taking the divergence from the gradient of one or another field, leads to the Laplace equation, which is symmetric with respect to the variables included in it.

However, in an anisotropic medium, one of the varieties of which is a gyrotropic medium, the same operation leads to an equation that is asymmetric with respect to variables, since the variables behave differently with respect to different coordinate axes.

In the magnetostatic approximation, that is, the position of the electric field equal to zero, such an asymmetric equation for the magnetization is called the Walker equation, which we will consider further.

2.3.1. Walker's equation with an arbitrary susceptibility tensor

Walker's equation is obtained from the classical equations of electrodynamics, taken in the magnetostatic approximation, that is, from the following system:

$$\operatorname{rot} \vec{H} = 0 \qquad (2.78)$$

$$\operatorname{div} \vec{B} = 0 \qquad (2.79)$$

Assuming the field and induction as a sum of constant and variable terms, and taking into account that the derivatives of constant values are equal to zero, we obtain the same equations for variable fields:

$$\operatorname{rot} \vec{h} = 0 \, ; \qquad (2.80)$$

$$\operatorname{div} \vec{b} = 0 \qquad (2.81)$$

where:

$$\vec{b} = \vec{h} + 4\pi\vec{m} \qquad (2.82)$$

According to the well-known vector analysis relation

$$\mathrm{rot\,grad}\, a = 0, \qquad (2.83)$$

where a is an arbitrary scalar function, it follows that the first equation admits the possibility of introducing a scalar potential such that:

$$\vec{h} = \mathrm{grad}\,\psi \qquad (2.84)$$

In this case, the first equation is satisfied identically.
From the second equation we get:

$$div\,\vec{b} = \mathrm{div}\,(\vec{h} + 4\pi\vec{m}) = \mathrm{div\,grad}\,\psi + 4\pi\,\mathrm{div}\,\vec{m} \qquad (2.85)$$

Considering the following relations from vector analysis:

$$\mathrm{div\,grad}\,\psi = \Delta\psi, \qquad (2.86)$$

where is the Laplace operator, we get:

$$\Delta\psi + 4\pi\,\mathrm{div}\,\vec{m} = 0 \qquad (2.87)$$

This is the Walker equation in general form, which is an analogue of the Laplace equation for a medium with a nonzero magnetization included in the expression for induction (2.82).

Variable magnetization is due to the action of an alternating field, so they are related to each other through the susceptibility by the following relationship:

$$\vec{m} = \ddot{\chi}\vec{h}, \qquad (2.88)$$

where, due to the anisotropy or gyrotropy of the medium, it has a tensor character.

We will assume that the tensor in the Cartesian coordinate system has the form:

$$\ddot{\chi} = \begin{pmatrix} a_{xx} & a_{xy} & a_{xz} \\ a_{yx} & a_{yy} & a_{yz} \\ a_{zx} & a_{zy} & a_{zz} \end{pmatrix}, \qquad (2.89)$$

that is, the expression for magnetization through the field is:

$$\begin{pmatrix} m_x \\ m_y \\ m_z \end{pmatrix} = \begin{pmatrix} a_{xx} & a_{xy} & a_{xz} \\ a_{yx} & a_{yy} & a_{yz} \\ a_{zx} & a_{zv} & a_{zz} \end{pmatrix} \cdot \begin{pmatrix} h_x \\ h_y \\ h_z \end{pmatrix}, \quad (2.90)$$

so that:

$$m_x = a_{xx} h_x + a_{xy} h_y + a_{xz} h_z \quad (2.91)$$

$$m_y = a_{yx} h_x + a_{yy} h_y + a_{yz} h_z \quad (2.92)$$

$$m_z = a_{zx} h_x + a_{zy} h_y + a_{zz} h_z \quad (2.93)$$

Differentiating (2.91)–(2.93) by the coordinates of the same name, and also taking into account that it follows from relation (2.84):

$$h_x = \frac{\partial \psi}{\partial x} \quad (2.94)$$

$$h_y = \frac{\partial \psi}{\partial y} \quad (2.95)$$

$$h_z = \frac{\partial \psi}{\partial z} \quad (2.96)$$

we get:

$$\frac{\partial m_x}{\partial x} = a_{xx} \frac{\partial^2 \psi}{\partial x^2} + a_{xy} \frac{\partial^2 \psi}{\partial x \partial y} + a_{xz} \frac{\partial^2 \psi}{\partial x \partial z} \quad (2.97)$$

$$\frac{\partial m_y}{\partial y} = a_{yx} \frac{\partial^2 \psi}{\partial y \partial x} + a_{yy} \frac{\partial^2 \psi}{\partial y^2} + a_{yz} \frac{\partial^2 \psi}{\partial y \partial z} \quad (2.98)$$

$$\frac{\partial m_z}{\partial z} = a_{zx} \frac{\partial^2 \psi}{\partial z \partial x} + a_{zy} \frac{\partial^2 \psi}{\partial z \partial y} + a_{zz} \frac{\partial^2 \psi}{\partial z^2} \quad (2.99)$$

Writing equation (2.87) in coordinates, that is, expanding the expressions for the Laplace operator and divergence, we get:

$$\frac{\partial^2 \psi}{\partial x^2} + \frac{\partial^2 \psi}{\partial y^2} + \frac{\partial^2 \psi}{\partial z^2} + 4\pi \left(\frac{\partial m_x}{\partial x} + \frac{\partial m_y}{\partial y} + \frac{\partial m_z}{\partial z} \right) \quad (2.100)$$

Substituting (2.97)–(2.99) into this expression, we get:

$$\frac{\partial^2 \psi}{\partial x^2} + \frac{\partial^2 \psi}{\partial y^2} + \frac{\partial^2 \psi}{\partial z^2} + 4\pi \left[a_{xx} \frac{\partial^2 \psi}{\partial x^2} + a_{yy} \frac{\partial^2 \psi}{\partial y^2} + a_{zz} \frac{\partial^2 \psi}{\partial z^2} + \right. \quad (2.101)$$

$$+ \left(a_{xy}+a_{yx}\right)\frac{\partial^2\psi}{\partial x\partial y} + \left(a_{xz}+a_{zx}\right)\frac{\partial^2\psi}{\partial x\partial z} + \left(a_{yz}+a_{zy}\right)\frac{\partial^2\psi}{\partial y\partial z}\Bigg] = 0$$

Bringing similar terms and writing the coordinates in ascending order, we get:

$$\left(1+4\pi a_{xx}\right)\frac{\partial^2\psi}{\partial x^2} + \left(1+4\pi a_{yy}\right)\frac{\partial^2\psi}{\partial y^2} + \left(1+4\pi a_{zz}\right)\frac{\partial^2\psi}{\partial z^2} +$$

$$+4\pi\left(a_{xy}+a_{yx}\right)\frac{\partial^2\psi}{\partial x\partial y} + 4\pi\left(a_{xz}+a_{zx}\right)\frac{\partial^2\psi}{\partial x\partial z} + 4\pi\left(a_{yz}+a_{zy}\right)\frac{\partial^2\psi}{\partial y\partial z} = 0$$

(2.102)

This is the required Walker equation written in the Cartesian coordinate system for an arbitrary susceptibility tensor (2.89). The conditions of gyrotropy and anisotropy impose certain relations on the tensor components, which can be obtained from the solution of the equation of motion for magnetization (Landau–Lifshitz equation). In the general case, the equation of motion for the magnetization can be solved in any other Cartesian system, oriented in an arbitrary way relative to the one adopted here, where the Walker equation (2.102) is written. In this case, the components of the susceptibility tensor obtained as a result of such a solution should be transformed into the system adopted here using the corresponding rotation matrices.

2.3.2. Walker equation in the Damon–Eshbach problem

Let us show how the Walker equation used in the Damon–Eshbach problem is obtained from the general equation (2.102).

In the section devoted to the susceptibility, taking into account the notation changes and, relations (2.58), (2.59) were obtained, which we write in the form:

$$4\pi m_x = \kappa h_x + i\nu h_y, \qquad (2.103)$$

$$4\pi m_y = -i\nu h_x + \kappa h_y, \qquad (2.104)$$

where, according to the notation (2.68), (2.69) adopted in the same place:

$$\kappa = \frac{\Omega_H}{\Omega_H^2 - \Omega^2} \qquad (2.105)$$

$$\nu = \frac{\Omega}{\Omega_H^2 - \Omega^2} \qquad (2.106)$$

Dividing (2.103)–(2.104) by 4π, we get:

$$m_x = \frac{\kappa}{4\pi} h_x + \frac{i\nu}{4\pi} h_y \quad ; \tag{2.107}$$

$$m_y = -\frac{i\nu}{4\pi} h_x + \frac{\kappa}{4\pi} h_y \tag{2.108}$$

From these expressions we obtain the components of the susceptibility tensor:

$$a_{xx} = \frac{\kappa}{4\pi} \quad a_{xy} = \frac{i\nu}{4\pi} \quad a_{xz} = 0 \tag{2.109}$$

$$a_{yy} = \frac{\kappa}{4\pi} \quad a_{yy} = \frac{\kappa}{4\pi} \quad a_{yz} = 0 \tag{2.110}$$

$$a_{zx} = 0; \; ; \; a_{zy} = 0; \; a_{zz} \tag{2.111}$$

Substituting these components into the general Walker equation (2.102), we obtain:

$$\left(1 + 4\pi\frac{\kappa}{4\pi}\right)\frac{\partial^2 \psi}{\partial x^2} + \left(1 + 4\pi\frac{\kappa}{4\pi}\right)\frac{\partial^2 \psi}{\partial y^2} + (1+0)\frac{\partial^2 \psi}{\partial z^2} +$$
$$+ 4\pi\left(\frac{i\nu}{4\pi} - \frac{i\nu}{4\pi}\right)\frac{\partial^2 \psi}{\partial x \partial y} + 4\pi(0+0)\frac{\partial^2 \psi}{\partial x \partial z} + 4\pi(0+0)\frac{\partial^2 \psi}{\partial y \partial z} = 0 \tag{2.112}$$

Reducing fractions and bringing similar terms, we get:

$$k_x^e = \sqrt{\mu_0^2 \frac{\mu-1}{\mu} \lambda_9^2} \tag{2.113}$$

Introducing the notation adopted for permeability:

$$k_x^i = i \cdot \mu_0, \tag{2.114}$$

we get:

$$k_y = \sqrt{\mu_0^2 - \frac{\lambda_0^2}{\mu}} \tag{2.115}$$

which is the Walker equation used below in the Damon–Eshbach problem (Section 2.4).

2.4. Dispersion ratio for magnetic plate with free surface Damon–Eshbach's solution

Let us now consider magnetostatic waves in a tangentially magnetized plate with free surfaces. The dispersion relation for such waves was first obtained by Damon and Eshbach in [118]. This problem was solved using the expressions for the susceptibility and the Walker equation obtained in the previous sections. In view of the importance of this problem, we present these calculations once again in a form as close as possible to the original work [118].

2.4.1. Basic equations

Magnetostatic equations:

$$\frac{d\vec{M}}{dt} = -\gamma \cdot \left[\vec{M} \times \vec{H} \right] \tag{2.116}$$

$$div\vec{B} = 0. \tag{2.117}$$

Here:

$$\vec{B} = \vec{H} + 4\pi\vec{M} \tag{2.118}$$

Let's assume that the field and magnetization are of the form:

$$\vec{H} = H_i \cdot \vec{n}_z + \vec{h} \cdot e^{i\omega t} \tag{2.119}$$

$$\vec{M} = M_0 \cdot \vec{n}_z + \vec{m} \cdot e^{i\omega t} \tag{2.120}$$

Here: H_t internal constant field; $h - m$, $H_t \sim M_0$.

Substituting (2.119), (2.120) into (2.116), (2.117), and taking into account that, as well as (since,), we obtain:

$$rot\,\vec{h} = 0; \tag{2.121}$$

$$div\left(\vec{h} + 4\pi\vec{m}\right) = 0 \tag{2.122}$$

Equation (2.121) implies the possibility of introducing an alternating field potential such that:

$$\vec{h} = grad\,\psi = \left\{ \frac{\partial\psi}{\partial x};\ \frac{\partial\psi}{\partial y};\ \frac{\partial\psi}{\partial z} \right\} \tag{2.123}$$

In this case, equation (2.122) takes the form:

$$\text{div}\left(\text{grad}\,\psi + 4\pi\vec{m}\right) = 0 \quad (2.124)$$

or:

$$\Delta\psi + 4\pi\,\text{div}\,\vec{m} = 0 \quad (2.125)$$

This equation is the basic equation for the potential in a medium with magnetization. It includes variable magnetization $\vec{m}$, which is determined from the equation of motion of the magnetization vector (Landau–Lifshitz) through an alternating field using the susceptibility tensor $\ddot{\chi}$. The components of the alternating field are expressed in terms of the potential ψ in accordance with (2.123). In this way, the components of the variable magnetization $\vec{m}$ can be expressed in terms of the potential ψ, after which the obtained expressions can be substituted into (2.124). In this case, we obtain the equation for the potential in its pure form ψ. This will be the Walker equation, the form of which will be determined by the explicit form of the susceptibility tensor $\ddot{\chi}$.

Let us derive the Walker equation for a homogeneous isotropic medium in which the internal field is equal to H_i.

The equation of motion of the magnetization vector (Landau-Lifshitz) has the form:

$$\frac{d\vec{M}}{dt} = -\gamma \cdot \left[\vec{M} \times \vec{H}\right] \quad (2.126)$$

Here, $\psi > 0$ while the free precession is right.

In the Cartesian coordinate system, the cross product is:

$$\left[\vec{M} \times \vec{H}\right] = \vec{i}\left(M_y H_z - M_z H_y\right) + \vec{j}\left(M_y H_z - M_z H_y\right) + \vec{k}\left(M_y H_z - M_z H_y\right) \quad (2.127)$$

Since the constant field is directed along the axis (2.119), and in what follows, linearization is assumed, it can be assumed that the alternating field along this axis is absent and confined to the components of variable magnetization only along the axes and:

$$\dot{M}_x = -\gamma\left(M_y H_z - M_z H_y\right) \quad (2.128)$$

$$\dot{M}_y = -\gamma\left(M_z H_x - M_x H_z\right) \quad (2.129)$$

Substituting and in the form (2.119) and (2.120), as well as performing linearization, we obtain a system of equations for and:

$$i\omega m_x + \gamma H_i m_y = \gamma M_0 h_y \tag{2.130}$$

$$\gamma H_i m_x - i\omega m_y = \gamma M_0 h_x \tag{2.131}$$

Solving this system, we find:

$$m_x = \frac{\gamma^2 M_0 H_i}{(\gamma H_i)^2 - \omega^2} \cdot h_x + \frac{i\omega \gamma M_0}{(\gamma H_i)^2 - \omega^2} \cdot h_y \tag{2.132}$$

$$m_y = -\frac{i\omega \gamma M_0}{(\gamma H_i)^2 - \omega^2} \cdot h_x + \frac{\gamma^2 M_0 H_i}{(\gamma H_i)^2 - \omega^2} \cdot h_y \tag{2.133}$$

Let's introduce the notation:

$$\omega_H = \gamma H_i \tag{2.134}$$

$$\omega_M = 4\pi \gamma M_0 \tag{2.135}$$

$$\Omega = \frac{\omega}{\omega_M} = \frac{\omega}{4\pi \gamma M_0} \tag{2.136}$$

$$\Omega_H = \frac{\omega_H}{\omega_M} = \frac{H_i}{4\pi M_0} \tag{2.137}$$

$$\Omega_H = \frac{\omega_H}{\omega_M} = \frac{H_i}{4\pi M_0} \tag{2.138}$$

$$\frac{\gamma^2 M_0 H_i}{(\gamma H_i)^2 - \omega^2} = \frac{1}{4\pi} \cdot \frac{\omega_M \cdot \omega_H}{\omega_H^2 - \omega^2} = \frac{1}{4\pi} \cdot \frac{\Omega_H}{\Omega_H^2 - \Omega^2} = \frac{1}{4\pi} \cdot \kappa \tag{2.139}$$

With these notation we get:

$$\frac{\gamma^2 M_0 H_i}{(\gamma H_i)^2 - \omega^2} = \frac{1}{4\pi} \cdot \frac{\omega_M \cdot \omega_H}{\omega_H^2 - \omega^2} = \frac{1}{4\pi} \cdot \frac{\Omega_H}{\Omega_H^2 - \Omega^2} = \frac{1}{4\pi} \cdot \kappa \tag{2.140}$$

$$\frac{i\omega \gamma M_0}{(\gamma H_i)^2 - \omega^2} = \frac{1}{4\pi} \cdot \frac{i\omega \cdot \omega_M}{\omega_H^2 - \omega^2} = \frac{1}{4\pi} \cdot \frac{i \cdot \Omega}{\Omega_H^2 - \Omega^2} = \frac{1}{4\pi} \cdot i\nu \tag{2.141}$$

Substituting in (2.140) and (2.141) in (2.132) and (2.133), we obtain:

$$4\pi m_x = \kappa h_x + i\nu h_y \tag{2.142}$$

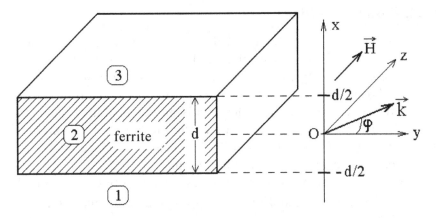

Fig. 2.2. Geometry of the problem.

$$4\pi m_y = -i\nu h_x + \kappa h_y \tag{2.143}$$

Substituting fields in (2.142) and (2.143) and in accordance with (2.123), we obtain:

$$4\pi m_x = \kappa \frac{\partial \psi}{\partial x} + i\nu \frac{\partial \psi}{\partial y} \tag{2.144}$$

$$4\pi m_y = -i\nu \frac{\partial \psi}{\partial x} + \kappa \frac{\partial \psi}{\partial y} \tag{2.145}$$

Substituting (2.144) and (2.145) into (2.125) and introducing the notation:

$$\mu = 1 + \kappa = 1 + \frac{\Omega_H}{\Omega_H^2 - \Omega^2} \tag{2.146}$$

we obtain an equation for the potential inside the medium – Walker's equation in the form:

$$\mu \cdot \left(\frac{\partial^2 \psi^i}{\partial x^2} + \frac{\partial^2 \psi^i}{\partial y^2} \right) + \frac{\partial^2 \psi^i}{\partial z^2} = 0 \tag{2.147}$$

In this case, the equation for the potential outside the medium at (so that, according to (2.132)–(2.133), also), takes the form of the Laplace equation:

$$\frac{\partial^2 \psi^e}{\partial x^2} + \frac{\partial^2 \psi^e}{\partial y^2} + \frac{\partial^2 \psi^e}{\partial z^2} = 0 \tag{2.148}$$

2.4.2. Border conditions

Consider the geometry of the problem illustrated in Fig. 2.2. The structure is an infinite ferrite plate 2 thickness d, on both sides of which there are empty half-spaces 1 and 3. The coordinate system $Oxyz$ is chosen in such a way that its plane Oyz is parallel to the planes of the ferrite plate, and the axis is perpendicular to them. In this case, the axis Oz is oriented along the direction of the external field. The origin O is in the middle between the surface planes of the plate, the coordinates of which are equal to $\pm d/2$.

Boundary conditions on plate surfaces – continuity of the normal induction component $\vec{B}$ and the tangential field component $\vec{H}$:

$$B^i_x \Big|_{x=\pm\frac{d}{2}} = B^e_x \Big|_{x=\pm\frac{d}{2}} ; \tag{2.149}$$

$$H^i_{y,z} \Big|_{x=\pm\frac{d}{2}} = H^e_{y,z} \Big|_{x=\pm\frac{d}{2}} . \tag{2.150}$$

Calculating the individual components $\vec{B}$ and, $\vec{H}$ we get:

$$B^i_x = h^i_x + 4\pi m_x = \frac{\partial \psi^i}{\partial x} + \kappa \frac{\partial \psi^i}{\partial x} + i\nu \frac{\partial \psi^i}{\partial y} = \mu \frac{\partial \psi^i}{\partial x} + i\nu \frac{\partial \psi^i}{\partial y} \tag{2.151}$$

$$B^e_x = h^e_x = \frac{\partial \psi^e}{\partial x} \tag{2.152}$$

$$H^i_y = h^i_y = \frac{\partial \psi^i}{\partial x} \tag{2.153}$$

$$H^e_y = h^e_y = \frac{\partial \psi^e}{\partial x} \tag{2.154}$$

$$H^i_z = h^i_z = \frac{\partial \psi^i}{\partial z} \tag{2.155}$$

$$H^e_z = h^e_z = \frac{\partial \psi^e}{\partial z} \tag{2.156}$$

Boundary conditions take the form (4 boundary conditions in total):

$$\mu \frac{\partial \psi^i}{\partial x} + i\nu \frac{\partial \psi^i}{\partial y} \Big|_{x=\pm\frac{d}{2}} = \frac{\partial \psi^e}{\partial x} \Big|_{x=\pm\frac{d}{2}} \tag{2.157}$$

$$\psi^i \Big|_{x=\pm\frac{d}{2}} = \psi^e \Big|_{x=\pm\frac{d}{2}} \tag{2.158}$$

The second pair of boundary conditions is obtained from the equality of derivatives, that is, it is fulfilled up to an arbitrary constant added to the potential (which is the result of integration). In the work of Damon–Eshbach, this constant is assumed to be zero, since all fields are calculated through the derivatives of the potential, and the derivative of a constant is equal to zero.

2.4.3. Complete problem statement

Thus, we get the complete problem in the form:

Equations:

region 1 (outside ferrite):

$$\frac{\partial^2 \psi_1}{\partial x^2} + \frac{\partial^2 \psi_1}{\partial y^2} + \frac{\partial^2 \psi_1}{\partial z^2} = 0 \tag{2.159}$$

region 2 (inside ferrite):

$$\mu \cdot \left(\frac{\partial^2 \psi_2}{\partial x^2} + \frac{\partial^2 \psi_2}{\partial y^2} \right) + \frac{\partial^2 \psi_2}{\partial z^2} = 0 \tag{2.160}$$

region 3 (outside ferrite):

$$\frac{\partial^2 \psi_3}{\partial x^2} + \frac{\partial^2 \psi_3}{\partial y^2} + \frac{\partial^2 \psi_3}{\partial z^2} = 0 \tag{2.161}$$

Boundary conditions on the bottom and top surfaces:

$$\left. \mu \frac{\partial \psi_2}{\partial x} + i v \frac{\partial \psi_2}{\partial y} \right|_{x=-\frac{d}{2}} = \left. \frac{\partial \psi_1}{\partial x} \right|_{x=-\frac{d}{2}} \tag{2.162}$$

$$\left. \mu \frac{\partial \psi_2}{\partial x} + i v \frac{\partial \psi_2}{\partial y} \right|_{x=\frac{d}{2}} = \left. \frac{\partial \psi_3}{\partial x} \right|_{x=\frac{d}{2}} \tag{2.163}$$

$$\left. \psi_2 \right|_{x=-\frac{d}{2}} = \left. \psi_1 \right|_{x=-\frac{d}{2}} \tag{2.164}$$

$$\left. \psi_2 \right|_{x=\frac{d}{2}} = \left. \psi_3 \right|_{x=\frac{d}{2}} \tag{2.165}$$

2.4.4. Solving equations without boundary conditions

Let's consider the solution in areas 1–3 separately.

AREA 1 (outside ferrite).

The equation for the potential ψ_1 in this region has the form (2.159):

$$\frac{\partial^2 \psi_1}{\partial x^2} + \frac{\partial^2 \psi_1}{\partial y^2} + \frac{\partial^2 \psi_1}{\partial z^2} = 0 \tag{2.166}$$

Let us solve equation (2.166) by the method of separation of variables. To do this, assume that the solution has the form:

$$\psi_1 = X_1(x) \cdot Y_1(y) \cdot Z_1(z) \tag{2.167}$$

Substitute solution (2.167) into equation (2.166) and divide everything by $X_1 \cdot Y_1 \cdot Z_1$. Let us introduce further separation constants ψ_1 and μ_1. Hereinafter, no assumptions are made about the reality or imaginary of these constants; signs and squares are used for convenience.

Substituting (2.167) into (2.166), we obtain:

$$\frac{X_1''}{X_1} + \frac{Y_1''}{Y_1} + \frac{Z_1''}{Z_1} = 0 \tag{2.168}$$

Let's introduce the separation constant:

$$\frac{X_1''}{X_1} + \frac{Y_1''}{Y_1} = -\frac{Z_1''}{Z_1} = \lambda_1^2 \tag{2.169}$$

From (2.169) we obtain the equation for:

$$Z_1'' + \lambda_1^2 Z_1 = 0 \tag{2.170}$$

Its solution looks like:

$$Z_1 = G_1 e^{i\lambda_1 z} + H_1 e^{-i\lambda_1 z} \tag{2.171}$$

Here and below, the letters A, B, C, D, G, H with the corresponding indices will denote arbitrary constants, which will be further determined from the boundary conditions.

Similarly, we find Y_1 and X_1. From (2.169) we obtain:

$$\frac{X_1''}{X_1} + \frac{Y_1''}{Y_1} = \lambda_1^2 \tag{2.172}$$

Introducing another separation constant, we get:

$$\frac{X_1''}{X_1} = -\frac{Y_1''}{Y_1} + \lambda_1^2 = \mu_1^2 \tag{2.173}$$

From (2.173) we obtain the equation for:

$$X_1'' - \mu_1^2 X_1 = 0 \tag{2.174}$$

Its solution looks like:

$$X_1 = A_1 e^{\mu_1 x} + B_1 e^{-\mu_1 x} \tag{2.175}$$

Similarly, from (2.173) we obtain the equation for:

$$Y_1'' + \left(\mu_1^2 - \lambda_1^2\right) Y_1 = 0 \tag{2.176}$$

His solution looks like:

$$Y_1 = C_1 e^{i\sqrt{\mu_1^2 - \lambda_1^2}\, y} + D_1 e^{-i\sqrt{\mu_1^2 - \lambda_1^2}\, y} \tag{2.177}$$

AREA 2 (inside the ferrite).

The equation for the potential in this region has the form (2.160):

$$\mu \cdot \left(\frac{\partial^2 \psi_2}{\partial x^2} + \frac{\partial^2 \psi_2}{\partial y^2}\right) + \frac{\partial^2 \psi_2}{\partial z^2} = 0 \tag{2.178}$$

Let us solve equation (2.178) by the method of separation of variables. To do this, assume that the solution has the form:

$$\psi_2 = X_2(x) \cdot Y_2(y) \cdot Z_2(z) \tag{2.179}$$

We substitute the solution (2.179) into the equation (2.178) and divide everything by. Let us introduce further separation constants and. Here, too, no assumptions are made about the reality or imaginary of these constants, signs and squares are used for convenience.

Substituting (2.179) into (2.178), we obtain:

$$\mu \cdot \left(\frac{X_2''}{X_2} + \frac{Y_2''}{Y_2} \right) + \frac{Z_2''}{Z_2} = 0 \tag{2.180}$$

Let's introduce the separation constant:

$$\mu \cdot \left(\frac{X_2''}{X_2} + \frac{Y_2''}{Y_2} \right) = -\frac{Z_2''}{Z_2} = \lambda_2^2 \tag{2.181}$$

From (2.181) we obtain the equation for:

$$Z_2'' + \lambda_2^2 Z_2 = 0 \tag{2.182}$$

Its solution looks like:

$$Z_2 = G_2 e^{i\lambda_2 z} + H_2 e^{-i\lambda_2 z} \tag{2.183}$$

Similarly, we find Y_2 and X_2. From (2.181) we obtain:

$$\mu \cdot \left(\frac{X_2''}{X_2} + \frac{Y_2''}{Y_2} \right) = \lambda_2^2 \tag{2.184}$$

Introducing another separation constant, we get:

$$\frac{X_2''}{X_2} = -\frac{Y_2''}{Y_2} + \frac{\lambda_2^2}{\mu} = \mu_2^2 \tag{2.185}$$

From (2.185) we obtain the equation for:

$$X_2'' - \mu_2^2 X_2 = 0 \tag{2.186}$$

Its solution looks like:

$$X_2 = A_2 e^{\mu_2 x} + B_2 e^{-\mu_2 x} \tag{2.187}$$

Similarly, from (2.185) we obtain the equation for:

$$Y_2'' + \left(\mu_2^2 - \frac{\lambda_2^2}{\mu}\right) Y_2 = 0 \tag{2.188}$$

Its solution looks like:

$$Y_2 = C_2 e^{i\sqrt{\mu_2^2 - \frac{\lambda_2^2}{\mu}}\, y} + D_2 e^{-i\sqrt{\mu_2^2 - \frac{\lambda_2^2}{\mu}}\, y} \tag{2.189}$$

AREA 3 (outside ferrite).

The solution in this area is completely similar to the solution in area 1 with the replacement of the index '1' by the index '3'.

The COMPLETE SOLUTION has the form (formulas (2.173), (2.175), (2.177), (2.183), (2.187), (2.189)):

Area 1:
$$X_1 = A_1 e^{\mu_1 x} + B_1 e^{-\mu_1 x} \tag{2.190}$$

$$Y_1 = C_1 e^{i\sqrt{\mu_1^2 - \lambda_1^2}\, y} + D_1 e^{-i\sqrt{\mu_1^2 - \lambda_1^2}\, y} \tag{2.191}$$

$$Z_1 = G_1 e^{i\lambda_1 z} + H_1 e^{-i\lambda_1 z} \tag{2.192}$$

Area 2:
$$X_2 = A_2 e^{\mu_2 x} + B_2 e^{-\mu_2 x} \tag{2.193}$$

$$Y_2 = C_2 e^{i\sqrt{\mu_2^2 - \frac{\lambda_2^2}{\mu}}\, y} + D_2 e^{-i\sqrt{\mu_2^2 - \frac{\lambda_2^2}{\mu}}\, y} \tag{2.194}$$

$$Z_2 = G_2 e^{i\lambda_2 z} + H_2 e^{-i\lambda_2 z} \tag{2.195}$$

Area 3:
$$X_3 = A_3 e^{\mu_3 x} + B_3 e^{-\mu_3 x} \tag{2.196}$$

$$Y_3 = C_3 e^{i\sqrt{\mu_3^2 - \lambda_3^2}\, y} + D_3 e^{-i\sqrt{\mu_3^2 - \lambda_3^2}\, y} \tag{2.197}$$

$$Z_3 = G_3 e^{i\lambda_3 z} + H_3 e^{-i\lambda_3 z} \qquad (2.198)$$

Suppose the wave propagates in a plane. In this case, the dependences of all three solutions on and should coincide. From this condition we get:

$$\sqrt{\mu_1^2 - \lambda_1^2} = \sqrt{\mu_2^2 - \frac{\lambda_2^2}{\mu}} = \sqrt{\mu_3^2 - \lambda_3^2} \qquad (2.199)$$

$$\lambda_1 = \lambda_2 = \lambda_3 \qquad (2.200)$$

Let us introduce the notation and from the conditions:

$$\mu_0 = \mu_2, \qquad (2.201)$$

$$\lambda_0 = \lambda_1 = \lambda_2 = \lambda_3 \qquad (2.202)$$

Substituting these notation into (2.199), we obtain:

$$\sqrt{\mu_1^2 - \lambda_0^2} = \sqrt{\mu_0^2 - \frac{\lambda_0^2}{\mu}} = \sqrt{\mu_3^2 - \lambda_0^2} \qquad (2.203)$$

Expression (2.203) can be regarded as a system of two equations for μ_1 and μ_3. Squaring all three components (2.203), then equating the first and third, and then the first and second, we get:
;

$$\mu_1 = \sqrt{\mu_0^2 + \frac{\mu-1}{\mu} \cdot \lambda_0^2} \qquad (2.204)$$

$$\mu_3 = \sqrt{\mu_0^2 + \frac{\mu-1}{\mu} \cdot \lambda_0^2} \qquad (2.205)$$

Since (according to the assumption made) the wave propagates in the plane Oyz, then they are periodic relative to $Y_{1,2,3}$ and $Z_{1,2,3}$. Moreover, since they are expressed through an exponent with an imaginary unit i in the exponent, then λ_1 and $\sqrt{\mu_0^2 - \frac{\lambda_0^2}{\mu}}$ must be real.

Suppose that the solution in region 1 decreases to zero at $x \to +\infty$, and in region 3 at $x \to -\infty$. Moreover, taking into account that $\mu_1 > 0$ and $\mu_3 > 0$, we get that $B_1 = 0$ and $A_3 = 0$..

For convenience, we introduce notation without indices: $A_1 = A$ and $B_3 = B$, and also $C_{1,2,3} = C$, $D_{1,2,3} = D$, $H_{1,2,3} = H$.

As a result, the COMPLETE SOLUTION takes the form:
part of the solution depending only on x:

$$X_1 = A e^{\sqrt{\mu_0^2 + \frac{\mu-1}{\mu}\lambda_0^2}\, x} \tag{2.206}$$

$$X_2 = A_2 e^{\mu_0 x} + B_2 e^{-\mu_0 x} \tag{2.207}$$

$$X_3 = B e^{-\sqrt{\mu_0^2 + \frac{\mu-1}{\mu}\lambda_0^2}\, x} \tag{2.208}$$

the part of the solution that depends only on y and z is the same in all areas:

$$Y = C e^{i\sqrt{\mu_0^2 - \frac{\lambda_0^2}{\mu}}\, y} + D e^{-i\sqrt{\mu_0^2 - \frac{\lambda_0^2}{\mu}}\, y} \tag{2.209}$$

$$\psi_2 = X_2 \cdot Y \cdot Z \tag{2.210}$$

In this solution there are 8 constant coefficients, and the boundary conditions by which they are to be determined, there are only 4. Therefore, 4 coefficients can be set arbitrarily. Suppose further that the waves propagate only in the positive directions of the axes Oy and Oz, that is, $C = 0$ and $G = 0$. Since we can arbitrarily set two more coefficients, we can put $D = 1$ and $H = 1$. In this case, the part of the solution that depends only on y and z takes the form:

$$Y = e^{-i\sqrt{\mu_0^2 - \frac{\lambda_0^2}{\mu}}\, y} \tag{2.211}$$

$$Z = e^{-i\lambda_0 z} \tag{2.212}$$

Thus, only 4 coefficients A, A_2, B_2, B remain in the complete solution, for the determination of which there are 4 boundary conditions (2.157), (2.158).

We now introduce into the solution the wave numbers in an explicit form using the notation:

the wavenumber along the coordinate x outside the plate:

$$k_x^e = \sqrt{\mu_0^2 - \frac{\mu-1}{\mu}\lambda_0^2}$$
(2.213)

the wavenumber along the coordinate inside the plate:

$$k_x^i = i \cdot \mu_0;$$
(2.214)

the wavenumber along the coordinate y in all space:

$$k_y = \sqrt{\mu_0^2 - \frac{\lambda_0^2}{\mu}};$$
(2.215)

the wavenumber along the coordinate z in all space:

$$k_z = \lambda_0$$
(2.216)

This system of expressions contains 6 values k_{ez}, k_{iz}, k_y, k_z, μ_0, λ_0. Since the four equations (2.213)–(2.216) contain six unknowns, two of them can be excluded, resulting in two equations with four unknowns. We exclude μ_0 and λ_0. To do this, we express them through k_z^i and k_x with the help of (2.214) and (2.216), substitute the obtained expressions in (2.213) and (2.215), whence, after squaring, we obtain:

$$(k_x^e)^2 = -(k_x^i)^2 + \frac{\mu-1}{\mu} \cdot k_z^2$$
(2.217)

$$\mu = 1 + \frac{\Omega_H}{\Omega_H^2 - \Omega^2}.$$
(2.218)

These expressions are no longer included. However, there are four wave numbers which are related by two equations. Therefore, two of them can be specified initially, and the other two can be expressed through them. Since there are no restrictions for wave propagation in the plane (geometrically, the plate is limited only along the axis), we will assume that and are given, and we will express through them. To do this, we will do the following. Subtracting (2.218) from (2.217), we obtain:

$$(k_x^e)^2 - k_y^2 - k_z^2 = 0$$
(2.219)

Finding from this expression and substituting in (2.217), we obtain:

$$\mu \cdot \left[\left(k_x^i\right)^2 + k_y^2 \right] + k_z^2 = 0 \qquad (2.220)$$

From (2.219) and (2.220), after simple transformations, we obtain:

$$k_x^e = \sqrt{k_y^2 + k_z^2} \qquad (2.221)$$

$$k_x^i = \sqrt{-\left(k_y^2 + \frac{k_z^2}{\mu}\right)} = i \cdot \sqrt{k_y^2 + \frac{k_z^2}{\mu}} \qquad (2.222)$$

The COMPLETE SOLUTION in the notation (2.213)–(2.216) takes the form:

part of the solution depending only on x:

$$X_1 = A e^{k_x^e x} \qquad (2.223)$$
$$X_2 = A_2 e^{i k_x^i x} + B_2 e^{-i k_x^i x} \qquad (2.224)$$
$$X_3 = B e^{-k_x^e x} \qquad (2.225)$$

the part of the solution that depends only on y and z takes the form:

$$Y = e^{-i k_y y} \qquad (2.226)$$
$$Z = e^{-i k_z z} \qquad (2.227)$$

Here k_y and k_z are still arbitrary real values (their validity is due to the absence of attenuation). In this case, it follows from (2.221) that it is also always real, that is, they exponentially decrease outside the plate. At the same time, from (2.222) it follows that it can be both real and imaginary, depending on the sign of the radical expression, which is determined by the relationship between the quantities and, as well as the quantity and sign depending on.

2.4.5. Frequency regions of body and surface waves

Let us find the frequency regions of bulk and surface waves, for which we consider the dependence determined by formula (2.146):

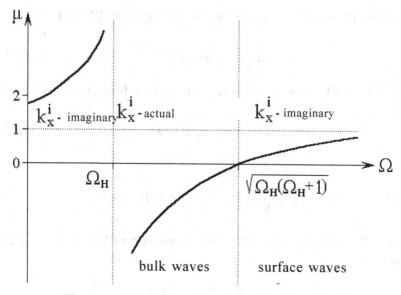

Fig. 2.3. Diagram of the dependence of μ on Ω.

$$\mu = 1 + \frac{\Omega_H}{\Omega_H^2 - \Omega^2} \qquad (2.228)$$

The following cases are possible:

1) $\Omega = 0$, and $\mu = 1 + \Omega_H^{-1}$; (2.229)

2) $0 < \Omega < \Omega_H$, while: $1 + \Omega_H^{-1} < \mu < +\infty$ (2.230)

3) $\Omega_H < \Omega < \sqrt{\Omega_H(\Omega_H + 1)}$, while: $-\infty < \mu < 0$; (2.231)

4) $\Omega = \sqrt{\Omega_H(\Omega_H + 1)}$, , while: $\mu < 0$ (2.232)

5) $\sqrt{\Omega_H(\Omega_H + 1)}$, , while:. $0 < \mu < 1$. (2.233)

The diagram of the dependence is shown in Fig. 2.3.

There is only one area 3: $\Omega_H < \Omega < \sqrt{\Omega_H(\Omega_H + 1)}$ where $\mu < 0$, that is k_x^i can be valid. This is the region of bulk waves, for which the dependence of the potential on the coordinate periodically oscillates. In this area, reality k_x^i is determined by the relationship between k_y and k_z, determined by the condition:

whence we get:
$$k_y^2 + \frac{k_z^2}{\mu} < 0 \qquad (2.234)$$

$$\frac{k_z}{k_y} > \sqrt{-\mu} \qquad (2.235)$$

which gives a sector that determines the possible limits of the propagation of body waves. It is convenient to determine its value in a cylindrical coordinate system using the cutoff angle.

In region 2: $0 < \Omega < \Omega_H$, where $\mu > 0$, the waves do not propagate, since in this case their frequency should have been lower than the frequency of a uniform FMR.

In area 5: $\sqrt{\Omega_H(\Omega_H + 1)} < \Omega < +\infty$, where $\mu > 0$, the value k_x^i is always imaginary. This is the region of surface waves, for which the dependence of the potential ψ_2 on the coordinate x decreases exponentially from one or another surface of the plate.

2.4.6. Derivation of the dispersion relation from the solution and boundary conditions

The solution looks like:
$$\psi_1 = X_1 \cdot Y \cdot Z; \qquad (2.236)$$
$$\psi_2 = X_2 \cdot Y \cdot Z; \qquad (2.237)$$
$$\psi_3 = X_3 \cdot Y \cdot Z. \qquad (2.238)$$

The boundary conditions on the lower $\left(x = -\frac{d}{2}\right)$ and upper $\left(x = \frac{d}{2}\right)$ surfaces have the form (2.162)–(2.165):

$$\left.\mu\frac{\partial \psi_2}{\partial x} + i\nu\frac{\partial \psi_2}{\partial y}\right|_{x=-\frac{d}{2}} = \left.\frac{\partial \psi_1}{\partial x}\right|_{x=-\frac{d}{2}} \qquad (2.239)$$

$$\left.\mu\frac{\partial \psi_2}{\partial x} + i\nu\frac{\partial \psi_2}{\partial y}\right|_{x=\frac{d}{2}} = \left.\frac{\partial \psi_3}{\partial x}\right|_{x=\frac{d}{2}} \qquad (2.240)$$

$$\left.\psi_2\right|_{x=-\frac{d}{2}} = \left.\psi_1\right|_{x=-\frac{d}{2}} \qquad (2.241)$$

$$\left.\psi_2\right|_{x=\frac{d}{2}} = \left.\psi_3\right|_{x=\frac{d}{2}} \qquad (2.242)$$

We substitute the solution (2.236)–(2.238) in the boundary conditions (2.239)–(2.242) and divide the first two equations by Z, and the second two equations by $Y \cdot Z$. In this case, we obtain the boundary conditions in the form:

$$\mu \frac{\partial X_2}{\partial x} \cdot Y + i v X_2 \frac{\partial Y}{\partial y} \bigg|_{x=-\frac{d}{2}} = \frac{\partial X_1}{\partial x} \cdot Y \bigg|_{x=-\frac{d}{2}} \qquad (2.243)$$

$$\mu \frac{\partial X_2}{\partial x} \cdot Y + i v X_2 \frac{\partial Y}{\partial y} \bigg|_{x=\frac{d}{2}} = \frac{\partial X_3}{\partial x} \cdot Y \bigg|_{x=\frac{d}{2}} \qquad (2.244)$$

$$X_2 \big|_{x=-\frac{d}{2}} = X_1 \big|_{x=-\frac{d}{2}} \qquad (2.245)$$

$$X_2 \big|_{x=\frac{d}{2}} = X_3 \big|_{x=\frac{d}{2}} \qquad (2.246)$$

Here, according to (2.223)–(2.227):

$$X_1 = A e^{k_x^e x} \qquad (2.247)$$

$$X_2 = A_2 e^{i k_x^i x} + B_2 e^{-i k_x^i x} \qquad (2.248)$$

$$X_3 = B e^{-k_x^e x} \qquad (2.249)$$

$$Y = e^{-i k_y y} \qquad (2.250)$$

Find the derivatives from the solution (2.247)–(2.250):

$$\frac{\partial X_1}{\partial x} = k_x^e \cdot A \cdot e^{k_x^e x} \qquad (2.251)$$

$$\frac{\partial X_2}{\partial x} = i k_x^i \cdot \left(A_2 e^{i k_x^i x} - B_2 e^{-i k_x^i x} \right) \qquad (2.252)$$

$$\frac{\partial X_3}{\partial x} = -k_x^e \cdot B \cdot e^{-k_x^e x} \qquad (2.253)$$

$$\frac{\partial Y}{\partial y} = -i k_y \cdot e^{-i k_y y} \qquad (2.254)$$

Let's introduce the notation:

$$\rho = \frac{k_x^i d}{2} \qquad (2.255)$$

$$\delta = \frac{k_x^e d}{2} \qquad (2.256)$$

The derivatives and the solution at the boundaries take the form:

$$\left.\frac{\partial X_1}{\partial x}\right|_{x=-\frac{d}{2}} = k_x^e \cdot A \cdot e^{-\delta} \qquad (2.257)$$

$$\left.\frac{\partial X_2}{\partial x}\right|_{x=-\frac{d}{2}} = i k_x^i \cdot \left(A_2 e^{-i\rho} - B_2 e^{i\rho}\right) \qquad (2.258)$$

$$\left.\frac{\partial X_2}{\partial x}\right|_{x=\frac{d}{2}} = i k_x^i \cdot \left(A_2 e^{i\rho} - B_2 e^{-i\rho}\right) \qquad (2.259)$$

$$\left.\frac{\partial X_3}{\partial x}\right|_{x=\frac{d}{2}} = -k_x^e \cdot B \cdot e^{-\delta} \qquad (2.260)$$

$$\left.X_1\right|_{x=-\frac{d}{2}} = A \cdot e^{-\delta} \qquad (2.261)$$

$$\left.X_2\right|_{x=-\frac{d}{2}} = A_2 e^{-i\rho} + B_2 e^{i\rho} \qquad (2.262)$$

$$\left.X_2\right|_{x=\frac{d}{2}} = A_2 e^{i\rho} + B_2 e^{-i\rho} \qquad (2.263)$$

$$\left.X_3\right|_{x=\frac{d}{2}} = B \cdot e^{-\delta} \qquad (2.264)$$

Substituting (2.250), (2.254), (2.257)–(2.264) into the boundary conditions (2.243)–(2.246) and dividing by $e^{-ik_y y}$, we obtain a system of equations for A, A_2, B_2, B,:

$$\mu \cdot i k_x^i \cdot \left(A_2 e^{-i\rho} - B_2 e^{i\rho}\right) + \nu k_y \cdot \left(A_2 e^{-i\rho} + B_2 e^{i\rho}\right) = k_x^e \cdot A e^{-\delta} \qquad (2.265)$$

$$\mu \cdot i k_x^i \cdot \left(A_2 e^{i\rho} - B_2 e^{-i\rho}\right) + \nu k_y \cdot \left(A_2 e^{i\rho} + B_2 e^{-i\rho}\right) = -k_x^e \cdot B e^{-\delta} \qquad (2.266)$$

$$A_2 e^{-i\rho} + B_2 e^{i\rho} = A e^{-\delta} \qquad (2.267)$$

$$A_2 e^{i\rho} + B_2 e^{-i\rho} = B e^{-\delta} \qquad (2.268)$$

Let's introduce the notation:

$$\varepsilon_1 = i\mu k_x^i + \nu k_y \qquad (2.269)$$

$$\varepsilon_2 = -i\mu k_x^i + \nu k_y \qquad (2.270)$$

With this notation, from (2.265)–(2.268) we obtain:

$$A_2 \varepsilon_1 e^{-ip} + B_2 \varepsilon_2 e^{ip} - A k_x^e e^{-\delta} = 0 \qquad (2.271)$$

$$A_2 \varepsilon_1 e^{ip} + B_2 \varepsilon_2 e^{-ip} + B k_x^e e^{-\delta} = 0 \qquad (2.272)$$

$$A_2 e^{-ip} + B_2 e^{ip} - A e^{-\delta} = 0 \qquad (2.273)$$

$$A_2 e^{ip} + B_2 e^{-ip} - B e^{-\delta} = 0 \qquad (2.274)$$

This is a system of 4 equations with 4 unknowns A, A_2, B_2, B. The condition for the existence of a nonzero solution is the equality to zero of its determinant: $D_0 = 0$, which gives the dispersion relation. This determinant is:

$$D_0 = \begin{vmatrix} \varepsilon_1 e^{-ip} & \varepsilon_2 e^{ip} & -k_x^e e^{-\delta} & 0 \\ \varepsilon_1 e^{ip} & \varepsilon_2 e^{-ip} & 0 & k_x^e e^{-\delta} \\ e^{-ip} & e^{ip} & -e^{-\delta} & 0 \\ e^{ip} & e^{-ip} & 0 & -e^{-\delta} \end{vmatrix} \qquad (2.275)$$

To expand this determinant, we multiply the third row by $-k_x^e$ and add to the first, and then multiply the fourth row by k_x^e and add to the second. As a result, we get:

$$D_0 = \begin{vmatrix} (\varepsilon_1 - k_x^e) e^{-ip} & (\varepsilon_2 - k_x^e) e^{ip} & 0 & 0 \\ (\varepsilon_1 + k_x^e) e^{ip} & (\varepsilon_2 + k_x^e) e^{-ip} & 0 & 0 \\ e^{-ip} & e^{ip} & -e^{-\delta} & 0 \\ e^{ip} & e^{-ip} & 0 & -e^{-\delta} \end{vmatrix} \qquad (2.276)$$

We expand this determinant in the fourth column, and the one that remains in the third. Considering further what should be, we get an equation of the form:

$$\begin{vmatrix} (\varepsilon_1 - k_x^e) e^{-i\rho} & (\varepsilon_2 - k_x^e) e^{i\rho} \\ (\varepsilon_1 + k_x^e) e^{i\rho} & (\varepsilon_2 + k_x^e) e^{-i\rho} \end{vmatrix} = 0 \qquad (2.277)$$

Expanding the determinant in (2.277), we obtain:

$$(\varepsilon_1 - k_x^e)(\varepsilon_2 + k_x^e) e^{-2i\rho} - (\varepsilon_1 + k_x^e)(\varepsilon_2 - k_x^e) e^{2i\rho} = 0 \qquad (2.278)$$

whence we obtain the dispersion relation in the form:

$$\frac{e^{2i\rho} - e^{-2i\rho}}{e^{2i\rho} + e^{-2i\rho}} = \frac{k_x^e (\varepsilon_2 - \varepsilon_1)}{(k_x^e)^2 - \varepsilon_1 \varepsilon_2} \qquad (2.279)$$

Here, the quantities included in (2.277) are determined through wave numbers in accordance with the formulas (duplicated here for reference):

$$\rho = \frac{k_x^i d}{2} \qquad (2.280)$$

$$\varepsilon_1 = i\mu k_x^i + \nu k_y \qquad (2.281)$$

$$\varepsilon_2 = -i\mu k_x^i + \nu k_y \qquad (2.282)$$

$$k_x^e = \sqrt{k_y^2 + k_z^2} \qquad (2.283)$$

$$k_x^i = \sqrt{-\left(k_y^2 + \frac{k_z^2}{\mu}\right)} = i \cdot \sqrt{k_y^2 + \frac{k_z^2}{\mu}} \qquad (2.284)$$

Relation (2.279) was obtained in the Cartesian coordinate system. However, for a number of problems, including those related to the propagation of magnetostatic waves in nonuniformly magnetized structures, it is more convenient to use a polar coordinate system, the transition to which will be considered in the next section.

2.4.7. Transition to the polar coordinate system

The dispersion relation (2.279) includes four wave numbers: k_x^i, k_x^e, k_y, k_z. Two of them k_x^i, k_x^e can be eliminated using the expressions (2.283) and (2.284). In this case, only k_y and k_z will enter the dispersion relation. This means that if k_y is given, for example,

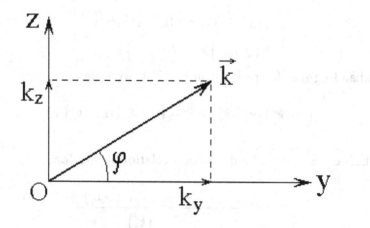

Fig. 2.4. Wave vector in the polar coordinate system

then the dispersion relation allows you to determine k_z through k_y. Such an exception can be conveniently made in the polar coordinate system shown in Figure 2.4.

Here:
$$k_y = k \cdot \cos\varphi \qquad (2.285)$$
$$k_z = k \cdot \sin\varphi \qquad (2.286)$$

In this coordinate system, wave numbers (2.283) and (2.284) take the form:
$$k_x^e = \sqrt{k_y^2 + k_z^2} = k \qquad (2.287)$$
$$k_x^i = i \cdot \sqrt{k_y^2 + \frac{k_z^2}{\mu}} = i \cdot k \cdot \sqrt{\cos^2\varphi + \frac{\sin^2\varphi}{\mu}} = ik\alpha \qquad (2.288)$$

where the notation is introduced:
$$\alpha = \sqrt{\cos^2\varphi + \frac{\sin^2\varphi}{\mu}} \qquad (2.289)$$

Let us calculate the intermediate expressions included in the dispersion relation (2.279):
$$2i\rho = -kd\alpha \qquad (2.290)$$
$$\varepsilon_2 - \varepsilon_1 = 2\mu k\alpha \qquad (2.291)$$

$$\varepsilon_1 \cdot \varepsilon_2 = k^2 \left[\left(v^2 - \mu^2 + \mu\right)\cos^2\phi - \mu \right] = k^2 \cdot \beta \tag{2.292}$$

where the notation is introduced:

$$\beta = \left(v^2 - \mu^2 + \mu\right)\cos^2\phi - \mu \tag{2.293}$$

Thus we get:

$$k_x^e \left(\varepsilon_2 - \varepsilon_1\right) = 2\mu k^2 \alpha \tag{2.294}$$

$$\left(k_x^e\right)^2 - \varepsilon_1 \varepsilon_2 = -k^2(\beta - 1) \tag{2.295}$$

Substituting (2.290), (2.294) and (2.295) in (2.279), we obtain:

$$\frac{e^{-kd\alpha} - e^{kd\alpha}}{e^{-kd\alpha} + e^{kd\alpha}} = -\frac{2\mu\alpha}{\beta - 1} \tag{2.296}$$

whence we get:

$$\frac{2\mu\alpha}{\beta - 1} = \frac{e^{kd\alpha} - e^{-kd\alpha}}{e^{kd\alpha} + e^{-kd\alpha}} \tag{2.297}$$

or:

$$\frac{2\mu\alpha}{\beta - 1} = th(kd\alpha) \tag{2.298}$$

Transforming (2.298) we obtain the basic DISPERSION RATIO for an isotropic ferrite plate with free surfaces magnetized in the plane:

$$\beta - 1 - 2\mu\,\alpha\,cth(kd\alpha) = 0 \tag{2.299}$$

Here, in accordance with (2.289), (2.293), (2.146), (2.136), (2.137), (2.139):

$$\alpha = \sqrt{\cos^2\phi + \frac{\sin^2\phi}{\mu}} \tag{2.300}$$

$$\beta = \left(v^2 - \mu^2 + \mu\right)\cos^2\phi - \mu \tag{2.301}$$

$$\mu = 1 + \frac{\Omega_H}{\Omega_H^2 - \Omega^2} \tag{2.302}$$

$$\nu = \frac{\Omega}{\Omega_H^2 - \Omega^2} \tag{2.303}$$

$$\Omega = \frac{\omega}{4\pi \gamma M_0} \tag{2.304}$$

$$\Omega_H = \frac{H_i}{4\pi M_0} \tag{2.305}$$

Dispersion relation (2.299) can be resolved relatively to k. To do this, it is convenient to use expression (2.297), namely: multiply the numerator and denominator of its right-hand side by $e^{kd\alpha}$, and then solve the resulting equation with respect to $e^{2kd\alpha}$, from where to express k using the logarithm. As a result, we obtain the dispersion relation in the form:

$$k = \frac{1}{2d\alpha} \cdot \ln \frac{\beta - 1 + 2\mu\alpha}{\beta - 1 - 2\mu\alpha} \tag{2.306}$$

2.4.8. Potentials

The system of equations for the coefficients A, A_2, B_2, B, determining the potentials ψ_1, ψ_2, ψ_3, in accordance with formulas (2.223)–(2.227) and (2.236)–(2.238), has the form (2.271)–(2.274):

$$A_2 \varepsilon_1 e^{-ip} + B_2 \varepsilon_2 e^{ip} - A k_x^e e^{-\delta} = 0 \tag{2.307}$$

$$A_2 \varepsilon_1 e^{ip} + B_2 \varepsilon_2 e^{-ip} + B k_x^e e^{-\delta} = 0 \tag{2.308}$$

$$A_2 e^{-ip} + B_2 e^{ip} - A e^{-\delta} = 0 \tag{2.309}$$

$$A_2 e^{ip} + B_2 e^{-ip} - B e^{-\delta} = 0 \tag{2.310}$$

This is a system of 4 equations with 4 unknowns, the determinant of which (2.275) is equal to zero. Therefore, three of the unknowns can be expressed through the fourth, for which it is convenient to choose the coefficient inside the plate, for example A_2. To do this, first, from (2.309) and (2.310), we express A and B:

$$A = \left(A_2 e^{-ip} + B_2 e^{ip} \right) \cdot e^{\delta} \tag{2.311}$$

$$B = \left(A_2 e^{i\rho} + B_2 e^{-i\rho}\right) \cdot e^{\delta} \tag{2.312}$$

Substituting (2.311) and (2.312) in (2.307) and (2.308), we obtain:

$$B_2 = -\frac{\varepsilon_1 - k_x^e}{\varepsilon_2 - k_x^e} \cdot e^{-2i\rho} \cdot A_2 \tag{2.313}$$

$$B_2 = -\frac{\varepsilon_1 + k_x^e}{\varepsilon_2 + k_x^e} \cdot e^{2i\rho} \cdot A_2 \tag{2.314}$$

Equating expressions (2.313) and (2.314), we obtain (2.278), which is the dispersion relation. Thus, relations (2.313) and (2.314) are equivalent, that is, for further work, you can choose any of them, based on convenience. Let us further take expression (2.314), since in it the numerator and denominator of the fraction contain summation. According to (2.287) and (2.288):

$$k_x^e = k \tag{2.315}$$

$$k_x^i = i k \alpha \tag{2.316}$$

and also, according to (2.290):

$$-2i\rho = k d \alpha \tag{2.317}$$

From (2.281)–(2.282):

$$\varepsilon_1 = i\mu k_x^i + \nu k_y \tag{2.318}$$

$$\varepsilon_2 = -i\mu k_x^i + \nu k_y \tag{2.319}$$

whence, taking into account that according to (2.285):

$$k_y = k \cdot \cos\varphi \tag{2.320}$$

taking into account (2.316), we obtain:

$$\varepsilon_1 = -\mu k\alpha + \nu k \cos\varphi \tag{2.321}$$

$$\varepsilon_2 = \mu k\alpha + \nu k \cos\phi \tag{2.322}$$

Substituting (2.315), (2.317), (2.321), (2.322) into (2.314), we obtain:

$$B_2 = \frac{\mu\alpha - \nu\cos\phi - 1}{\mu\alpha + \nu\cos\phi + 1} \cdot e^{-kd\alpha} \cdot A_2 \tag{2.323}$$

Let's introduce the notation:

$$g = \frac{\mu\alpha - v\cos\phi - 1}{\mu\alpha + v\cos\phi + 1} \qquad (2.324)$$

Moreover, from (2.232) we obtain:

$$B_2 = A_2 \cdot g \cdot e^{-kd\alpha} \qquad (2.325)$$

Substituting B_2 by (2.325) into (2.311) and (2.312), we obtain the coefficients A and B, expressed in terms of A_2:

$$A = A_2 \cdot \left(1 + g \cdot e^{-2kd\alpha}\right) \cdot e^{\frac{kd}{2}(1+\alpha)} \qquad (2.326)$$

$$B = A_2 \cdot (1+g) \cdot e^{\frac{kd}{2}(1-\alpha)} \qquad (2.327)$$

According to (2.223)–(2.225), the part of the solution that depends only on has the form:

$$X_1 = A e^{k_x^e x} \qquad (2.328)$$

$$X_2 = A_2 e^{ik_x^i x} + B_2 e^{-ik_x^i x} \qquad (2.329)$$

$$X_3 = B e^{-k_x^e x} \qquad (2.330)$$

whence, substituting (2.325)–(2.327), taking into account (2.329) and (2.330), we obtain:

$$X_1 = A_2 \cdot \left(1 + g \cdot e^{-2kd\alpha}\right) \cdot e^{\frac{kd}{2}(1+\alpha)} \cdot e^{kx} \qquad (2.331)$$

$$X_2 = A_2 \cdot \left[e^{-k\alpha x} + g \cdot e^{-k\alpha d} \cdot e^{k\alpha x}\right] \qquad (2.332)$$

$$X_3 = A_2 \cdot (1+g) \cdot e^{\frac{kd}{2}(1-\alpha)} \cdot e^{-kx} \qquad (2.333)$$

Taking into account (2.285) and (2.286):

$$k_y = k \cdot \cos\phi \qquad (2.334)$$

$$k_z = k \cdot \sin\phi \qquad (2.335)$$

the part of the solution depending only on y and z (2.226)–(2.227) takes the form:

$$Y = e^{-i(k \cdot \cos\phi) y} \qquad (2.336)$$

$$Z = e^{-i(k \cdot \sin\phi) z} \qquad (2.337)$$

Assuming, we get potentials in the form:

$$\psi_1 = \left(1 + g \cdot e^{-2kd\alpha}\right) \cdot e^{\frac{kd}{2}(1+\alpha)} \cdot e^{kx} \cdot e^{-i(k\cdot\cos\phi)y} \cdot e^{-i(k\cdot\sin\phi)z} \quad (2.338)$$

$$\psi_2 = \left[e^{-k\alpha x} + g \cdot e^{-k\alpha d} \cdot e^{k\alpha x}\right] \cdot e^{-i(k\cdot\cos\phi)y} \cdot e^{-i(k\cdot\sin\phi)z} \quad (2.339)$$

$$\psi_3 = (1+g) \cdot e^{\frac{kd}{2}(1-\alpha)} \cdot e^{-kx} \cdot e^{-i(k\cdot\cos\phi)y} \cdot e^{-i(k\cdot\sin\phi)z} \quad (2.340)$$

where and are determined by expressions (2.289) and (2.324):

$$\alpha = \sqrt{\cos^2\phi + \frac{\sin^2\phi}{\mu}} \quad (2.341)$$

$$g = \frac{\mu\alpha - \nu\cos\phi - 1}{\mu\alpha + \nu\cos\phi + 1} \quad (2.342)$$

By direct substitution, it is easy to verify that (2.339) satisfies the Walker equation (2.147), and (2.338) and (2.340) satisfy the Laplace equation (2.148).

For the convenience of further notation, we introduce the notation:

$$\psi_{yz} = e^{-i(k\cdot\cos\phi)y} \cdot e^{-i(k\cdot\sin\phi)z} \quad (2.343)$$

In this case, the potentials (2.338)–(2.340) take the form:

$$\psi_1 = \left(1 + g \cdot e^{-2kd\alpha}\right) \cdot e^{\frac{kd}{2}(1+\alpha)} \cdot e^{kx} \cdot \psi_{yz} \quad (2.344)$$

$$\psi_2 = \left[e^{-k\alpha x} + g \cdot e^{-k\alpha d} \cdot e^{k\alpha x}\right] \cdot \psi_{yz} \quad (2.345)$$

$$\vec{h} = \text{grad}\psi = \left\{\frac{\partial \psi}{\partial x}; \frac{\partial \psi}{\partial y}; \frac{\partial \psi}{\partial z}\right\} \quad (2.346)$$

2.4.9. Fields

Let us now find the components of the wave fields in all three regions in accordance with the formula (2.123):

$$\vec{h} = \text{grad}\psi = \left\{\frac{\partial \psi}{\partial x}; \frac{\partial \psi}{\partial y}; \frac{\partial \psi}{\partial z}\right\} \quad (2.347)$$

To simplify the notation, we will use expressions (2.344)–(2.346):

in area 1:

$$h_{1x} = k \cdot \left(1 + g \cdot e^{-2kd\alpha}\right) \cdot e^{\frac{kd}{2}(1+\alpha)} \cdot e^{kx} \cdot \psi_{yz} \qquad (2.348)$$

$$h_{1y} = -ik\cos\phi \cdot \left(1 + g \cdot e^{-2kd\alpha}\right) \cdot e^{\frac{kd}{2}(1+\alpha)} \cdot e^{kx} \cdot \psi_{yz} \qquad (2.349)$$

$$h_{1z} = -ik\sin\phi \cdot \left(1 + g \cdot e^{-2kd\alpha}\right) \cdot e^{\frac{kd}{2}(1+\alpha)} \cdot e^{kx} \cdot \psi_{yz} \qquad (2.350)$$

in area 2:

$$h_{2x} = -k\alpha \cdot \left[e^{-k\alpha x} - g \cdot e^{-k\alpha d} \cdot e^{k\alpha x}\right] \cdot \psi_{yz} \qquad (2.351)$$

$$h_{2y} = -ik\cos\phi \cdot \left[e^{-k\alpha x} + g \cdot e^{-k\alpha d} \cdot e^{k\alpha x}\right] \cdot \psi_{yz} \qquad (2.352)$$

$$h_{2z} = -ik\sin\phi \cdot \left[e^{-k\alpha x} + g \cdot e^{-k\alpha d} \cdot e^{k\alpha x}\right] \cdot \psi_{yz} \qquad (2.353)$$

in area 3:

$$h_{1x} = -k \cdot (1+g) \cdot e^{\frac{kd}{2}(1+\alpha)} \cdot e^{-kx} \cdot \psi_{yz} \qquad (2.354)$$

$$h_{1y} = -ik\cos\phi \cdot (1+g) \cdot e^{\frac{kd}{2}(1+\alpha)} \cdot e^{-kx} \cdot \psi_{yz} \qquad (2.355)$$

$$h_{1z} = -ik\sin\phi \cdot (1+g) \cdot e^{\frac{kd}{2}(1+\alpha)} \cdot e^{-kx} \cdot \psi_{yz} \qquad (2.356)$$

2.4.10. Magnetization

Let us now find the magnetization. Outside the magnetic medium in the regions 1 and 3, the magnetizations are zero. Inside the magnetic medium in region 2, the magnetizations are determined by relations (2.144) and (2.145):

$$4\pi m_x = \kappa \frac{\partial \psi}{\partial x} + iv \frac{\partial \psi}{\partial y} \qquad (2.357)$$

$$4\pi m_y = -iv \frac{\partial \psi}{\partial x} + \kappa \frac{\partial \psi}{\partial y} \qquad (2.358)$$

from which, taking into account (2.146), we obtain:

$$m_x = \frac{\mu - 1}{4\pi} \cdot \frac{\partial \psi}{\partial x} + \frac{iv}{4\pi} \cdot \frac{\partial \psi}{\partial y} \qquad (2.359)$$

$$m_y = -\frac{i\nu}{4\pi} \cdot \frac{\partial \psi}{\partial x} + \frac{\mu-1}{4\pi} \cdot \frac{\partial \psi}{\partial y} \qquad (2.360)$$

In this case, in the state of saturation, the variable magnetization is zero.

Differentiating the potential (2.345), we obtain:

$$m_{2x} = \frac{k}{4\pi} \cdot \Big\{ \big[-(\mu-1)\alpha + \nu\cos\phi \big] \cdot e^{-k\alpha x} +$$
$$+ g \cdot \big[(\mu-1)\alpha + \nu\cos\phi \big] \cdot e^{-k\alpha d} \cdot e^{k\alpha x} \Big\} \cdot \psi_{yz} \qquad (2.361)$$

$$m_{2y} = i\frac{k}{4\pi} \cdot \Big\{ \big[-(\mu-1)\cos\phi + \nu\alpha \big] \cdot e^{-k\alpha x} -$$
$$- g \cdot \big[(\mu-1)\cos\phi + \nu\alpha \big] \cdot e^{-k\alpha d} \cdot e^{k\alpha x} \Big\} \cdot \psi_{yz} \qquad (2.362)$$

where according to (2.343), (2.289), (2.324), (2.146), (2.139), (2.136), (2.137):

$$\psi_{yz} = e^{-i(k\cos\phi)y} \cdot e^{-i(k\sin\phi)z} \qquad (2.363)$$

$$\alpha = \sqrt{\cos^2\phi + \frac{\sin^2\phi}{\mu}} \qquad (2.364)$$

$$g = \frac{\mu\alpha - \nu\cos\phi - 1}{\mu\alpha + \nu\cos\phi + 1} \qquad (2.365)$$

$$\mu = 1 + \frac{\Omega_H}{\Omega_H^2 - \Omega^2} \qquad (2.366)$$

$$\nu = \frac{\Omega}{\Omega_H^2 - \Omega^2} \qquad (2.367)$$

$$\Omega = \frac{\omega}{4\pi\gamma M_0} \qquad (2.368)$$

$$\Omega_H = \frac{H_i}{4\pi M_0} \qquad (2.369)$$

and the wave number is determined from the dispersion relation (2.299):

$$\beta - 1 - 2\mu\,\alpha\,cth(k\,d\,\alpha) = 0 \qquad (2.370)$$

or, according to (2.306):

$$k = \frac{1}{2d\alpha} \cdot \ln\frac{\beta - 1 + 2\mu\,\alpha}{\beta - 1 - 2\mu\,\alpha} \qquad (2.371)$$

where in accordance with (2.301):

$$\beta = (v^2 - \mu^2 + \mu)\cos^2\phi - \mu \qquad (2.372)$$

2.4.11. Cutoff angle for the Damon–Eshbach ratio

As noted in Section 2.4, the propagation of surface magnetostatic waves is possible only in a certain range of angles between the directions of the wave vector and the applied constant field. The extreme values of this range of angles are called 'cutoff angles'. Due to the symmetry of the dispersion relation with respect to the normal to the direction of the field, the positive and negative values of the cutoff angle are equal in absolute value to each other; therefore, it is sufficient to determine only one of these values, namely the positive one. Let's proceed to this definition.

The cutoff angle is found from the dispersion relation in the polar coordinate system (2.299):

$$\beta - 1 - 2\mu\,\alpha\,cth(k\,d\,\alpha) = 0 \qquad (2.373)$$

where:

$$\alpha = \sqrt{\cos^2\phi + \frac{\sin^2\phi}{\mu}} \qquad (2.374)$$

$$\beta = (v^2 - \mu^2 + \mu)\cos^2\phi - \mu \qquad (2.375)$$

$$\mu = 1 + \frac{\Omega_H}{\Omega_H^2 - \Omega^2} \qquad (2.376)$$

$$v = \frac{\Omega}{\Omega_H^2 - \Omega^2} \qquad (2.377)$$

and Ω and Ω_H represent the normalized frequencies:

$$\Omega = \frac{\omega}{4\pi\gamma M_0} \tag{2.378}$$

$$\Omega_H = \frac{H_0}{4\pi M_0} \tag{2.379}$$

linking frequency and field with material parameters (magnetization). We will assume that the cutoff corresponds to the wave number tending to infinity, that is. Moreover, so that (3.373) takes the form:

$$\beta - 1 - 2\mu\alpha = 0 \tag{2.380}$$

Substituting (2.375) and (2.374) into this expression, we obtain:

$$\left(v^2 - \mu^2 + \mu\right)\cos^2\phi - \mu - 1 - 2\mu\sqrt{\cos^2\phi + \frac{\sin^2\phi}{\mu}} = 0 \tag{2.381}$$

Expressing the sine in the last term through the cosine and transferring the root to the right side, we get:

$$\left(v^2 - \mu^2 + \mu\right)\cos^2\phi - (\mu+1) = 2\sqrt{\mu\left[1+(\mu-1)\cos^2\phi\right]} = 0 \tag{2.382}$$

By squaring and quoting similar terms, we get:

$$\left(v^2 - \mu^2 + \mu\right)^2 \cos^4\phi -$$
$$-2\left[(\mu+1)\left(v^2 - \mu^2 + \mu\right) + 2\mu(\mu-1)\right]\cdot\cos^2\phi + (\mu-1)^2 = 0 \tag{2.383}$$

Let us introduce auxiliary notation:

$$A = \left(v^2 - \mu^2 + \mu\right)^2 \tag{2.384}$$

$$B = -2\left[(\mu+1)\left(v^2 - \mu^2 + \mu\right) + 2\mu(\mu-1)\right] \tag{2.385}$$

$$C = (\mu-1)^2 \tag{2.386}$$

With these notations (2.383) takes the form:

$$A\cdot\cos^4\phi + B\cdot\cos^2\mu + C = 0 \tag{2.387}$$

This is a biquadratic equation for $\cos\phi$. Its solution (at 90°) has the form:

$$\cos\phi = \sqrt{\frac{-B+\sqrt{B^2-4AC}}{2A}} \quad (2.388)$$

whence we get the cutoff angle in the form:

$$\phi_c = \pm\arccos\left(\sqrt{\frac{-B+\sqrt{B^2-4AC}}{2A}}\right) \quad (2.389)$$

The value of the cutoff angle at specific values of the field and material parameters is discussed in more detail below.

2.4.12. Damon–Eshbach dispersion relation in the Cartesian coordinate system

In the previous section, the dispersion relation was obtained in a cylindrical coordinate system. However, for some tasks it is desirable to have this relationship in Cartesian coordinates. In this case, the expression (2.279) obtained in the same place is not quite convenient due to the fact that it includes the wave numbers k_x^i and k_x^e. These wave numbers can be eliminated using the expressions (2.280)–(2.284), which will be done below.

So, we proceed from the Damon–Eshbach dispersion relation written in the form (2.279):

where:
$$\frac{e^{i2\rho}-e^{-i2\rho}}{e^{i2\rho}+e^{-i2\rho}} = \frac{k_x^e(\xi_2-\xi_1)}{(k_x^e)^2-\xi_1\xi_2} \quad (2.390)$$

$$\rho = k_x^i \cdot \frac{d}{2} \quad (2.391)$$

$$k_x^i = ik\alpha \quad (2.392)$$

$$k_x^e = k \quad (2.393)$$

$$\xi_1 = i\mu k_x^i + \nu k_y \quad (2.394)$$

$$\xi_2 = -i\mu k_x^i + \nu k_y \quad (2.395)$$

moreover:

$$k = \sqrt{k_y^2 + k_z^2} \quad (2.396)$$

and the parameter, being introduced in the polar coordinate system

in the form:
$$\alpha = \sqrt{\cos^2\varphi + \frac{\sin^2\varphi}{\mu}} \qquad (2.397)$$

in the Cartesian system takes the form:
$$\alpha = \sqrt{\frac{\mu k_y^2 + k_z^2}{\mu\left(k_y^2 + k_z^2\right)}} \qquad (2.398)$$

where it is taken into account that:
$$k_y = k\cdot\cos\varphi \qquad (2.399)$$
$$k_z = k\cdot\sin\varphi \qquad (2.400)$$

i.e:
$$\cos\phi = \frac{k_y}{k} \qquad (2.401)$$
$$\sin\phi = \frac{k_z}{k} \qquad (2.402)$$

as well as:
$$\phi = \mathrm{arctg}\left(\frac{k_z}{k_y}\right) \qquad (2.403)$$

Let us find the intermediate relations included in (2.390):

$$i2\rho = -kd\alpha = -d\cdot\sqrt{\frac{\mu k_y^2 + k_z^2}{\mu}} \qquad (2.404)$$

$$\xi_2 - \xi_1 = 2\sqrt{\mu\cdot\left(\mu k_y^2 + k_z^2\right)} \qquad (2.405)$$

$$\xi_1\xi_2 = \left(v^2 - \mu^2\right)k_y^2 - \mu k_z^2 \qquad (2.406)$$

In this case, we obtain the numerator and denominator of the fraction on the right-hand side of (2.390) in the form:

$$k_x^e\left(\xi_2 - \xi_1\right) = 2\sqrt{\left(\mu k_y^2 + k_z^2\right)\cdot\left(k_y^2 + k_z^2\right)} \qquad (2.407)$$

$$\left(k_x^e\right)^2 - \xi_1\xi_2 = -\left(v^2 - \mu^2 - 1\right)k_y^2 + \left(1 + \mu\right)k_z^2 \qquad (2.408)$$

Substituting (2.404), (2.407) and (2.408) into (2.390) and transferring everything to the left side, we obtain the desired dispersion relation in the form:

$$\frac{2\sqrt{\left(\mu k_y^2 + k_z^2\right)\cdot\left(k_y^2 + k_z^2\right)}}{\left(v^2 - \mu^2 - 1\right)k_y^2 - \left(1+\mu\right)k_z^2} - \text{th}\left\{d\sqrt{\frac{\mu k_y^2 + k_z^2}{\mu}}\right\} = 0 \quad (2.409)$$

where, as before:

$$\mu = 1 + \frac{\Omega_H}{\Omega_H^2 - \Omega^2} \quad (2.410)$$

$$v = \frac{\Omega}{\Omega_H^2 - \Omega^2} \quad (2.411)$$

Ω and Ω_H represent the normalized frequencies:

$$\Omega = \frac{\omega}{4\pi\gamma M_0} \quad (2.412)$$

$$\Omega_H = \frac{H_i}{4\pi M_0} \quad (2.413)$$

linking frequency and field with material parameters (magnetization). Let us now bring the obtained relation to a form containing only and. For this, we first write (2.379) in a more convenient form:

$$\alpha = \sqrt{\frac{(\mu-1)\cos^2\phi + 1}{\mu}} \quad (2.414)$$

Using this expression, we find the exponent:

$$i2\rho = -k\,d\,\vartheta = -k\,d\cdot\sqrt{\frac{(\mu-1)\cos^2\phi + 1}{\mu}} \quad (2.415)$$

Further, taking into account (2.393)–(2.395), we obtain:

$$k_x^e\left(\xi_2 - \xi_1\right) = 2k^2\sqrt{\mu\left[(\mu-1)\cos^2\phi + 1\right]} \quad (2.416)$$

$$\left(k_x^e\right)^2 - \xi_1\xi_2 = -k^2\left[\left(v^2 - \mu^2 + \mu\right)\cos^2\phi - (1+\mu)\right] \quad (2.417)$$

Substituting all the obtained expressions in (2.390), cancelling the left side by and transferring everything to the left side, we obtain the dispersion relation in the form:

$$\frac{2\sqrt{\mu\left[(\mu-1)\cos^2\phi + 1\right]}}{\left(v^2 - \mu^2 + \mu\right)\cos^2\phi - (1+\mu)} - \text{th}\left\{k\,d\sqrt{\frac{(\mu-1)\cos^2\phi + 1}{\mu}}\right\} = 0 \quad (2.418)$$

It can be seen that, up to the notation, this relationship turns into the classic Damon–Eshbach relationship written in the polar coordinate system:

$$\beta - 1 - 2\mu\alpha \, cth \, (k \, d\alpha) = 0 \qquad (2.419)$$

where:

$$\alpha = \sqrt{\cos^2\phi + \frac{\sin^2\phi}{\mu}} \qquad (2.420)$$

$$\beta = (\nu^2 - \mu^2 + \mu)\cos^2\phi - \mu \qquad (2.421)$$

2.5. Dispersion ratio for metal–dielectric–ferrite–metal (MDFDM) structure and its particular cases

In a number of cases, the ferrite plate (film) considered in the previous sections is used as an integral part of more complex structures. As a fairly general example, consider a structure consisting of a plane-parallel ferrite plate enclosed between two metal planes and separated from both of these planes by dielectric layers. In the magnetostatic approximation, there are no electric fields in the wave; therefore, the dielectric constant of the dielectric does not participate in the formation of the propagating wave, that is, the dielectric can be replaced simply by a vacuum gap with a permeability equal to unity. So it is most logical to call the structure under consideration

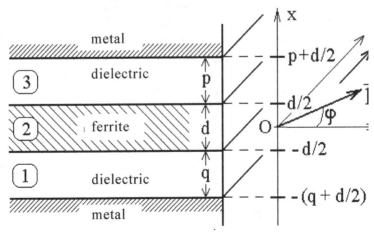

Fig. 2.5. Diagram of the structure of MDFDM.

'metal-gap–ferrite–gap–metal" (MGFGM), however, due to the established tradition, we will call it with the preservation of the term 'dielectric', that is, 'metal–dielectric–ferrite–dielectric-metal' (MDFDM).

2.5.1. General derivation of the dispersion relation

A diagram of the structure and the associated coordinate system is shown in Fig. 2.5.

Similarly to the Damon–Eshbach problem, we introduce potentials in all three media, for which we obtain the following equations: medium 1 is a dielectric, Laplace's equation:

$$\frac{\partial^2 \psi_1}{\partial x^2} + \frac{\partial^2 \psi_1}{\partial y^2} + \frac{\partial^2 \psi_1}{\partial z^2} = 0 \tag{2.422}$$

medium 2 - ferrite, Walker's equation:

$$(1+\kappa)\left(\frac{\partial^2 \psi_2}{\partial x^2} + \frac{\partial^2 \psi_2}{\partial y^2}\right) + \frac{\partial^2 \psi_2}{\partial z^2} = 0 \tag{2.423}$$

medium 3 – Laplace's equation:

$$\frac{\partial^2 \psi_3}{\partial x^2} + \frac{\partial^2 \psi_3}{\partial y^2} + \frac{\partial^2 \psi_3}{\partial z^2} = 0 \tag{2.424}$$

Boundary conditions on surfaces – continuity of tangent field components and normal induction components. Similarly to the Damon–Eshbach problem, we assume that the potentials on the boundaries at are equal. We will also assume that the magnetic field does not penetrate into the metal, the derivative of the potential along the coordinate (determining the field) is equal to zero. As a result, we obtain the boundary conditions in the form:

$$\left.(1+\kappa)\frac{\partial \psi_2}{\partial x} + i\nu\frac{\partial \psi_2}{\partial y}\right|_{x=-d/2} = \left.\frac{\partial \psi_1}{\partial x}\right|_{x=-d/2} \tag{2.425}$$

$$\left.(1+\kappa)\frac{\partial \psi_2}{\partial x} + i\nu\frac{\partial \psi_2}{\partial y}\right|_{x=d/2} = \left.\frac{\partial \psi_3}{\partial x}\right|_{x=-d/2} \tag{2.426}$$

$$\left.\psi_1\right|_{x=-d/2} = \left.\psi_2\right|_{x=-d/2} \tag{2.427}$$

$$\psi_2\big|_{x=d/2} = \psi_3\big|_{x=d/2} \tag{2.428}$$

$$\frac{\partial \psi_1}{\partial x}\bigg|_{x=-(q+d/2)} = 0 \tag{2.429}$$

$$\frac{\partial \psi_3}{\partial x}\bigg|_{x=p+d/2} = 0 \tag{2.430}$$

In these formulas, the parameters κ and ν are determined by the expressions:

$$\kappa = \frac{\Omega_H}{\Omega_H^2 - \Omega^2} \tag{2.431}$$

$$\nu = \frac{\Omega}{\Omega_H^2 - \Omega^2} \tag{2.432}$$

where:

$$\Omega = \frac{\omega}{\omega_M} = \frac{\omega}{4\pi\gamma M_0} \tag{2.433}$$

$$\Omega_H = \frac{\omega_H}{\omega_M} = \frac{H_i}{4\pi M_0} \tag{2.434}$$

and:

$$\omega_H = \gamma H_i \tag{2.435}$$

$$\omega_M = 4\pi\gamma M_0 \tag{2.436}$$

Equations (2.422)–(2.424) in all three areas are solved by the method of separation of variables. Assuming that the waves propagate in the plane Oyz, so that the dependences on y and z in all three regions are the same, similarly to the Damon–Eshbach problem, we obtain a solution in the form:

$$X_1 = A_1 \cdot e^{\left(\sqrt{\mu^2 + \frac{\kappa}{1+\kappa}\lambda^2}\right)\cdot x} + B_1 \cdot e^{-\left(\sqrt{\mu^2 + \frac{\kappa}{1+\kappa}\lambda^2}\right)\cdot x} \tag{2.437}$$

$$Y_1 = C_1 \cdot e^{\left(i\sqrt{\mu^2 - \frac{\lambda^2}{1+\kappa}}\right) \cdot y} + D_1 \cdot e^{-\left(i\sqrt{\mu^2 - \frac{\lambda^2}{1+\kappa}}\right) \cdot y} \tag{2.438}$$

$$Z_1 = G_1 \cdot e^{i\lambda z} + H_1 \cdot e^{-i\lambda z} \tag{2.439}$$

$$X_2 = A_2 \cdot e^{\mu x} + B_2 \cdot e^{-\mu x} \tag{2.440}$$

$$Y_2 = C_2 \cdot e^{\left(i\sqrt{\mu^2 - \frac{\lambda^2}{1+\kappa}}\right) \cdot y} + D_2 \cdot e^{-\left(i\sqrt{\mu^2 - \frac{\lambda^2}{1+\kappa}}\right) \cdot y} \tag{2.441}$$

$$Z_2 = G_2 \cdot e^{i\lambda z} + H_2 \cdot e^{-i\lambda z} \tag{2.442}$$

$$X_3 = A_3 \cdot e^{\left(\sqrt{\mu^2 + \frac{\kappa}{1+\kappa}\lambda^2}\right) \cdot x} + B_3 \cdot e^{-\left(\sqrt{\mu^2 + \frac{\kappa}{1+\kappa}\lambda^2}\right) \cdot x} \tag{2.443}$$

$$Y_3 = C_3 \cdot e^{\left(i\sqrt{\mu^2 - \frac{\lambda^2}{1+\kappa}}\right) \cdot y} + D_3 \cdot e^{-\left(i\sqrt{\mu^2 - \frac{\lambda^2}{1+\kappa}}\right) \cdot y} \tag{2.444}$$

$$Z_3 = G_3 \cdot e^{i\lambda z} + H_3 \cdot e^{-i\lambda z} \tag{2.445}$$

Here μ and λ are permanent divisions.

Let's introduce the notation:

$$k_x^e = \sqrt{\mu^2 + \frac{\kappa}{1+\kappa}\lambda^2} \tag{2.446}$$

$$k_x^i = i\mu \tag{2.447}$$

$$k_y = \sqrt{\mu^2 - \frac{\lambda^2}{1+\kappa}} \tag{2.448}$$

$$k_z = \lambda \tag{2.449}$$

excluding from which μ and λ, we obtain auxiliary relations between the components of the wave numbers:

$$(1+\kappa)\left[(k_x^i)^2 + k_y^2\right] + k_z^2 = 0 \tag{2.450}$$

$$\left(k_x^e\right)^2 - k_y^2 - k_z^2 = 0 \qquad (2.451)$$

For simplicity of analytical calculations, at this stage, we will consider waves only with positive exponents in Y_i and Z_i, that is, at $D_{1,2,3} = 0$ and $H_{1,2,3} = 0$, and also assume that $C_{1,2,3} = 1$ and $G_{1,2,3} = 1$. In this case, writing down the exponents through k_x^i and k_x^e, we get the potentials in the form:

$$\psi_1 = X_1 \cdot Y \cdot Z \qquad (2.452)$$

$$\psi_2 = X_2 \cdot Y \cdot Z \qquad (2.453)$$

$$\psi_2 = X_2 \cdot Y \cdot Z \qquad (2.454)$$

where:

$$X_1 = A_1 \cdot e^{k_x^e x} + B_1 \cdot e^{-k_x^e x} \qquad (2.455)$$

$$X_2 = B_2 \cdot e^{i k_x^i x} + A_2 \cdot e^{-i k_x^i x} \qquad (2.456)$$

$$X_3 = A_3 \cdot e^{k_x^e x} + B_3 \cdot e^{-k_x^e x} \qquad (2.457)$$

and:

$$Y = e^{i k_y y} \qquad (2.458)$$

$$Z = e^{i k_z z} \qquad (2.459)$$

The boundary conditions take the form:

$$\left[(1+\kappa)\frac{\partial X_2}{\partial x} - \frac{\partial X_1}{\partial X}\right] \cdot Y + i \nu X_2 \frac{\partial Y}{\partial y} \bigg|_{x=-d/2} = 0 \qquad (2.460)$$

$$\left[(1+\kappa)\frac{\partial X_2}{\partial x} - \frac{\partial X_3}{\partial X}\right] \cdot Y + i \nu X_2 \frac{\partial Y}{\partial y} \bigg|_{x=d/2} = 0 \qquad (2.461)$$

$$X_2 - X_1\big|_{x=-d/2} = 0 \tag{2.462}$$

$$X_2 - X_3\big|_{x=d/2} = 0 \tag{2.463}$$

$$\frac{\partial X_1}{\partial x}\bigg|_{x=-(q+d/2)} = 0 \tag{2.464}$$

$$\frac{\partial X_3}{\partial x}\bigg|_{x=p+d/2} = 0 \tag{2.465}$$

Let us introduce auxiliary notation:

$$\beta = \frac{k_x^i d}{2} \tag{2.466}$$

$$\delta = \frac{k_x^e d}{2} \tag{2.467}$$

$$\xi = k_x^e (q + d/2) \tag{2.468}$$

$$\eta = k_x^e (p + d/2) \tag{2.469}$$

$$\varepsilon_1 = i k_x^i (1+\kappa) - \nu k_y \tag{2.470}$$

$$\varepsilon_2 = -\left[i k_x^i (1+\kappa) + \nu k_y \right] \tag{2.471}$$

Substituting the solution into the boundary conditions, we obtain the system of equations:

$$-A_1 k_x^e e^{-\delta} + B_1 k_x^e e^{\delta} + A_2 \varepsilon_2 e^{i\beta} + B_2 \varepsilon_1 e^{-i\beta} = 0 \tag{2.472}$$

$$A_2 \varepsilon_2 e^{-i\beta} + B_2 \varepsilon_1 e^{i\beta} - A_3 k_x^e e^{\delta} + B_3 k_x^e e^{-\delta} = 0 \tag{2.473}$$

$$A_1 e^{-\delta} - B_1 e^{\delta} + A_2 e^{i\beta} + B_2 e^{-i\beta} = 0 \tag{2.474}$$

$$A_2 e^{-i\beta} + B_2 e^{i\beta} - A_3 e^{\delta} - B_3 e^{-\delta} = 0 \tag{2.475}$$

$$A_1 e^{-\xi} - B_1 e^{\xi} = 0 \qquad (2.476)$$

$$A_3 e^{\eta} - B_3 e^{-\eta} = 0 \qquad (2.477)$$

This is a system of six equations with six unknowns A_1, B_1, A_2, B_2, A_3, B_3. The condition for a nonzero solution of this system is the equality to zero of its determinant:

$$\begin{vmatrix} -k_x^e e^{-\delta} & k_x^e e^{\delta} & \varepsilon_2 e^{i\beta} & \varepsilon_2 e^{-i\beta} & 0 & 0 \\ 0 & 0 & \varepsilon_2 e^{-i\beta} & \varepsilon_1 e^{i\beta} & -k_x^e e^{\delta} & k_x^e e^{-\delta} \\ e^{-\delta} & -e^{\delta} & e^{i\beta} & e^{-i\beta} & 0 & 0 \\ 0 & 0 & e^{-i\beta} & e^{i\beta} & -e^{\delta} & -e^{-\delta} \\ e^{-\xi} & -e^{\xi} & 0 & 0 & 0 & 0 \\ 0 & 0 & 0 & 0 & e^{\eta} & -e^{-\eta} \end{vmatrix} = 0 \qquad (2.478)$$

From this expression, taking into account the definition of hyperbolic functions, we obtain the dispersion relation in the form:

$$\operatorname{ch}\left[k_x^e(p-q)\right] \cdot \left[\left(k_x^e\right)^2 + \varepsilon_1 \varepsilon_2\right] +$$

$$+ \operatorname{sh}\left[k_x^e(p-q)\right] \cdot \left(k_x^e\right) \cdot (\varepsilon_1 + \varepsilon_2) -$$

$$- \operatorname{ch}\left[k_x^e(p+q)\right] \cdot \left[\left(k_x^e\right)^2 - \varepsilon_1 \varepsilon_2\right] -$$

$$- \operatorname{sh}\left[k_x^e(p+q)\right] \cdot \left(k_x^e\right) \cdot (\varepsilon_1 - \varepsilon_2) \cdot \operatorname{cth}(2i\beta) = 0 \qquad (2.479)$$

This dispersion relation connects and, moreover, the expressions included in it, containing $\varepsilon_{1,2}$, taking into account (2.470) and (2.471), are determined through k_y and k_z as follows:

$$\varepsilon_1 \varepsilon_2 = \left(k_x^i\right)^2 (1+\kappa)^2 + v^2 k_y^2 \qquad (2.480)$$

$$\varepsilon_1 + \varepsilon_2 = -2\nu k_y \tag{2.481}$$

$$\varepsilon_1 - \varepsilon_2 = 2i k_x^i (1+\kappa) \tag{2.482}$$

For practice, it is of interest to find the dispersion relation connecting k_y and k_z, since it is these quantities that are available for determination in experiment by probing the wave field in the plane of the structure Oyz.

In order to express both k_x^i and k_x^e in terms of k_y and k_z, we use the relations (2.450) and (2.451) obtained above, from which we obtain:

$$k_x^i = i\sqrt{k_y^2 + \frac{k_z^2}{1+\kappa}} \tag{2.483}$$

$$k_x^e = \sqrt{k_y^2 + k_z^2} \tag{2.484}$$

Substituting these expressions in (2.479) taking into account (2.480)-(2.482) and bringing similar terms, we obtain the dispersion relation in the form:

$$\operatorname{ch}\left[(p-q)\cdot\sqrt{k_y^2 + k_z^2}\right]\cdot\left[(\nu^2 - \kappa^2 - 2\kappa)k_y^2 - \kappa k_z^2\right] -$$

$$-\operatorname{ch}\left[(p+q)\cdot\sqrt{k_y^2 + k_z^2}\right]\cdot\left[(-\nu^2 + \kappa^2 + 2\kappa + 2)k_y^2 + (2+\kappa)k_z^2\right] \times$$

$$\times \operatorname{cth}\left(d\sqrt{k_y^2 + \frac{k_z^2}{1+\kappa}}\right) = 0 \tag{2.485}$$

This is the dispersion relation for the MDFDM structure in the Cartesian coordinate system. For given values of the field and magnetization, it connects the components of the wave number and with the frequency included in and in accordance with expressions (2.431)–(2.436). Note, however, that its form is associated with a certain direction of propagation of the wave phase front, so that it is not general enough. Let us consider the generalization of this relation to an arbitrary direction of propagation of the phase front.

2.5.2. Dispersion relation for an arbitrary direction of propagation of the phase front

When deriving the dispersion relation (2.485), for simplicity of calculations in expressions (2.438), (2.439), (2.441), (2.442), (2.444), (2.445), only terms with positive exponents were left. This means that only waves of the form were considered: $e^{i(\omega t + k_y y)}$ and $e^{i(\omega t + k_z z)}$. At positive wavenumbers and the phase fronts of such waves propagate in the negative directions of the Oy and Oz axes.

To introduce into consideration the waves, the phase fronts of which propagate in the positive directions of the same axes, while maintaining the positivity k_y and k_z the exponents must have the form: $e^{i(\omega t - k_y y)}$ and $e^{i(\omega t - k_z z)}$. The same can be achieved without changing the signs in the exponents, but assuming k_y and k_z negative ones. If we wish to preserve the positiveness of k_y and k_z, then in expression (64) we must change the signs in front of these wave numbers.

From the structure of expression (2.485), one can see that it is symmetric about the axis, since everywhere it enters only in the square, so that when the sign changes, nothing changes. At the same time, this expression with respect to the axis is asymmetric, since the second term is included in the first power, so that when the sign changes, this term changes sign.

Thus, in order to obtain the dispersion relation for waves whose phase fronts propagate in the positive direction of the Oz xis, nothing needs to be changed in expression (2.485), and to obtain the dispersion relation for waves whose phase fronts propagate in the positive direction of the Oy axis under the condition of positiveness of k_y, it is necessary in expression (2.485) change the sign in front of its second term.

In practical use, the condition of positivity k_y can be realized by taking the module $|k_y|$. In this case, it is convenient to introduce an auxiliary parameter S such that $S = 1$ corresponds to the propagation of the phase front of the wave in the positive direction of the Oy axis, and $S = 1$ corresponds to the propagation of the phase front of the wave in the negative direction of the same axis.

As a result, the dispersion relation (2.485) takes the form:

$$\text{ch}\left[(p-q)\cdot\sqrt{k_y^2 + k_z^2}\right]\cdot\left[(v^2 - \kappa^2 - 2\kappa)k_y^2 - \kappa k_z^2\right] +$$

$$+2 \cdot S \cdot \operatorname{sh}\left[(p-q) \cdot \sqrt{k_y^2 + k_z^2}\right] \cdot \nu \left|k_y\right| \sqrt{k_y^2 + k_z^2} -$$
$$-\operatorname{ch}\left[(p+q) \cdot \sqrt{k_y^2 + k_z^2}\right] \cdot \left[\left(-\nu^2 + \kappa^2 + 2\kappa + 2\right) k_y^2 + (2+\kappa)k_z^2\right] -$$
$$-2\operatorname{sh}\left[(p+q) \cdot \sqrt{k_y^2 + k_z^2}\right] \cdot \sqrt{(1+\kappa)\left(k_y^2 + k_z^2\right)\left[(1+\kappa)k_y^2 + k_z^2\right]} \times$$
(2.486)

For $k_y > 0$ and $S = 1$, this relation describes waves whose phase fronts propagate in the positive direction of the Oy axis, for $k_y < 0$ and $S = -1$, in the negative direction.

Thus, this relationship describes the dispersion of magnetostatic waves in the structure of an MDFDM in the most general case.

Comment. Within the limits of this section, the discussion was about the propagation of the phase front of the wave, that is, about its phase velocity. The propagation of the wave energy flux is determined by the group velocity and, in the general case, can either coincide or occur at any angle to the phase velocity vector, up to the opposite direction to it, as is the case for backward waves. The mutual orientation of the phase and group velocity vectors is determined by the wave dispersion and is considered in more detail in subsequent sections.

2.5.3. Transition to the polar coordinate system

The resulting relation (2.486) is rather cumbersome. On the other hand, the dispersion relation for the Damon–Eshbach problem has the simplest form not in Cartesian, but in cylindrical coordinates. Therefore, here it makes sense to go over from the Cartesian components of the wave number k_y and k_z and to the cylindrical variables and to k and φ.

Here:

$$k_y = k\cos\varphi \qquad (2.487)$$

$$k_z = k\sin\varphi \qquad (2.488)$$

With this setting, the value k is always assumed to be positive, and can take any values, both positive and negative. In this case, due to the quadratic nature of the dispersion relation in k_z, as well as the parity of the cosine function, the change in sign φ does not change anything, that is, the dispersion relation is symmetric with respect to the angle φ, so it is sufficient to consider only in the range from 0 to 180°.

The interval $0° \leq \varphi \leq 90°$ corresponds to the propagation of the phase front in the positive direction of the Oy axis, so we must take $S = 1$. In this case, the product $S \cdot |k_y|$ included in the second term of expression (2.486) is positive, but also k_y, according to (2.487), is also positive. In this case, the product $S \cdot |k_y|$ is simply equivalent to the value k_y.

The interval $90°0° \leq \varphi \leq 180°$ corresponds to the propagation of the phase front in the negative direction of the Oy axis, so we must take $S = -1$. In this case, the product $S \cdot |k_y|$ is negative. But according to (2.487), k_y is also negative, therefore the mentioned product is again equivalent to the quantity k_y.

Therefore, when transforming expression (2.486) into a cylindrical coordinate system, for any direction of propagation of the wave front, it is possible to set S = 1, and the module $|k_y|$ can be opened to k_y, determined by the relation (2.487), that is, to put:

$$S \cdot |k_y| \rightarrow k_y = k \cdot \cos\phi \qquad (2.489)$$

To further simplify (2.486), we now calculate some auxiliary expressions:

$$\sqrt{k_y^2 + k_z^2} = k \qquad (2.490)$$

$$\sqrt{k_y^2 + \frac{k_z^2}{1+\kappa}} = k \cdot \sqrt{\cos^2\phi + \frac{\sin^2\phi}{1+\kappa}} \qquad (2.491)$$

To simplify the last expression, we introduce a notation similar to that in the Damon–Eshbach problem:

$$\alpha = \sqrt{\frac{\sin^2\phi}{1+\kappa} + \cos^2\phi} \qquad (2.492)$$

In this case, the hyperbolic cotangent in expression (2.486) takes the form:

$$\operatorname{cth}\left(d\sqrt{k_y^2 + \frac{k_z^2}{1+\kappa}}\right) = cth(\alpha k d) \tag{2.493}$$

similar to a similar cotangent in the Damon–Eshbach dispersion relation.

Substituting these expressions in (2.486) under the condition (2.459) and dividing everything into, we get:

$$\operatorname{ch}[k(p-q)] \cdot \left[(v^2 - \kappa^2 - 2\kappa)\cos^2\phi - \kappa\sin^2\phi\right] + 2v\operatorname{sh}[k(p-q)]\cos\phi - $$
$$-\operatorname{ch}[k(p+q)] \cdot \left[(-v^2 + \kappa^2 + 2\kappa + 2)\cos^2\phi + (2+\kappa)\sin^2\phi\right] - $$
$$- 2\operatorname{sh}[k(p+q)] \cdot (1+\kappa)\alpha \cdot \operatorname{cth}(\alpha k d) = 0 \tag{2.494}$$

For further simplification, we introduce a notation similar to that used in the Damon–Eshbach problem:

$$\mu = 1 + \kappa \tag{2.495}$$

that is, taking into account (2.431):

$$\mu = 1 + \frac{\Omega_H}{\Omega_H^2 - \Omega^2} \tag{2.496}$$

Moreover, from (2.492) we obtain:

$$\alpha = \sqrt{\frac{\sin^2\phi}{\mu} + \cos^2\phi} \tag{2.497}$$

Let us introduce one more notation similar to that in the Damon–Eshbach problem:

$$\beta = \left(v^2 - \mu^2 + \mu\right)\cos^2\phi - \mu \tag{2.498}$$

To exclude explicitly the angle from the dispersion relation, we introduce an additional auxiliary notation:

$$\delta = \cos\phi \tag{2.499}$$

Substituting these expressions in (2.494), we obtain the dispersion relation in the final form:

$$(\beta+1)\cdot \text{ch}\left[k(p-q)\right] + 2\nu\delta\cdot \text{sh}\left[k(p-q)\right] + \\ + (\beta-1)\cdot \text{ch}\left[k(p+q)\right] - 2\mu\alpha\cdot \text{sh}\left[k(p+q)\right]\cdot \text{cth}(\alpha k d) = 0 \quad (2.500)$$

Here β it is determined by the expression (2.498), ν by the expression (2.432), δ by the expression (2.499), μ by the expression (2.496), α by the expression (2.497), and the normalized frequencies Ω and Ω_H included in these expressions – by the relations (2.433) and (2.434).

2.5.4. Passage to the limit for dispersion relations for other structures

The obtained dispersion relation (2.500) is quite general and includes, as special cases, dispersion relations for a number of simpler structures. To pass to such structures, it is enough to set and tend to zero or infinity in various combinations. Consider such transitions for the hyperbolic functions included in (2.500):

$$\lim_{q\to\infty} \text{ch}\left[k(p-q)\right] = \frac{1}{2}\cdot e^{-kp}\cdot \lim_{q\to\infty} e^{kq} \quad (2.501)$$

$$\lim_{q\to\infty} \text{sh}\left[k(p-q)\right] = -\frac{1}{2}\cdot e^{-kp}\cdot \lim_{q\to\infty} e^{kq} \quad (2.502)$$

$$\lim_{q\to\infty} \text{ch}\left[k(p+q)\right] = \frac{1}{2}\cdot e^{kp}\cdot \lim_{q\to\infty} e^{kq} \quad (2.503)$$

$$\lim_{q\to\infty} \text{sh}\left[k(p-q)\right] = \frac{1}{2}\cdot e^{kp}\cdot \lim_{q\to\infty} e^{kq} \quad (2.504)$$

$$\lim_{p\to\infty} \text{ch}\left[k(p-q)\right] = \frac{1}{2}\cdot e^{-kq}\cdot \lim_{p\to\infty} e^{kp} \quad (2.505)$$

$$\lim_{p\to\infty} \text{sh}\left[k(p-q)\right] = \frac{1}{2}\cdot e^{-kq}\cdot \lim_{p\to\infty} e^{kp} \quad (2.506)$$

$$\lim_{p\to\infty} \text{ch}\left[k(p+q)\right] = \frac{1}{2}\cdot e^{kq} \cdot \lim_{p\to\infty} e^{kp} \qquad (2.507)$$

$$\lim_{p\to\infty} \text{sh}\left[k(p+q)\right] = \frac{1}{2}\cdot e^{kq} \cdot \lim_{p\to\infty} e^{kp} \qquad (2.508)$$

$$\lim_{q\to 0} \text{ch}\left[k(p-q)\right] = \text{ch}(kp) \qquad (2.509)$$

$$\lim_{q\to 0} \text{sh}\left[k(p-q)\right] = \text{sh}(kp) \qquad (2.510)$$

$$\lim_{q\to 0} \text{ch}\left[k(p+q)\right] = \text{ch}(kp) \qquad (2.511)$$

$$\lim_{q\to 0} \text{sh}\left[k(p+q)\right] = \text{sh}(kp) \qquad (2.512)$$

$$\lim_{p\to 0} \text{ch}\left[k(p-q)\right] = \text{ch}(kq) \qquad (2.513)$$

$$\lim_{p\to 0} \text{sh}\left[k(p-q)\right] = -\text{sh}(kq) \qquad (2.514)$$

$$\lim_{p\to 0} \text{ch}\left[k(p+q)\right] = \text{ch}(kq) \qquad (2.515)$$

$$\lim_{p\to 0} \text{sh}\left[k(p+q)\right] = \text{sh}(kq) \qquad (2.516)$$

Using these expressions, we obtain the dispersion relations for various structures.

Ferrite–dielectric–metal structure – a ferrite plate, one surface of which is free, and a layer of metal is located parallel to the second, separated from the ferrite by a certain gap.

Assuming (the lower surface is free, metal is located above the upper one with a gap), we get:

$$\beta - 1 - 2\mu\alpha \cdot \text{cth}(\alpha k d) = 0 \qquad (2.517)$$

In this form, this relation coincides with that obtained when solving the Damon–Eshbach problem (2.299).

The ferrite–metal (FM) structure is a ferrite plate, one surface of which is free, and the other is covered with a closely adjacent metal layer.

Assuming $q \to \infty$, $p \to \infty$ (the lower surface is free, the upper one is covered with metal), we get:

$$\beta - \nu\delta - \mu\alpha \cdot \operatorname{cth}(\alpha k d) = 0 \qquad (2.518)$$

When $q \to \infty$, $p \to \infty$, (the lower surface is covered with metal, the upper one is free), we get:

$$\beta + \nu\delta - \mu\alpha \cdot \operatorname{cth}(\alpha k d) = 0 \qquad (2.519)$$

The ferrite–dielectric–metal (FDM) structure is a ferrite plate, one surface of which is free, and parallel to the second there is a layer of metal, separated from the ferrite by a certain gap.

Assuming $q \to \infty$ (the lower surface is free, the metal is located above the upper one with a gap q), we get:

$$\beta - 1 - 2\mu\alpha \cdot \operatorname{cth}(\alpha k d) + (\beta + 1 - 2\nu\delta) \cdot e^{-2kp} = 0 \qquad (2.520)$$

When (under the lower surface with a gap there is metal, the upper surface is free), we get:

$$\beta - 1 - 2\mu\alpha \cdot \operatorname{cth}(\alpha k d) + (\beta + 1 + 2\nu\delta) \cdot e^{-2kq} = 0 \qquad (2.521)$$

In these expressions, the first group of terms (up to the exponential) in the structure coincides with the Damon–Eshbach dispersion relation for a free plate (2.517), and the second (with an exponential) reflects the action of the metal layer.

It can be seen that, for $p \to \infty$ or $q \to \infty$, the relations (2.520) and (2.521) in both cases go over into (2.517), that is, into the relation for a free Damon–Eshbach plate, and for $q \to 0$ or $p \to 0$ into relations (2.518) or (2.519), respectively, that is, in one kind or another of the 'ferrite–metal' structure.

Recall that in expressions (2.520) and (2.521) the wave number k is always only positive, and the direction of propagation of the wave front relative to the Oy axis is determined by the interval in which the angle value φ is located. So the interval $0° \leq \varphi \leq 90°$

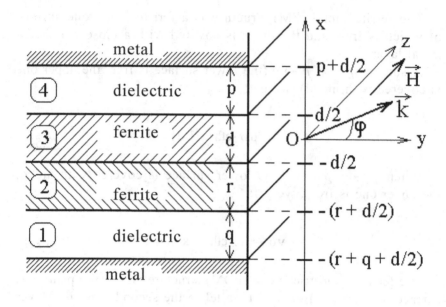

Fig. 2.6. General geometry of the problem.

corresponds to the propagation of the phase front in the positive direction of the Oy axis, and the interval $90° \leq \varphi \leq 180°$ corresponds to the propagation of the phase front in the negative direction of the same axis.

2.6. Dispersion ration for metal–dielectric–ferrite–ferrite–dielectric–metal structure (MDFFDM)

Let us now consider an even more general case when, instead of one ferrite layer, the structure has two folded together. This case is important for analyzing the propagation of magnetostatic waves in two-layer ferrite films. So, let us consider the relatively magnetized metal–dielectric–ferrite–ferrite–dielectric-metal (MDFFDM) structure under the condition of an arbitrary direction of MSW propagation into its structure. For the first time such a relation was given in [544]; here we will describe it in somewhat more detail.

2.6.1. General conclusion and character of the dispersion relation

The general geometry of the problem is shown in Fig. 2.6. The structure consists of two ferrite layers stacked together, adjacent to the outside are two dielectric layers coated with metal layers. The thicknesses of the ferrite layers are equal to d and r, the thicknesses of the dielectric layers are equal to p and q. The magnetization of the thickness layer is equal to, of the thickness layer d. The origin of the coordinate system is in the middle of the ferrite layer of thickness, the plane is parallel to the plane of this layer, and the axis is perpendicular to it. The field is applied along the plane of the structure parallel to the Oz axis. MSWs propagate in the plane of the structure parallel to the plane Oyz. The angle between the direction of the wave vector of the MCW $\vec{k}$ and the axis is indicated by φ.

In the magnetostatic approximation, the problem is reduced to the Walker and Laplace equations of the type (2.422)–(2.424) with boundary conditions of the type (2.425)–(2.430) and similar conditions for the metal surface and the interface between the ferrite layers [2–4, 134], and is solved by the method of separation of variables completely analogously to [118]. Time dependence is assumed in the form $e^{i\omega t}$.

Omitting the intermediate calculations due to their excessive cumbersomeness, we present the final results of calculating the dispersion relation in a cylindrical coordinate system.

Since the magnetizations of the layers No. 2 and No. 3 are different, it is convenient to take as a basis the magnetization of layer No. 3, equal to M_{d0}, and for the magnetization of layer No. 2, equal to M_{r0}, enter the normalization parameter:

$$g = \frac{M_{r0}}{M_{d0}} \tag{2.522}$$

In this case, the recording of parameters, and in comparison with the recording of parameters, is somewhat more complicated, however, the recording of all subsequent much more complex expressions for both layers becomes uniform.

Thus, if the parameters of layer 3 are:

$$\alpha_d = \sqrt{\frac{\sin^2 \varphi}{\mu_d} + \cos^2 \varphi}; \tag{2.523}$$

$$\mu_d = 1 + \frac{\Omega_{dH}}{\Omega_{dH}^2 - \Omega_d^2}$$
(2.524)

$$v_d = \frac{\Omega_d}{\Omega_{dH}^2 - \Omega_d^2}$$
(2.525)

then the parameters of layer 4, taking into account the normalization factor, take the form:

$$\alpha_r = \sqrt{\frac{\sin^2 \varphi}{\mu_d g + 1 - g} + \cos^2 \varphi};$$
(2.526)

$$\mu_r = \mu_d g + 1 - g$$
(2.527)

$$v_r = v_d g$$
(2.528)

$$\Omega_{dH} = \frac{H_0}{4\pi M_{d0}};$$
(2.529)

$$\Omega_d = \frac{\omega}{4\pi \gamma M_{d0}}$$
(2.530)

With this recording, the magnetostatic potential has the form:

$$\Psi_i = X_i Y Z$$
(2.531)

where $i = 1, 2, 3, 4$, and also:

$$X_1 = A_1 e^{kx} + B_1 e^{-kx}$$
(2.532)

$$X_2 = A_2 e^{k\alpha_d x} + B_2 e^{-k\alpha_d x}$$
(2.533)

$$X_3 = A_3 e^{k\alpha_r x} + B_3 e^{-k\alpha_r x}$$
(2.534)

$$X_4 = A_4 e^{kx} + B_4 e^{-kx}$$
(2.535)

$$Y = e^{-i(k\cos\phi)y}; \tag{2.536}$$

$$Z = e^{i(k\sin\phi)z} \tag{2.537}$$

Here k is the wave number of the MSW.

Substitution of the magnetostatic potential in the boundary conditions with the subsequent equating to zero the determinant of the resulting system of linear equations of the eighth order for the constants A_2, B_2, gives the dispersion relation in the Cartesian coordinate system. The transition to the polar coordinate system is carried out similarly to the one done in Section 2.5, as a result of which we obtain the dispersion relation in the form:

$$S_1 e^{k(p+q)} + S_2 e^{-k(p+q)} + S_3 e^{k(p-q)} + S_4 e^{-k(p+q)} = 0, \tag{2.538}$$

where:

$$S_1 = -(1-\beta_3) M_1 e^{k r \alpha_r} + (1-\beta_4) M_2 e^{-k r \alpha_r} \tag{2.539}$$

$$S_2 = -(1+\beta_3) M_3 e^{k r \alpha_r} + (1+\beta_4) M_4 e^{-k r \alpha_r} \tag{2.540}$$

$$S_3 = (1+\beta_3) M_1 e^{k r \alpha_r} - (1+\beta_4) M_2 e^{-k r \alpha_r} \tag{2.541}$$

$$S_4 = (1-\beta_3) M_3 e^{k r \alpha_r} - (1-\beta_4) M_4 e^{-k r \alpha_r} \tag{2.542}$$

and:

$$M_1 = (1+\beta_2)(\beta_1 - \beta_4) e^{k d \alpha_d} - (1+\beta_1)(\beta_2 - \beta_4) e^{-k d \alpha_d}; \tag{2.543}$$

$$M_2 = (1+\beta_2)(\beta_1 - \beta_3) e^{k d \alpha_d} - (1+\beta_1)(\beta_2 - \beta_3) e^{-k d \alpha_d} \tag{2.544}$$

$$M_3 = (1-\beta_2)(\beta_1 - \beta_4) e^{k d \alpha_d} - (1-\beta_1)(\beta_2 - \beta_4) e^{-k d \alpha_d} \tag{2.545}$$

$$M_4 = (1-\beta_2)(\beta_1 - \beta_3) e^{k d \alpha_d} - (1-\beta_1)(\beta_2 - \beta_3) e^{-k d \alpha_d}; \tag{2.546}$$

where:

$$\beta_1 = -\mu_d \alpha_d + \nu_d \cos\phi \tag{2.547}$$

$$\beta_2 = \mu_d \alpha_d + v_d \cos\phi \qquad (2.548)$$

$$\beta_3 = -\mu_r \alpha_r + v_r \cos\phi \qquad (2.549)$$

$$\beta_4 = \mu_r \alpha_r + v_r \cos\phi \qquad (2.550)$$

An analysis of the obtained dispersion relation shows that the full spectrum of MSWs contains eight groups of branches, four of which are bulk (multimode, corresponding to imaginary) and four surface (single-mode, corresponding to real). At, the bulk and surface groups of branches are combined in pairs, and the surface branches always lie higher than the corresponding bulk ones. The lowest frequency of body waves for all four groups of branches is the same, it corresponds and is equal. The highest frequencies of the body waves are equal to the lowest frequencies of the surface waves and at are equal to and. The highest frequencies of surface waves are reached at intermediate values and are determined by the thicknesses of the dielectric layers. For one pair of surface waves, the frequencies are

$$\Omega_{dH} + \frac{1}{2} \qquad (2.551)$$

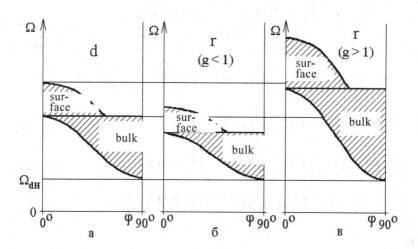

Fig. 2.7. Boundaries of the spectra of body and surface waves for the structure of MDFDM.

$$\Omega_{dH} + \frac{g}{2} \tag{2.552}$$

and for another pair:

$$\frac{1}{4}\left\{-(1-g)+\sqrt{(1-g)^2+8\Omega_{dH}\left(2\Omega_{dH}+g+1\right)}\right\}. \tag{2.553}$$

The boundaries of the spectra of all branches strongly depend on the angle φ. These dependences for four groups of body wave branches and two external surface waves at $k \to 0$ and $k \to \infty$ are shown in Fig. 2.7. Case a corresponds to a layer of thickness d. Cases b and c to a layer of thickness r at $g < 1$ and $g > 1$. In a structure with two ferrite layers, two cases are realized simultaneously: a and b or a and c. Depending on the magnitude g (that is, the ratio of the magnetizations of the ferrite layers), one or another overlap in frequency of the spectra of body and surface waves corresponding to individual layers can be realized. In this case, the surface wave in the layer propagates along the upper (according to Fig.2.6) surface of this layer, and the surface wave in the layer r – along the lower (according to Fig.2.6) surface of this layer. Along with the external, there are two branches of internal surface waves (not shown in Fig. 2.7), propagating along the adjacent surface of two ferrite layers. At a given frequency, both surface and bulk waves can propagate only in a certain range of angles φ between cutoff angles that are different for different branches of the spectrum. For bulk and external surface waves, the cutoff angles are found from the quadratic equation with respect to cos φ, for internal surface waves – from the biquadratic equation.

2.6.2. Passage to the limit for dispersion relations for other structures

The obtained dispersion relation (2.538) contains, as special cases, all the dispersion relations found earlier for a ferrite film with free surfaces (FFS), as well as for ferrite–metal (FM), ferrite–dielectric-metal (FDM) structures, metal–ferrite–metal (MFM), ferrite–ferrite (FF) and others. Here are some of them that are important for further consideration.

Free ferrite films (FFF)

The case of a ferrite film with free surfaces corresponds to $p \to \infty$ and $q \to \infty$, as well as to $r = 0$ and $M_{r0} = 0$. In this case, (2.538) turns into $S1 = 0$, that after substitution (2.539)–(2.550) and separation in an explicit form of the dependence $(kd\alpha)$ on takes the form:

$$\beta - 1 - 2\mu\alpha \, cth(k\,d\alpha) = 0 \qquad (2.554)$$

Here:

$$\alpha = \sqrt{\cos^2\phi + \frac{\sin^2\phi}{\mu}} \qquad (2.555)$$

$$\beta = (v^2 - \mu^2 + \mu)\cos^2\phi - \mu \qquad (2.556)$$

and μ and v are defined by the formulas (2.524), (2.525). The resulting expression (2.554) is completely identical to (2.299), and (2.555), (2.556) coincide with (2.300), (2.301).

Ferrite–metal (FM) structure

The MDFFDM structure transforms into a ferrite–metal (FM) structure at $p = 0$, $q \to \infty$, and also $r = 0$ and $M_{r0} = 0$. In this case, (2.538) turns into that after substitution (2.539)–(2.550) and separation in an explicit form of dependence on $(kd\alpha)$ takes the form:

$$\beta - v\delta - \mu\alpha \cdot cth(\alpha kd) = 0 \qquad (2.557)$$

where α, μ, v andnβ are determined by formulas (2.523)–(2.525) and (2.556), and $\delta = \cos\phi$.

Ferrite–dielectric–metal (FDM) structure

The structure of MDFDM transforms into a ferrite–dielectric–metal (FDM) structure at $q \to \infty$, and also r - 0 and $M_{r0} = 0$. The value p that determines the thickness of the dielectric layer is specified by the specific conditions of the problem. In this case, (2.538) turns into $S_1 + S_2 e^{-2kp} = 0$, that after substitution (2.539)–(2.550) and selection in an explicit form, the dependence on takes the form:

$$\beta - 1 - 2\mu\alpha \cdot cth(\alpha k d) + (\beta + 1 - 2\nu\delta) \cdot e^{-2kp} = 0 \quad (2.558)$$

where all designations coincide with those introduced above.

By successive transformations, one can make sure that (2.558) at $q \to \infty$ goes into (2.556), and at $p = 0$ into (2.557).

Ferrite–ferrite (FF) structure

The structure of MDFFDM transforms into a ferrite–ferrite (FF) structure at $p \to \infty$ and $q \to \infty$, and r and M_{r0} both along with d and M_{d0} are determined by the specific conditions of the problem. In this case, (2.538) turns into $S_1 = 0$, however, in contrast to a film with free surfaces, the expressions (2.539)–(2.550) remain more complicated. The dispersion relation takes the form:

$$-2\nu_d\nu_r \cos^2\phi + \left[\nu_d(\beta_r + 1) - \nu_r(\beta_d + 1)\right]\cos\phi + (\beta_d + \beta_r) +$$

$$+\mu_d\alpha_d(\beta_r - 1) cth(k d \alpha_d) + \mu_r\alpha_r(\beta_d - 1) cth(k d \alpha_r) -$$

$$-2\mu_d\mu_r\alpha_d\alpha_r cth(k d \alpha_d) cth(k d \alpha_r) = 0 \quad (2.559)$$

where α_{dr}, μ_{dr}, ν_{dr}, are defined by the formulas (2.523)–(2.530), and

$$\beta_d = \left(\nu_d^2 - \mu_d^2 + \mu_d\right)\cos^2\phi - \mu_d \quad (2.560)$$

$$\beta_r = \left(\nu_r^2 - \mu_r^2 + \mu_r\right)\cos^2\phi - \mu_r \quad (2.561)$$

By successive transformations, one can make sure that for $r = 0$ and $M_{r0} = 0$ or for d and $M_{d0} = 0$ the expression (2.559) turns into (2.554) for a layer of thickness d or r, respectively.

2.7. Phase and group velocities, phase rise and delay time of wave beams SMSW

The main subject of consideration in this work is the propagation of

SMSWs in ferrite films and structures based on them under conditions of various inhomogeneities. In this case, along with the trajectories of the wave beams, the most important local characteristics of the wave are the phase and group velocities, and the integral ones are the corresponding phase incursion and the delay time, which are determined both by the properties of the medium and by the wave dispersion law. Let's consider these characteristics in more detail.

2.7.1. Phase and group velocities

In the general case, a wave of the type $A \cdot e^{i(\omega t - kx)}$ is characterized by two parameters – amplitude A and phase $(\omega t - kx)$. In this case, the speed of movement in space of the constant value of the phase – the phase front may not coincide with the speed of movement of the constant value of the amplitude – the amplitude front. To accurately characterize the movement of both fronts, the concept of phase v_p and group v_g velocities is introduced. In a dispersionless medium, the phase and group velocities are equal. The difference between the phase and group velocities occurs when the medium in which the wave propagates has dispersive properties, that is, waves with different frequencies have different phase velocities.

In the absence of dispersion, frequency and wavenumber are linearly related:

$$\omega = v_p \cdot k \tag{2.562}$$

where is $v_p = const$:

$$v_p = \frac{\omega}{k} \tag{2.563}$$

In a dispersive medium, there is a nonlinear relationship:

$$\omega = v_p \cdot k(\omega) \tag{2.564}$$

so the phase velocity, defined as (2.563), is frequency dependent. This dependence is determined by the dispersion relation:

$$F(\omega, k) = 0 \tag{2.565}$$

In some cases, the dispersion relation can be resolved relatively, but when such resolution is not possible, it contains an implicit dependence. The solvable relationship is the Damon–Eshbach relationship (2.299), the insoluble one is the relationship for the ferrite–dielectric–metal structure (2.520).

Following [545], we obtain a formula for determining the group velocity according to the dispersion law. Consider the wave process as a set of two waves and, the frequencies and wavenumbers of which differ from the central values and in both directions by small additions and, so that $\omega_1 = \omega_0 - \delta\omega$; $\omega_2 = \omega_0 - \delta\omega$; $k_1 = k_0 - \delta k$; $k_2 = k_0 + \delta k$.

The total wave is beats of the form:

$$y = y_1 + y_2 = 2\hat{A} \cdot \cos(t \cdot \delta\omega - x \cdot \delta k) \cdot \sin(\omega_0 t - k_0 x), \qquad (2.566)$$

where the second factor determines the phase of the total wave, and the first one determines its slowly varying amplitude.

The phase constancy condition $\omega_0 t = k_0 x = const$, which determines the phase velocity, after differentiation gives the relation:

$$v_p = \frac{dx}{dt} = \frac{\omega}{k} \qquad (2.567)$$

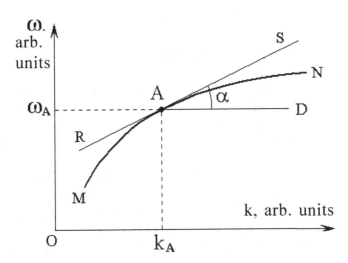

Fig. 2.8. Illustration of the formation of the absolute values of the phase and group velocities on the dispersion curve.

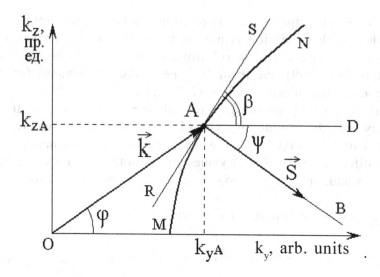

Fig. 2.9. Illustration of the formation of vectors of phase and group velocities on the isofrequency curve.

The condition for the constancy of the amplitude $t\cdot\delta\omega - x\delta k$ which determines the group velocity, after a similar differentiation gives the relation:

$$v_g = \frac{dx}{dt} = \frac{d\omega}{dk}$$

(2.568)

When studying the phase and group velocities of magnetostatic waves, two types of dependences are of great importance.

The first one is the actual dispersion relation, as the dependence of the wave frequency on its wave number. This type of dependence is illustrated in Fig.2.8, where the dispersion dependence $\omega(k)$ is shown by a thickened curve *MN*. The wavenumber k_A corresponds to a given frequency ω_A. point *A* corresponds to this frequency and wave number on the dispersion curve. At this point, the phase velocity is $v_p = \omega_A/k_A$. A tangent *RS* is drawn through the point *AD* to the dispersion curve, making an angle α with the horizontal equal to. According to the property of the tangent to the curve $\omega(k)$, the tangent of the angle α is equal to $d\omega/dk$, that is, this is the group velocity v_g.

Let us now turn to another type of dependence, determined again by the dispersion relation. In a Cartesian space $Ok_yk_z\omega$, that is, with a vertical coordinate ω and horizontal k_y and k_z, the section of the dispersion surface $\omega = F(k_y, k_z)$ by a plane $\omega = const.$ is an 'isofrequency' curve, an approximate view of which in the first quarter $Ok_y k_z$ of the plane is illustrated in Fig.2.9.

In this figure, the isofrequency curve itself is shown by a thickened line MV. In the sense of construction, this curve is the locus of points with coordinates k_y and k_z, so that a vector OA drawn from the origin O to any desired point A of the curve MN has a length $k = \sqrt{k_{yA}^2 + k_{zA}^2}$ and makes an angle $\varphi = \operatorname{arctg}(k_{zA}/k_{yA})$ with the axis OK_y, that is, it has components $k_{yA} = k \cdot \cos \varphi$ and $k_{zA} = k \cdot \sin \varphi$, and is a wave vector $\vec{k}$.

Thus, the components of the phase velocity vector in the general case are equal:

$$v_{py} = \frac{\omega}{k_y} = \frac{\omega}{k \cos \phi} \qquad (2.569)$$

$$v_{pz} = \frac{\omega}{k_z} = \frac{\omega}{k \sin \phi}, \qquad (2.570)$$

Consider now the group velocity. Draw a tangent to the curve through to point A and construct a perpendicular AB to it. Straight-line AS makes an angle β with the horizontal AD so tg $\beta = dk_z/dk_y$. Let's denote the angle between the horizontal AD and the straight line AB through ψ. By virtue of the equality $\beta + \psi = 90°$, we obtain:

$$tg\psi = 1/tg\beta = dk_y/dk_z \qquad (2.571)$$

In this last equality, we multiply the numerator and denominator by $1/\partial\omega$ and, passing from total derivatives to partial derivatives (since ω depends on two variables k_y and k_z), we write tg ψ in the form of a quotient from division of fractions:

$$tg\psi = \frac{dk_y}{dk_z} = \frac{\partial k_y/\partial \omega}{\partial k_z/\partial \omega} = \frac{\partial \omega/\partial k_z}{\partial \omega/\partial k_y} \qquad (2.572)$$

On the other hand, according to the definition of group speed:

$$\vec{V}_g = \vec{n}_y \cdot V_{gy} + \vec{n}_z \cdot v_{gz} = \vec{n}_y \cdot \frac{\partial \omega}{\partial k_y} + \vec{n}_z \cdot \frac{\partial \omega}{\partial k_z},$$
(2.573)

so one can see that:

$$tg\psi = \frac{V_{gz}}{V_{gy}}$$
(2.574)

This means that the angle ψ determines the direction of the group velocity vector $\vec{v}_g$, indicated in Fig. 2.9 through $\vec{s}$. According to common terminology [545], the vector $\vec{s}$ is called a 'beam' vector. The components of this vector are equal respectively:

$$s_y = v_{gy} = v_g \cdot \cos\psi$$
(2.575)

$$s_z = v_{gz} = v_g \cdot \sin\psi$$
(2.576)

Let us find the components of the group velocity vector, taking into account that ω, in accordance with the dispersion relation depends on k and φ, where:

$$k = \sqrt{k_y^2 + k_z^2}$$
(2.577)

$$\phi = \operatorname{arctg}\frac{k_z}{k_y}$$
(2.578)

Thus we get:

$$v_{gy} = \frac{\partial \omega}{\partial k_y} = \frac{\partial \omega}{\partial k} \cdot \frac{\partial k}{\partial k_y} + \frac{\partial \omega}{\partial \phi} \cdot \frac{\partial \phi}{\partial k_y}$$
(2.579)

$$v_{gy} = \frac{\partial \omega}{\partial k_y} = \frac{\partial \omega}{\partial k} \cdot \frac{\partial k}{\partial k_y} + \frac{\partial \omega}{\partial \phi} \cdot \frac{\partial \phi}{\partial k_y}$$
(2.580)

Differentiating (2.577) and (2.578) with respect to k_y and k_z, and also taking into account that $k_y = k \cos\varphi$ and $k_z = k \sin\varphi$, we obtain:

$$v_{gy} = \frac{\partial \omega}{\partial k} \cdot \cos\phi - \frac{\partial \omega}{\partial \phi} \cdot \frac{\sin\phi}{k} \tag{2.581}$$

$$v_{gz} = \frac{\partial \omega}{\partial k} \cdot \sin\phi + \frac{\partial \omega}{\partial \phi} \cdot \frac{\cos\phi}{k} \tag{2.582}$$

Here, according to (2.68), ω is related to the normalized frequency Ω by the relation:

$$\omega = 4\pi\gamma M_0 \cdot \Omega \tag{2.583}$$

so (2.581) and (2.582) can be written as:

$$v_{gy} = 4\pi\gamma M_0 \cdot \left(\frac{\partial \Omega}{\partial k} \cdot \cos\phi - \frac{\partial \Omega}{\partial \phi} \cdot \frac{\sin\phi}{k} \right) \tag{2.584}$$

$$v_{gz} = 4\pi\gamma M_0 \cdot \left(\frac{\partial \Omega}{\partial k} \cdot \sin\phi + \frac{\partial \Omega}{\partial \phi} \cdot \frac{\cos\phi}{k} \right) \tag{2.585}$$

The derivatives $\frac{\partial \Omega}{\partial k}$ and $\frac{\partial \Omega}{\partial \phi}$ are determined by the specific form of the dispersion relation and are discussed in more detail in Section 2.9 below.

Figure 2.9 shows that the mutual orientation of the wave and ray vectors is determined by the slope of the isofrequency curve at the point corresponding to the end of the wave vector. In this case, the sign of the projection of the ray vector onto the direction of the wave vector can be considered as a criterion for the forward or reverse nature of the wave. So, if the projection of the group velocity vector onto the phase velocity vector is positive, then the wave is direct, if negative, it is reverse. In the language of vector algebra, we can say that a wave is direct if the sign of the scalar product of the wave and ray vectors is positive and inverse if this sign is negative. Within the framework of this work, such a definition of the nature of the wave is quite sufficient, but for a more detailed acquaintance with this issue, one can refer to [564].

2.7.2. Phase run and delay time

The propagation of a wave between some two points in space with a uniform dependence on time at all its points is accompanied by a phase incursion, as well as a delay of the amplitude in time.

Let us turn to the diagram of the section of the trajectory of the wave beam shown in Fig. 2.10. The wave propagates from point 1 to point 2, traveling the distance ΔS. The phase incursion in this section is equal to:

$$\Delta\phi_{1\to2} = k \cdot \Delta S \qquad (2.586)$$

Considering that:

$$\Delta S = \sqrt{(\Delta y)^2 + (\Delta z)^2} \qquad (2.587)$$

and also passing to the differentials, we get:

$$dS = \sqrt{(dy)^2 + (dz)^2} = \sqrt{1 + \left(\frac{dz}{dy}\right)^2} \cdot dy \qquad (2.588)$$

Integrating over the full trajectory of the wave beam propagation, we obtain:

$$\Delta\phi = \int_{y_b}^{y_e} k \cdot \sqrt{1 + \left(\frac{dz}{dy}\right)^2} \cdot dy \qquad (2.589)$$

where y_b and y_e are the y-coordinates of the starting and ending points of the trajectory.

The delay time between points 1 and 2 in Fig. 2.10 is determined by dividing the path traveled by the wave by its group velocity:

$$\Delta\tau_{1\to2} = \frac{\Delta S}{v_g} \qquad (2.590)$$

Passing to the differentials and writing v_g through the components, we get:

$$d\tau = \frac{\sqrt{1+\left(\dfrac{dz}{dy}\right)^2}}{\sqrt{v_{gy}^2 + v_{gz}^2}} \cdot dy \qquad (2.591)$$

Considering that $dy = v_{gy}\, dt$, $dz = v_{gz}\, dt$, we get:

$$\frac{dz}{dy} = \frac{v_{gz}}{v_{gy}} \qquad (2.592)$$

Substituting this expression in (2.591), we get:

$$d\tau = \frac{\sqrt{1+\left(\dfrac{v_{gz}}{v_{gy}}\right)^2}}{\sqrt{v_{gy}^2 + v_{gz}^2}} \cdot dy = \frac{dy}{v_{gy}} \qquad (2.593)$$

Integrating over the entire path of propagation of the wave beam, we get:

$$\Delta\tau = \int_{y_b}^{y_e} \frac{dy}{v_{gy}} \qquad (2.594)$$

where y_b and y_e are the coordinates of the starting and ending points of the trajectory.

2.8. System of equations for the Hamilton–Auld method

One of the most important tools for calculating the propagation of MSW wave beams in inhomogeneously magnetized structures is the Hamilton–Auld method, which was apparently first used for MSWs in [274] and somewhat later in [275]. Let us dwell briefly on the mathematical apparatus used here. We restrict ourselves to the case of wave propagation in a plane Oyz using a polar coordinate system. We note right away that these two restrictions are not unconditional, so the calculation method presented below can be easily generalized to a three-dimensional space using a spherical coordinate system, the use of a Cartesian coordinate system is also possible, however, for the main purposes of this work, such generalizations are not required, so we leave them as material for further research.

2.8.1. General derivation of the Hamilton–Auld equations

So, let us first consider the Cartesian coordinate system $Oxyz$, the Oyz plane of which coincides with the plane of the structure under study. It is in this plane that the wave will propagate. The Hamilton–Auld method is based on the analogy between wave propagation in an inhomogeneous medium and the motion of a particle in a field determined by the inhomogeneity of the structure. In this case, the analogue of the energy (Hamiltonian) of the particle is the dispersion relation for the wave.

We will proceed from the Hamilton equations in general form [546]:
;

$$\frac{dx_i}{dt} = \frac{\partial H}{\partial p_i} \qquad \frac{dp_i}{dt} = -\frac{\partial H}{\partial x_i} \qquad (2.595)$$

where H is the Hamilton operator, according to the mechanical definition, for a material point freely moving in a potential field, equal to the sum of potential and kinetic energies [546].

For two coordinates y, z, and impulses p_y and p_z, we get:

$$\frac{dy}{dt} = \frac{\partial H}{\partial p_y} \qquad (2.596)$$

$$\frac{dz}{dt} = \frac{\partial H}{\partial p_z} \qquad (2.597)$$

$$\frac{dp_y}{dt} = -\frac{\partial H}{\partial y} \qquad (2.598)$$

$$\frac{dp_z}{dt} = -\frac{\partial H}{\partial z} \qquad (2.599)$$

We exclude from these equations t by dividing (2.597)–(2.599) by (2.596). We replace H by Ω, where $\Omega = \Omega\ (\omega,\ k_y,\ k_z)$ is the dispersion relation, and also p_i by k_i (replacement made by Auld in [274]). As a result, from (2.597)–(2.599) we obtain three equations:

$$\frac{dz}{dy} = \frac{\frac{\partial \Omega}{\partial k_z}}{\frac{\partial \Omega}{\partial k_y}} \qquad (2.600)$$

$$\frac{dk_y}{dy} = -\frac{\frac{\partial \Omega}{\partial y}}{\frac{\partial \Omega}{\partial k_y}} \qquad (2.601)$$

$$\frac{dk_z}{dy} = -\frac{\frac{\partial \Omega}{\partial z}}{\frac{\partial \Omega}{\partial k_z}} \qquad (2.602)$$

Equations (2.600)–(2.602) allow us to find the trajectory of the wave and the dependences of both components of the wave number on in the Cartesian coordinate system. With a known dispersion relation and to describe a one-dimensional wave at a given point, it is necessary to know its coordinate and length, or coordinate and wavenumber. In this case, one of the coordinates can be specified initially, and the other and both components of the wave number are determined by the obtained equations (2.600)–(2.602).

Here, we will consider the given coordinate y, while the equations (2.600)–(2.602) allow us to find another coordinate z and components of the wave number k_y, k_z.

2.8.2. Transition to the polar coordinate system

To obtain the characteristics of wave propagation in a polar coordinate system, equations (2.600)–(2.602) must be transformed into such a system. The wave vector in the polar coordinate system is shown in Fig. 2.4 (Section 2.4.7), and its components are:

$$k_y = k \cdot \cos\phi \qquad (2.603)$$

$$k_z = k \cdot \sin\phi \qquad (2.604)$$

whence we get:

$$k = \sqrt{k_y^2 + k_z^2} \qquad (2.605)$$

$$\phi = \operatorname{arctg}\frac{k_z}{k_y} \qquad (2.606)$$

Consider equation (2.600). This equation includes derivatives $\dfrac{\partial\Omega}{\partial k_y}$ and $\dfrac{\partial\Omega}{\partial k_z}$. Let's find these derivatives, taking into account that $\Omega = \Omega(k, \varphi)$:

$$\frac{\partial\Omega}{\partial k_y} = \frac{\partial\Omega}{\partial k}\cdot\frac{\partial k}{\partial k_y} + \frac{\partial\Omega}{\partial\phi}\cdot\frac{\partial\phi}{\partial k_y} \qquad (2.607)$$

$$\frac{\partial\Omega}{\partial k_z} = \frac{\partial\Omega}{\partial k}\cdot\frac{\partial k}{\partial k_z} + \frac{\partial\Omega}{\partial\phi}\cdot\frac{\partial\phi}{\partial k_z} \qquad (2.608)$$

These expressions include the derivatives $\dfrac{\partial\Omega}{\partial k}$ and $\dfrac{\partial\Omega}{\partial\phi}$, which are determined by the form of the dispersion relation; therefore, we will not calculate them at this stage, but will do them later (in Section 2.9). Other derivatives included in these expressions, such as $\dfrac{\partial k}{\partial k_y}, \dfrac{\partial k}{\partial k_z}, \dfrac{\partial\phi}{\partial k_y}, \dfrac{\partial\phi}{\partial k_z}$, are determined by the coordinate system, therefore, based on (2.605), (2.606), taking into account (2.603), (2.604), take the form:

$$\frac{\partial k}{\partial k_y} = \frac{\partial}{\partial k_y}\left(\sqrt{k_y^2 + k_z^2}\right) = \frac{k_y}{k} = \cos\phi \qquad (2.609)$$

$$\frac{\partial k}{\partial k_z} = \frac{\partial}{\partial k_z}\left(\sqrt{k_y^2 + k_z^2}\right) = \frac{k_z}{k} = \sin\phi \qquad (2.610)$$

$$\frac{\partial \phi}{\partial k_y} = \frac{\partial}{\partial k_y}\left(arctg\frac{k_z}{k_y}\right) = \frac{-\dfrac{k_z}{k_y^2}}{1+\left(\dfrac{k_z}{k_y}\right)^2} = \frac{-k_z}{k_y^2+k_z^2} = -\frac{\sin\phi}{k}$$

(2.611)

$$\frac{\partial \phi}{\partial k_z} = \frac{\partial}{\partial k_z}\left(arctg\frac{k_z}{k_y}\right) = \frac{\dfrac{1}{k_y}}{1+\left(\dfrac{k_z}{k_y}\right)^2} = \frac{k_y}{k_y^2+k_z^2} = \frac{\cos\phi}{k}$$

(2.612)

Thus, we get:

$$\frac{\partial k}{\partial k_y} = \cos\phi$$

(2.613)

$$\frac{\partial k}{\partial k_z} = \sin\phi$$

(2.614)

$$\frac{\partial \phi}{\partial k_y} = -\frac{\sin\phi}{k}$$

(2.615)

$$\frac{\partial \phi}{\partial k_z} = \frac{\cos\phi}{k}$$

(2.616)

Substituting (2.613)–(2.616) into (2.607)–(2.608), we obtain:

$$\frac{\partial \Omega}{\partial k_y} = \frac{\partial \Omega}{\partial k}\cdot\cos\phi - \frac{\partial \Omega}{\partial \phi}\cdot\frac{\sin\phi}{k}$$

(2.617)

$$\frac{\partial \Omega}{\partial k_z} = \frac{\partial \Omega}{\partial k}\cdot\sin\phi + \frac{\partial \Omega}{\partial \phi}\cdot\frac{\cos\phi}{k}$$

(2.618)

Substituting these expressions into (2.600), we get:

$$\frac{dz}{dy} = \frac{\frac{\partial \Omega}{\partial k}\cdot \sin\phi + \frac{\partial \Omega}{\partial \phi}\cdot\frac{\cos\phi}{k}}{\frac{\partial \Omega}{\partial k}\cdot \cos\phi - \frac{\partial \Omega}{\partial \phi}\cdot\frac{\sin\phi}{k}} \qquad (2.619)$$

Multiplying the numerator and denominator of the fraction on the right side of this expression by k, we get:

$$\frac{dz}{dy} = \frac{\frac{\partial \Omega}{\partial k}\cdot k\cdot \sin\phi + \frac{\partial \Omega}{\partial \phi}\cdot \cos\phi}{\frac{\partial \Omega}{\partial k}\cdot k\cdot \cos\phi - \frac{\partial \Omega}{\partial \phi}\cdot \sin\phi} \qquad (2.620)$$

Consider now equations (2.601) and (2.602). They include derivatives $\frac{dk_y}{dy}$ and $\frac{dk_z}{dy}$. Let us find these derivatives using (2.603) and (2.604):

$$\frac{dk_y}{dy} = \frac{d}{dy}(k\cdot\cos\phi) = \cos\phi\cdot\frac{dk}{dy} - k\cdot\sin\phi\cdot\frac{d\phi}{dy} \qquad (2.621)$$

$$\frac{dk_z}{dy} = \frac{d}{dy}(k\cdot\sin\phi) = \sin\phi\cdot\frac{dk}{dy} + k\cdot\cos\phi\cdot\frac{d\phi}{dy} \qquad (2.622)$$

Substituting these expressions in (2.601) and (2.602), we get:

$$\cos\phi\cdot\frac{dk}{dy} - k\cdot\sin\phi\cdot\frac{d\phi}{dy} = -\frac{\frac{\partial \Omega}{\partial y}}{\frac{\partial \Omega}{\partial k_y}} \qquad (2.623)$$

$$\sin\phi\cdot\frac{dk}{dy} + k\cdot\cos\phi\cdot\frac{d\phi}{dy} = -\frac{\frac{\partial \Omega}{\partial z}}{\frac{\partial \Omega}{\partial k_y}} \qquad (2.624)$$

Here the derivatives $\frac{\partial \Omega}{\partial y}$ and $\frac{\partial \Omega}{\partial z}$ are determined by the dispersion relation, and the derivative $\frac{\partial \Omega}{\partial k_y}$ by expression (2.617). The final problem is to find the derivatives d_k/d_y and $d\varphi/d_y$, for which it is necessary to solve the system of equations (2.623) and (2.624) with respect to these derivatives.

Let us introduce auxiliary notations:

$$A = -\frac{\dfrac{\partial \Omega}{\partial y}}{\dfrac{\partial \Omega}{\partial k_y}}$$
(2.625)

$$B = -\frac{\dfrac{\partial \Omega}{\partial z}}{\dfrac{\partial \Omega}{\partial k_y}}$$
(2.626)

With these designations, the equations (2.623) and (2.624) take the form:

$$\cos\phi \cdot \frac{dk}{dy} - k \cdot \sin\phi \cdot \frac{d\phi}{dy} = A$$
(2.627)

$$\sin\phi \cdot \frac{dk}{dy} + k \cdot \cos\phi \cdot \frac{d\phi}{dy} = B$$
(2.628)

Let us solve this system with respect to $\dfrac{dk}{dy}$ and $\dfrac{d\phi}{dy}$.

Its determinant is:

$$D = \begin{vmatrix} \cos\phi & -k \cdot \sin\phi \\ \sin\phi & k \cdot \cos\phi \end{vmatrix} = k \cdot \cos^2\phi + k \cdot \sin^2\phi = k$$
(2.629)

Find unknowns:

$$\frac{dk}{dy} = \frac{1}{k} \cdot \begin{vmatrix} A & -k \cdot \sin\phi \\ B & k \cdot \cos\phi \end{vmatrix} = A \cdot \cos\phi + B \cdot \sin\phi$$
(2.630)

$$\frac{\partial F}{\partial \Omega} = 2\mu \left[\frac{k\, d\alpha}{sh^2(kd\alpha)} - cth(kd\alpha) \right] \cdot \frac{\partial \alpha}{\partial \Omega} + \left(1 + e^{-2kp}\right) \cdot \frac{\partial \beta}{\partial \Omega}$$
(2.631)

Substituting into the right-hand sides of these expressions A and B, defined by formulas (2.625), (2.626), we obtain:

$$\frac{dk}{dy} = -\frac{\dfrac{\partial \Omega}{\partial y} \cdot \cos\phi + \dfrac{\partial \Omega}{\partial z} \cdot \sin\phi}{\dfrac{\partial \Omega}{\partial k_y}} \sigma_x \qquad (2.632)$$

$$\frac{d\phi}{dy} = \frac{1}{k} \cdot \frac{-\dfrac{\partial \Omega}{\partial z} \cdot \cos\phi + \dfrac{\partial \Omega}{\partial y} \cdot \sin\phi}{\dfrac{\partial \Omega}{\partial k_y}} \qquad (2.633)$$

Substituting into these equations the derivative $d\Omega/dk_y$ defined by formula (2.617), we obtain:

$$\frac{dk}{dy} = -\frac{\dfrac{\partial \Omega}{\partial y} \cdot \cos\phi + \dfrac{\partial \Omega}{\partial z} \cdot \sin\phi}{\dfrac{\partial \Omega}{\partial k} \cdot \cos\phi - \dfrac{\partial \Omega}{\partial \phi} \cdot \dfrac{\sin\phi}{k}} \qquad (2.634)$$

$$\frac{\partial F}{\partial y} = 2\mu \left[\frac{k\,d\alpha}{sh^2(kd\alpha)} - cth(kd\alpha) \right] \cdot \frac{\partial \alpha}{\partial y} + \left(1 + e^{-2kp}\right) \cdot \frac{\partial \beta}{\partial y} \qquad (2.635)$$

Multiplying the numerators and denominators of the fractions on the right-hand sides of these expressions and combining them with (2.620), we obtain a complete system of equations that determines the wave trajectory and its parameters in the polar coordinate system:

$$\frac{dz}{dy} = \frac{\dfrac{\partial \Omega}{\partial k} \cdot k \cdot \sin\phi + \dfrac{\partial \Omega}{\partial \phi} \cdot \cos\phi}{\dfrac{\partial \Omega}{\partial k} \cdot k \cdot \cos\phi - \dfrac{\partial \Omega}{\partial \phi} \cdot \sin\phi} \qquad (2.636)$$

$$\frac{dk}{dy} = -k \cdot \frac{\dfrac{\partial \Omega}{\partial y} \cdot \cos\phi + \dfrac{\partial \Omega}{\partial z} \cdot \sin\phi}{\dfrac{\partial \Omega}{\partial k} \cdot k \cdot \cos\phi - \dfrac{\partial \Omega}{\partial \phi} \cdot \sin\phi} \qquad (2.637)$$

$$\frac{d\phi}{dy} = \frac{-\dfrac{\partial \Omega}{\partial z} \cdot \cos\phi + \dfrac{\partial \Omega}{\partial y} \cdot \sin\phi}{\dfrac{\partial \Omega}{\partial k} \cdot k \cdot \cos\phi - \dfrac{\partial \Omega}{\partial \phi} \cdot \sin\phi} \qquad (2.638)$$

These expressions include derivatives of the dispersion relation, which has the form $\Omega = \Omega(\omega, y, z.k, \varphi)$ Let us express the derivatives of Ω through the derivatives of the wave number k in the same coordinates, that is, through $\dfrac{\partial k}{\partial y}$, $\dfrac{\partial k}{\partial z}$ and $\dfrac{\partial k}{\partial \phi}$ using the formula for the derivative of an implicitly given function $F(x, y)$:

$$\frac{\partial y}{\partial x} = -\frac{\dfrac{\partial F}{\partial x}}{\dfrac{\partial F}{\partial y}} \qquad (2.639)$$

Using this formula, we get:

$$\frac{\partial k}{\partial y} = -\frac{\dfrac{\partial \Omega}{\partial y}}{\dfrac{\partial \Omega}{\partial k}} \qquad (2.640)$$

$$\frac{\partial k}{\partial z} = -\frac{\dfrac{\partial \Omega}{\partial z}}{\dfrac{\partial \Omega}{\partial k}} \qquad (2.641)$$

$$\frac{\partial k}{\partial \phi} = -\frac{\dfrac{\partial \Omega}{\partial \phi}}{\dfrac{\partial \Omega}{\partial k}} \qquad (2.642)$$

From these expressions we find the required derivatives:

$$\frac{\partial \Omega}{\partial y} = -\frac{\partial k}{\partial y} \cdot \frac{\partial \Omega}{\partial k} \tag{2.643}$$

$$\frac{\partial \Omega}{\partial z} = -\frac{\partial k}{\partial z} \cdot \frac{\partial \Omega}{\partial k} \tag{2.644}$$

$$\frac{\partial \Omega}{\partial \phi} = -\frac{\partial k}{\partial \phi} \cdot \frac{\partial \Omega}{\partial k} \tag{2.645}$$

Substituting these derivatives in (2.636)–(2.638) and dividing the numerators and denominators of the fractions by $\frac{\partial \Omega}{\partial k}$, we obtain the complete equations that determine the wave trajectory and its parameters in the polar coordinate system:

$$\frac{dz}{dy} = \frac{k \cdot \sin\phi - \frac{\partial k}{\partial \phi} \cdot \cos\phi}{k \cdot \cos\phi + \frac{\partial k}{\partial \phi} \cdot \sin\phi} \tag{2.646}$$

$$\frac{dk}{dy} = k \cdot \frac{\frac{\partial k}{\partial y} \cdot \cos\phi + \frac{\partial k}{\partial z} \cdot \sin\phi}{k \cdot \cos\phi + \frac{\partial k}{\partial \phi} \cdot \sin\phi} \tag{2.647}$$

$$\frac{d\phi}{dy} = -\frac{\frac{\partial k}{\partial y} \cdot \sin\phi + \frac{\partial k}{\partial z} \cdot \cos\phi}{k \cdot \cos\phi + \frac{\partial k}{\partial \phi} \cdot \sin\phi} \tag{2.648}$$

This is the basic system of the Hamilton–Auld equations used to calculate the trajectories of propagation of magnetostatic waves in nonuniformly magnetized structures.

In the literature, the same system is often found written in a slightly different equivalent form:

$$\frac{dk}{dy} = k\left(\frac{\partial k}{\partial y}\cos\phi + \frac{\partial k}{\partial z}\sin\phi\right) \cdot \left(k\cos\phi + \frac{\partial k}{\partial \phi}\sin\phi\right)^{-1}$$
(2.649)

$$\frac{d\phi}{dy} = -\left(\frac{\partial k}{\partial y}\sin\phi - \frac{\partial k}{\partial z}\cos\phi\right) \cdot \left(k\cos\phi + \frac{\partial k}{\partial \phi}\sin\phi\right)^{-1}$$
(2.650)

$$\frac{dz}{dy} = \left(k\sin\phi - \frac{\partial k}{\partial \phi}\cos\phi\right) \cdot \left(k\cos\phi + \frac{\partial k}{\partial \phi}\sin\phi\right)^{-1}$$
(2.651)

2.9. Derivatives from the dispersion relationship for the ferritic–dielectric–metal structure

The formulas given in the previous sections for calculating the phase and group velocities, phase incursion, delay time, and wave trajectories in an inhomogeneous medium include the derivatives $\frac{\partial \Omega}{\partial k}, \frac{\partial \Omega}{\partial \phi}, \frac{\partial k}{\partial \phi}, \frac{\partial k}{\partial y}, \frac{\partial k}{\partial z}$. Let us present them here for the ferrite-dielectric-metal (FDM) structure most often considered below, the dispersion relation in which has the form:

$$F(\Omega, k, \phi) = \beta - 1 - 2\mu\alpha \cdot cth(kd\alpha) + (\beta + 1 - 2\nu\delta) \cdot e^{-2kp} = 0$$
, (2.652)

where the field is included as a parameter.

Differentiating (2.652) as an implicit function, we get:

$$\frac{\partial \Omega}{\partial k} = -\frac{\partial F/\partial k}{\partial F/\partial \Omega}$$
(2.653)

$$\frac{\partial \Omega}{\partial \phi} = -\frac{\partial F/\partial \phi}{\partial F/\partial \Omega}$$
(2.654)

$$\frac{\partial k}{\partial \phi} = -\frac{\partial F/\partial \phi}{\partial F/\partial k}$$
(2.655)

$$\frac{\partial k}{\partial y} = -\frac{\partial F/\partial y}{\partial F/\partial k} \qquad (2.656)$$

$$\frac{\partial k}{\partial z} = -\frac{\partial F/\partial z}{\partial F/\partial k} \qquad (2.657)$$

These formulas include derivatives $\partial F/\partial k$, $\partial F/\partial \varphi$, $\partial F/\partial \Omega$, $\partial F/\partial y$, $\partial F/\partial z$. Taking into account (2.652), they take the form:

$$\frac{\partial F}{\partial k} = 2\left[\frac{\mu \, d\alpha^2}{sh^2(k\,d\alpha)} - p e^{-2kp}(\beta + 1 - 2\nu\delta)\right] \qquad (2.658)$$

$$\frac{\partial F}{\partial \varphi} = 2\mu\left[\frac{k\,d\alpha}{sh^2(k\,d\alpha)} - cth(k\,d\alpha)\right]\cdot\frac{\partial \alpha}{\partial \varphi} +$$

$$+ \left(1 + e^{-2kp}\right)\cdot\frac{\partial \beta}{\partial \varphi} - 2\nu e^{-2kp}\cdot\frac{\partial \delta}{\partial \varphi} \qquad (2.659)$$

$$\frac{\partial F}{\partial \Omega} = 2\mu\left[\frac{k\,d\alpha}{sh^2(k\,d\alpha)} - cth(k\,d\alpha)\right]\cdot\frac{\partial \alpha}{\partial \Omega} + \left(1 + e^{-2kp}\right)\cdot\frac{\partial \beta}{\partial \Omega} -$$

$$- 2\alpha\, cth(k\,d\alpha)\cdot\frac{\partial \mu}{\partial \Omega} - 2\delta e^{-2kp}\cdot\frac{\partial \nu}{\partial \Omega} \qquad (2.660)$$

$$\frac{\partial F}{\partial y} = 2\mu\left[\frac{k\,d\alpha}{sh^2(k\,d\alpha)} - cth(k\,d\alpha)\right]\cdot\frac{\partial \alpha}{\partial y} + \left(1 + e^{-2kp}\right)\frac{\partial \beta}{\partial y} -$$

$$-2\alpha\, cth(k\,d\alpha)\frac{\partial \mu}{\partial y} - 2\delta e^{-2kp}\frac{\partial \nu}{\partial y};$$

$$(2.661)$$

$$\frac{\partial F}{\partial z} = 2\mu\left[\frac{k\,d\alpha}{sh^2(k\,d\alpha)} - cth(k\,d\alpha)\right]\cdot\frac{\partial \alpha}{\partial z} + \left(1 + e^{-2kp}\right)\cdot\frac{\partial \beta}{\partial z} -$$

$$-2\alpha\, cth(k\,d\alpha)\frac{\partial \mu}{\partial z} - 2\delta e^{-2kp}\frac{\partial \nu}{\partial z}.$$

$$, \qquad (2.662)$$

and the derivatives entering (2.659)–(2.662) have the form

$$\frac{\partial \alpha}{\partial \phi} = \frac{1}{2\alpha}\left(\frac{1}{\mu} - 1\right) \cdot \sin 2\phi \qquad (2.663)$$

$$\frac{\partial \alpha}{\partial \Omega} = -\frac{\sin^2 \phi}{2\alpha\mu^2} \cdot \frac{\partial \mu}{\partial \Omega} \qquad (2.664)$$

$$\frac{\partial \alpha}{\partial y} = -\frac{\sin^2 \phi}{2\alpha\mu^2} \cdot \frac{\partial \mu}{\partial y} \qquad (2.665)$$

$$\frac{\partial \alpha}{\partial z} = -\frac{\sin^2 \phi}{2\alpha\mu^2} \cdot \frac{\partial \mu}{\partial z} \qquad (2.666)$$

$$\frac{\partial \beta}{\partial \phi} = \left(\mu^2 - v^2 - \mu\right) \cdot \sin 2\phi \qquad (2.667)$$

$$\frac{\partial \beta}{\partial \Omega} = 2v\cos^2 \phi \cdot \frac{\partial v}{\partial \Omega} - \left(\sin^2 \phi + 2\mu \cos^2 \phi\right) \cdot \frac{\partial \mu}{\partial \Omega} \qquad (2.668)$$

$$\frac{\partial \beta}{\partial y} = 2v\cos^2 \phi \cdot \frac{\partial v}{\partial y} - \left(\sin^2 \phi + 2\mu \cos^2 \phi\right) \cdot \frac{\partial \mu}{\partial y} \qquad (2.669)$$

$$\frac{\partial \beta}{\partial z} = 2v\cos^2 \phi \cdot \frac{\partial v}{\partial z} - \left(\sin^2 \phi + 2\mu \cos^2 \phi\right) \cdot \frac{\partial \mu}{\partial z} \qquad (2.670)$$

$$\frac{\partial \delta}{\partial \phi} = -\sin \phi \qquad (2.671)$$

$$\frac{\partial \mu}{\partial \Omega} = \frac{2\Omega_H \Omega}{\left(\Omega_H^2 - \Omega^2\right)} \qquad (2.672)$$

$$\frac{\partial \mu}{\partial y} = -\frac{\Omega_H^2 + \Omega^2}{\left(\Omega_H^2 - \Omega^2\right)^2} \cdot \frac{\partial \Omega_H}{\partial y} \qquad (2.673)$$

$$\frac{\partial \mu}{\partial z} = -\frac{\Omega_H^2 + \Omega^2}{\left(\Omega_H^2 - \Omega^2\right)^2} \cdot \frac{\partial \Omega_H}{\partial z} \qquad (2.674)$$

$$\frac{\partial \nu}{\partial \Omega} = \frac{\Omega_H^2 + \Omega^2}{\left(\Omega_H^2 - \Omega^2\right)^2} \qquad (2.675)$$

$$\frac{\partial \nu}{\partial y} = -\frac{2\Omega_H \Omega}{\left(\Omega_H^2 - \Omega^2\right)^2} \cdot \frac{\partial \Omega_H}{\partial y} \qquad (2.676)$$

$$\frac{\partial \nu}{\partial z} = -\frac{2\Omega_H \Omega}{\left(\Omega_H^2 - \Omega^2\right)^2} \cdot \frac{\partial \Omega_H}{\partial z} \qquad (2.677)$$

At the same time, as before:

$$\alpha = \sqrt{\frac{\sin^2 \phi}{\mu} + \cos^2 \phi} \qquad (2.678)$$

$$\beta = \left(\nu^2 - \mu^2 + \mu\right) \cdot \cos^2 \phi - \mu \qquad (2.679)$$

$$\delta = \cos \phi \qquad (2.680)$$

$$\mu = 1 + \frac{\Omega_H}{\Omega_H^2 - \Omega^2} \qquad (2.681)$$

$$\nu = \frac{\Omega}{\Omega_H^2 - \Omega^2} \qquad (2.682)$$

$$\Omega = \frac{\omega}{4\pi \gamma M_0} \qquad (2.683)$$

$$\Omega_H = \frac{H}{4\pi M_0} \qquad (2.684)$$

The derivatives presented here form the basis of the computing apparatus for finding the phase and group velocities, trajectories, phase incursion, SMSW delay time for the FDM structure. Formulas for other structures are obtained in a similar way. Similar derivatives for other structures can be obtained from the above formulas by passing to the limit. So for a ferrite plate with free surfaces, it is enough to put, and for a ferrite–metal structure – to put. From the point of view of numerical algorithmization, it is most convenient to immediately compose a program for the general case – the FDM structure according to the formulas given here, after which, if it is necessary to switch to a free plate, set a sufficiently large value (obviously much larger than the plate thickness), and to switch to a ferrite-metal structure – put $p = 0$.

2.10. Equivalence of different kinds of equations of dynamics in classical mechanics

In this work, the Hamilton–Auld method is traditionally used to calculate the trajectories of SMSW wave beams, one of the versions of which is described in Section 2.8. However, this method is not the only one possible, just as in mechanics there are several fundamentally different methods for solving similar problems. Let us dwell here briefly on the situation in classical mechanics, which is completely similar to that in the problems of the propagation of MSWs.

In mechanics, there are three main types of equations that describe the motion of a material point under the action of a force: Newton's equations, Lagrange's and Hamilton's equations [546].

Traditionally, it is believed that each type of these equations is designed to solve its own class of problems.

So Newton's equations are designed to solve problems in which a material point is free, and in each space current a known force acts on it, so that the point moves under the action of this force. This class of problems includes, for example, the motion of planets (a problem solved by Newton himself), the scattering of particles by stationary bodies or other particles.

Here, the force can be specified as a function of a point in space, and this function can be determined as a property of the space itself, and specified by other bodies, for example, as a result of attraction or repulsion. The integral of this force over space can be considered as a potential, that is, the point seems to be moving in the field of this potential.

Lagrange's equations are designed to solve problems with constraints, such in which a material point is not free, but can only move within the limits imposed by constraints. In this case, the force causing the movement of the point is also known at each point in space. This class of problems includes various vibration problems, for example, a pendulum (load on a string, on a rod, on a spring), the movement of a ball in a hole, and others.

Here the force can consist of two parts: the same force as in the first case, as well as one more force imposed by the bonds, that is, the reaction force of the bonds. In the general case, this total force can also be reduced to potential.

To characterize the motion of a point, a special function is defined here – 'Lagrangian', equal to the difference between kinetic and potential energy. In this case, the structure of the Lagrange equations is such that a minus sign is present in front of the second term describing the force. Apparently, another function is possible, equal to the sum of the kinetic and potential energies, then the plus sign must be taken in the equation. Note that the Lagrangian determined through the energy difference is an artificial formation, while the sum of the kinetic and potential energies is simply the total mechanical energy of a material point, which is much more natural.

Generally speaking, to solve vibration problems, one can use not only the Lagrange equations, but also the Newton equations. For this, it is necessary to define the force as the derivative of the potential along the coordinate, that is, a certain potential must be given. In the Lagrangian, the role of such a potential is played by the potential energy. In both cases, the same second-order equations are obtained, connecting the second time derivative of the point's coordinate (that is, the acceleration of its motion) with the force acting on the point (that is, everything is reduced to Newton's second law).

Hamilton's equations are designed to solve problems about the motion of a particle in some fields. In contrast to the Newton and Lagrange equations, which are one second-order equation with one change – the coordinate of the point, there are two first-order equations with two variables – the coordinate of the point and its momentum, that is, the product of mass and velocity. Like the Lagrange equations, there is also a definite function here – the 'Hamiltonian', which is the sum of kinetic and potential energies, that is, the total energy of a moving point.

From the mathematical side, here we have reduced one second-order equation to two first-order equations. The reception of such information consists in the fact that a new variable is found, equal to the time derivative of the main variable, so that the second derivative of the main function becomes the first derivative of the introduced one. Here, the first function is the coordinate of the point, and the second is its impulse, proportional (up to mass) to the point's velocity, that is, the time derivative of the first variable.

Thus, this method is applicable to any of the equations obtained by the Newton or Lagrange method. The field in which the particle moves can also be described using a potential.

The reduction of one second-order equation to two first-order equations can be convenient in the case when the problem of the trajectory of motion of a point in some field is solved, if the value and direction of the velocity of a material point are given at the initial point of motion. A similar problem in mathematics is the 'Cauchy problem'. The solution is found numerically by the Euler or Runge–Kutta method. Along the way, in addition to the trajectory of the particle, its speed is determined in magnitude and direction. In this case, the result of the solution with the help of two first-order Hamilton equations is completely equivalent to the solution with the help of one Newton or Lagrange equation of the second order, provided that it is reduced to two first-order equations using the aforementioned mathematical technique.

Returning to the MSW, we note that the application of the Hamilton–Auld method here is not the only possible one, but is dictated primarily by tradition, as well as considerations of convenience and minimization of the amount of computations of the mathematical apparatus used.

2.11. Cauchy's problem in the distribution of SMSW

The application of the Hamilton–Auld method consists in the step-by-step calculation of the trajectories of the MSSW wave beams in accordance with certain first-order differential equations for a given initial condition (wave parameters at the initial point of the trajectory) and also a given distribution of the medium inhomogeneity along the wave propagation path.

The problem posed in this way is in fact the Cauchy problem widely known in mathematical physics [547]. Let us briefly consider the application of this approach, its expediency and legality of use. The classical Cauchy problem is that it is required to find a function defined on the half-line that satisfies the ordinary differential equation of the first order:

$$\frac{du}{dx} = f(x,u) \tag{2.685}$$

where f is a given function, and the desired function $u(x)$ at $x = x_0$ takes on a value u_0, that is $u(x_0) = u_0$ (this is an initial condition or a condition at the initial point).

Thus, the value of the derivative of the required function at a certain point (No. 1, at) is set through the value of the function itself at this point and the parameters of the environment (included in the function), depending on the coordinate of the same point. These data allow us to find the value of the function at another point (No. 2, at), located close enough to the first, provided that the parameters of the environment between these points are kept constant, using the formula:

$$u(x_2) = u(x_1) + \left[\frac{du}{dx}(x_1)\right] \cdot (x_2 - x_1) \tag{2.686}$$

This is the classical Euler method for solving a first-order linear differential equation. The same problem can be solved more accurately using the Runge–Kutta method. This is the procedure that is performed when finding the development of oscillations in time. Thus, knowing the parameters of the problem at some starting point, we can sequentially move from the previous point to the next and in this way find the parameters of the desired function along the entire path of its propagation. In this case, the parameters of the desired function at a given point are completely determined by the history of the problem, that is, the parameters of the environment lying behind this point, which the sought function has already passed. That is, we can say that here the present is completely determined by the past.

This is the classical Cauchy problem in its one-dimensional version. Multidimensional generalization is not difficult (the classical solution of a system of first-order differential equations with initial conditions), and this principle of defining the present by the past is always preserved.

Now let's see how this approach can be applied to the problem of wave propagation through an inhomogeneous medium, that is, is it possible, knowing the parameters of a wave at one point, by successive steps to find its parameters at other subsequent points.

Immediately, we note that in each current of the medium there will be two waves propagating in opposite directions: the first is propagating in the direction of incidence of the initial wave, that is, as it were, falling, and the second, propagating in the opposite direction, that is, as if reflected

In this case, the wave field (disturbance) at this point will be the sum of the fields of both these waves.

For this field (total), on both sides of the point under consideration, the boundary conditions must be satisfied: the equality of the values of this field and the equality of its first derivatives with respect to the coordinate.

The necessity of the existence of two counterpropagating waves follows from two possible solutions of the second-order wave equation.

Thus, if one incident wave arrives at some point in space, which we want to follow further, then the reflected and transmitted waves relative to this point will be determined not only by the incident wave propagating in the forward direction, but also by the parameters of the wave arriving at this point from the opposite direction.

The parameters of this counterpropagating wave are not known to us in advance, since they are determined not only by the incident wave, but also by the conditions that the counterpropagating wave underwent when approaching the point under consideration from the opposite side.

Thus, the parameters of the wave at a given point are determined not only by the prehistory of its propagation from the beginning to this point, but also by the parameters of the medium lying in front of this point, information about which is carried by the counter wave.

That is, we can say that here the present is determined not only by the past, but also by the future (if it were possible, the Wells time machine would be feasible and we would meet Morlocks and Eloi on the street).

Thus, knowing the parameters of a wave at one point, it is generally not possible to determine its parameters at subsequent points by successive steps along these points. In the most general form, it is necessary to consider the problem of wave propagation in an inhomogeneous medium from two ends at once (fortunately, we are talking not about time, but about space, and, fortunately, both ends of space are available to us), that is, in the form of the fall of two opposite waves If, ultimately, we are interested in the propagation of only one wave from one end, then we must set the amplitude of the counterpropagating wave at the point of its entry into the medium equal to zero, but at any infinitely close to this point inside the medium, the counterpropagating wave must be considered.

The authors of this work are not aware of a formal mathematical treatment of such problems in general form. However, for the particular case of stepwise inhomogeneities in [548-552], the possibility of solving the problem of propagation of a one-dimensional and electromagnetic wave from two ends is shown and a completely adequate mathematical apparatus is proposed, which can be easily implemented in algorithmic form both analytically and numerically.

In [553–557], a number of other methods were proposed, and in [558–561], the distribution of wave amplitudes in the layers of a multilayer structure, including one containing magnetic inhomogeneities, excited from both ends, was considered.

Thus, the general problem of wave propagation in inhomogeneous media must be solved 'from both ends'. However, there are a number of situations that allow one to neglect the counterpropagating waves, that is, to reduce the consideration to the classical Cauchy problem. Let us dwell briefly on some special cases.

To be able to use the 'one-sided' approach, that is, to avoid meeting the future, this 'future' must be somehow eliminated, that is, to make sure that the wave reflected inside the environment from some discontinuities lying ahead of the considered point is not interfered with. Without pretending to be an exhaustive consideration, we note three possible options.

The first option is to consider a medium with a large attenuation, so large that the counterpropagating wave reflected from the ahead lying inhomogeneity is already strongly attenuated at the point under consideration. Here, apparently, it is necessary to introduce some characteristic parameter of inhomogeneity along the coordinate, for which to obtain the limiting value based on the required attenuation value (for example, by an order of magnitude or at least by a factor). Here it is necessary to evaluate how the resulting solution will depend on the degree of inhomogeneity, that is, find the dependence of the reflection (or transmission) on the attenuation value and see at what parameters it will tend to a stationary value.

The second option is to assume that the wave reflected from the unknown lying inhomogeneity in front goes somewhere to the side and does not contribute to the wave reflected at this point. This option is not suitable for a one-dimensional problem, but it is already quite suitable for a two-dimensional problem on a plane. It can also be used for three-dimensional problems of oblique incidence of a wave on a plane boundary. At least a slight attenuation will further improve the situation (however, in the first case, you can do without it). Here it is necessary to consider not plane waves, but wave beams limited in width, so that the wave reflected from a local inhomogeneity would immediately go beyond the width of the initial beam and go to the side. In this case, the initial beam will propagate along a curved trajectory, the shape of which will be determined by the distribution of inhomogeneities within the medium. We note right away that the inhomogeneities should be sufficiently smooth, that is, the parameters of the medium at the wavelength should not change much. Here, too, a critical parameter of inhomogeneity can be introduced, which will somehow be related to the wavelength and curvature of the trajectory. This approach is widely used in problems of wave propagation in a medium consisting of plane-parallel layers with smoothly varying parameters, primarily for the propagation of radio waves in the ionosphere (where there are charged layers) and acoustic waves in the ocean (where the water temperature changes with depth). It is in this way that many problems have been made on the propagation of MSWs in nonuniformly magnetized ferrite films (discussed in more detail below). In this case, the problem is reduced to solving the system of Hamilton–Auld equations (first order) for the coordinate and wave number components of the wave in the polar coordinate system. All problems are solved here exactly as the Cauchy problem, that is, all the results are obtained using the traditional Euler or Runge–Kutta methods.

The third option is to consider the development of the process in time, that is, the propagation of a step front of the incident wave in an inhomogeneous medium. Note that everything that has been said before this line applies only to the regimes established in time. In the process of wave front propagation, counterpropagating waves in front of this front will not be formed, since the wave has not yet reached there, that is, there is nothing to be reflected. Behind the front, it is already necessary to consider the counterpropagating wave, but the front itself will propagate in its pure form, that is, it will probably again be possible to turn to a Cauchy-type problem. Some similar tasks were also considered, and here the ideology is rather complicated, but it goes beyond the authors' field of vision. As an example, one can point to the reviews [562, 563] concerning the propagation of pulses in inhomogeneous media.

This so far exhausts the possibilities, seen by the authors, of solving problems of wave propagation in inhomogeneous media by more or less simple means, including by reducing them to the Cauchy problem.

2.12. Technique for calculating the trajectories of wave beams of MSW in an inhomogeneous field

The calculation of the trajectories of the MSW wave beams in an inhomogeneous field is performed by numerically solving the system of Hamilton–Auld equations, which is a system of three first-order differential equations. Variables are wave number, angle, coordinate, parameter - coordinate. Such a problem in mathematics is called the Cauchy problem [547].

For all variables at the starting point of the trajectory, initial values are set, at the same point the derivatives of the variables with respect to the parameter are calculated using the equations, the step of the parameter is set, after which, using the obtained values of the derivatives, the values of the variables are found at the subsequent point determined by this step. The procedure is repeated until until the entire required trajectory is calculated. This is, in general outline, the scheme of Euler's method [340]. Calculations are carried out in accordance with the following formulas:

system of equations:

$$\frac{dk}{dy} = f(k, \phi, z, y) \qquad (2.687)$$

$$\frac{d\phi}{dy} = g(k, \phi, z, y) \qquad (2.688)$$

$$\frac{dz}{dy} = h(k, \phi, z, y) \qquad (2.6\,89)$$

decision:

$$k_{p+1} = k_p + f(k_p, \phi_p, z_p, y_p) \cdot \Delta y \qquad (2.690)$$

$$\phi_{p+1} = \phi_p + g(k_p, \phi_p, z_p, y_p) \cdot \Delta y \qquad (2.691)$$

$$z_{p+1} = z_p + h(k_p, \phi_p, z_p, y_p) \cdot \Delta y \tag{2.692}$$

More accurate results are obtained by the Runge–Kutta method [340], in which the derivatives within one step are calculated not at one point, but at several points, distributed in one way or another along the step length. Calculations using the fourth-order Runge-Kutta method are performed using the following formulas:

system of equations:

$$\frac{dk}{dy} = f(k, \phi, z, y) \tag{2.693}$$

$$\frac{d\phi}{dy} = g(k, \phi, z, y) \tag{2.694}$$

$$\frac{dz}{dy} = h(k, \phi, z, y) \tag{2.695}$$

decision:

$$k_{p+1} = k_p + \frac{1}{6} \cdot (k_1 + 2k_2 + 2k_3 + k_4) \tag{2.696}$$

$$\phi_{p+1} = \phi_p + \frac{1}{6} \cdot (m_1 + 2m_2 + 2m_3 + m_4) \tag{2.697}$$

$$z_{p+1} = z_p + \frac{1}{6} \cdot (n_1 + 2n_2 + 2n_3 + n_4) \tag{2.698}$$

In these expressions:

$$k_1 = f(k_p, \phi_p, z_p, y_p) \cdot \Delta y \tag{2.699}$$

$$m_1 = g(k_p, \phi_p, z_p, y_p) \cdot \Delta y \tag{2.700}$$

$$n_1 = h(k_p, \phi_p, z_p, y_p) \cdot \Delta y \tag{2.701}$$

$$k_2 = f\left(k_p + \frac{k_1}{2},\ \phi_p + \frac{m_1}{2},\ z_p + \frac{n_1}{2},\ y_p + \frac{\Delta y}{2}\right)\cdot \Delta y \tag{2.702}$$

$$m_2 = g\left(k_p + \frac{k_1}{2},\ \phi_p + \frac{m_1}{2},\ z_p + \frac{n_1}{2},\ y_p + \frac{\Delta y}{2}\right)\cdot \Delta y \tag{2.703}$$

$$n_2 = h\left(k_p + \frac{k_1}{2},\ \phi_p + \frac{m_1}{2},\ z_p + \frac{n_1}{2},\ y_p + \frac{\Delta y}{2}\right)\cdot \Delta y \tag{2.704}$$

$$k_3 = f\left(k_p + \frac{k_2}{2},\ \phi_p + \frac{m_2}{2},\ z_p + \frac{n_2}{2},\ y_p + \frac{\Delta y}{2}\right)\cdot \Delta y \tag{2.705}$$

$$m_3 = g\left(k_p + \frac{k_2}{2},\ \phi_p + \frac{m_2}{2},\ z_p + \frac{n_2}{2},\ y_p + \frac{\Delta y}{2}\right)\cdot \Delta y \tag{2.706}$$

$$n_3 = h\left(k_p + \frac{k_2}{2},\ \phi_p + \frac{m_2}{2},\ z_p + \frac{n_2}{2},\ y_p + \frac{\Delta y}{2}\right)\cdot \Delta y \tag{2.707}$$

$$k_4 = f\left(k_p + k_3,\ \phi_p + m_3,\ z_p + n_3,\ y_p + \Delta y\right)\cdot \Delta y \tag{2.708}$$

$$m_4 = g\left(k_p + k_3,\ \phi_p + m_3,\ z_p + n_3,\ y_p + \Delta y\right)\cdot \Delta y \tag{2.709}$$

$$n_4 = h\left(k_p + k_3,\ \phi_p + m_3,\ z_p + n_3,\ y_p + \Delta y\right)\cdot \Delta y \tag{2.710}$$

It can be seen from the above formulas that the Euler method, from the point of view of numerical algorithms, is much simpler than the Runge–Kutta method. The accuracy of both methods increases with decreasing the step in the variable, so that with the simpler Euler method, you can get an accuracy similar to that of the Runge–Kutta method by simply decreasing the step. However, for example, when solving oscillatory problems when finding the sweep of oscillations in time or in space, Euler's method usually works more or less satisfactorily for no more than 4–5 oscillation periods, after which the error increases noticeably, while the Runge–Kutta method with that the same step gives an error of no more than a few percent for up to several tens and even hundreds of periods.

Comment. Looking ahead, we note that in practice, when calculating the SMSW trajectories by the Hamilton–Auld method, for fields of a linear or 'valley' type, where the trajectories are aperiodic, it is usually sufficient to use the Euler method, however, for fields of the 'shaft' type, where trajectories are close to oscillatory, including those with many periods, the Runge–Kutta method is more preferable.

Conclusions for chapter 2

The main issues discussed in this chapter are as follows.

1. A detailed review of the basic mathematical apparatus used in solving problems related to the propagation of magnetostatic waves in iron–garnet films is carried out. The derivation of the equation of motion of the Landau–Lifshitz magnetization is presented. For a medium with an arbitrary tensor of magnetic susceptibility, the derivation of the Walker equation is given in general form, and the transition to a similar equation used in the Damon–Eshbach problem is shown. The classical solution of the Damon–Eshbach problem is considered, which makes it possible to find the dispersion relation, potentials, and field components of the magnetization vector of a wave propagating in a tangentially magnetized ferrite plate with free surfaces. The dispersion relation is found in both polar and Cartesian coordinate systems. Dispersion relations were obtained for more complex structures: 'metal–dielectric–ferrite–dielectric-metal' and 'metal–dielectric–ferrite–ferrite–dielectric-metal', from which, as special cases, the relationships for the structures 'ferrite–metal', 'ferrite–dielectric–metal', 'ferrite–ferrite' were obtained.

2. It is shown for the MDFFDM structure that the full MSW spectrum contains eight branches, four of which are bulk and four are surface, two of the surface branches are external and two are internal. The boundaries of the spectrum are found for all branches, it is shown that they strongly depend on the angle between the wave vector of the wave and the direction of the field, and the corresponding cutoff angles are determined.

3. For the Damon–Eshbach relation, an interval of admissible directions of wave propagation is obtained, limited by the cutoff angle, general formulas are found for calculating the phase and group velocities of the wave. The Hamilton–Auld equations are derived, which make it possible to find the propagation trajectories

of magnetostatic waves in inhomogeneously magnetized ferrite films and structures based on them. For the ferrite–dielectric–metal structure, derivatives of the variables are obtained, which determine the phase and group velocities, as well as the trajectories of wave beams in inhomogeneous fields. The analogy of the apparatus under consideration with the main methods of classical mechanics, as well as the correctness of its reduction to the Cauchy problem of mathematical physics, are discussed. A scheme for solving the system of Hamilton–Auld equations based on the Euler and Runge–Kutta equations is presented.

3

Magnetostatic waves in homogenized magnetized ferrite films and structures on their basis

This chapter discusses the main issues of the propagation of MSWs in uniformly magnetized ferrite films and structures based on them, which are important for the future.

The dispersion properties of the SMSW in the FDM structure are considered and are important for further presentation. Isofrequency curves are plotted on the plane of wave vectors, and the regions of existence of forward and backward SMSWs are determined. The propagation of SMSWs in spatially inhomogeneous structures magnetized by a uniform field is considered, in particular, the laws of refraction and reflection of SMSWs at the interface between a ferrite film with a free surface and an FDM structure are studied. The dispersion of SMSW in two-layer ferrite films with free and fixed surface spins is investigated.

3.1. Conditions of existence and dispersion of SMSW (surface magnetostatic waves) in ferrite films and structures on their basis

Let us first consider the general conditions for the existence and dispersion properties of SMSWs. Let us investigate in more detail the dispersion properties of forward and reverse SMSWs in the ferrite–dielectric–metal (FDM) structure, after which we present the results of an experimental study of the dispersion of SMSWs in this structure. Let us identify the reasons for the difficulty of observing backward waves and discuss the possibility of its implementation. Basically, we will follow the works [4, 9, 292, 296, 297, 305, 564-579]. Links to other works are given in the text.

3.1.1. Dispersion properties of forward and backward SMSWs in the FDM structure

The propagation of MSWs in structures based on ferrite films is determined by two factors: the parameters of the waveguide structure and the dispersion law of the MSW. Let us first consider the properties of MSWs that follow from the dispersion law under the assumption that the medium is homogeneous. We restrict ourselves to the properties of only surface MSWs (SMSWs), as they are more interesting for practice due to their single-mode nature. As an example, let us take the ferrite–dielectric–metal (FDM) structure, which is important for further presentation, the dispersion relation for which has the form (2.520):

$$\beta - 1 - 2\mu\alpha \cdot \operatorname{cth}(\alpha k d) + (\beta + 1 - 2\nu\delta) \cdot e^{-2kp} = 0 \tag{3.1}$$

In such a structure, the lower frequency boundary of the SMSW spectrum does not depend on the thickness of the dielectric layer p and in the given values at $\Omega_H = 0.25$ is equal to: $\sqrt{\Omega_H(\Omega_H + 1)} = 0.559$ and the upper boundary is determined by the thickness of the dielectric layer p and lies between $\Omega_H + 1/2 = 0.75$ at $p \to \infty$ (ferrite film with a free surface) and $\Omega_H + 1 = 1.25$ at $p = 0$ (ferrite film with a metallized surface).

Typical dispersion curves (in reduced values) at $\varphi = 0$ for different values of the ratio of the dielectric layer thickness to the ferrite film thickness d are shown in Fig. 3.1. It can be seen that they are all enclosed between the curve 1 corresponding to $p = 0$ (ferrite film with a metallized surface), and the curve 8, corresponding to $p \to \infty$ (ferrite film with a free surface). All curves, except for 1, at $k \to \infty$, asymptotically tend to the horizontal straight line corresponding to $\Omega_H + 1/2$, and curve 1 to the value $\Omega_H + 1$.

Small ones are characterized by three main types of dispersion curves:

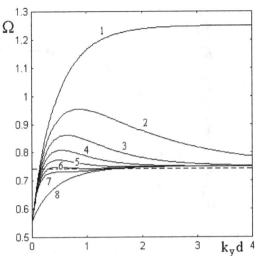

Fig. 3.1. Dispersion curves for $\varphi = 0$, corresponding to different values of the ratio of the thickness of the dielectric layer to the thickness of the ferrite film: $1 - p/d = 0$; $2 - p/d = 0.33$; $3 - p/d = 0.67$; $4 - p/d = 1.0$; $5 - p/d = 1.33$; $6 - p/d = 1.67$; $7 - p/d = 2.00$; $8 - p/d \to \infty$.

the first – with a characteristic maximum, which is more pronounced and corresponding to the higher values of the wavenumber, the smaller the value of the interval (curves 2, 3, 4, 5);

the second – with the appearance, along with the maximum, of a minimum at slightly larger values of the wavenumber than the corresponding maximum (curve 6);

the third – without extrema with monotonic growth and subsequent approximation to the horizontal asymptote (curves 7, 8).

In the frequency range where the dispersion curves have both maximum and minimum, their character depends on the frequency to a very significant extent. For clarity, Fig. 3.2 shows the same dispersion curves, but with smaller and smaller steps compared to Fig. 3.1. Dispersion curve 1 corresponds, curve 16 –. Curve 15 () corresponds to the transition from the maximum and minimum to the inflection. You can see that here the appearance of the minimum and its disappearance are much more pronounced than in Fig. 3.1. Curve 6 in Fig. 3.1 here corresponds to the intermediate region between the curves 11 and 12, and the minimum begins to appear already between the curves 8 and 9 (that is, at 5), and disappears between the curves 14 and 15 (that is, at $p/d \approx 1.35$).

The reason for this behaviour of the dispersion curves is that, at small values, the SMSW length is large, so that its field penetrates

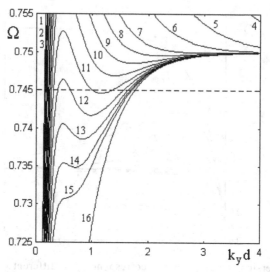

Fig. 3.2. Dispersion curves for different values: 1 – p/d = 0; 2 – p/d = 0.2; 3 – p/d = 0.4; 4 p/d = 0.6; 5 – p/d = 0.8; 6 – p/d = 1.0; 7 – p/d = 1.2; 8 – p/d = 1.3; 9 – p/d = 1.4; 10 – p/d = 1.5; 11 – p/d = 11.6; 12 – p/d = 1.7; 13 – p/d = 1.8; 14 – p/d = 1.9; 15 – 2.0; 16 – $p/d \to \infty$.

deep enough into the dielectric layer and strongly 'feels' the presence of the outer metal layer. In this case, the dispersion curve tends to the curve 11, which corresponds to the surface with metallization. As the depth of penetration of the wave into the dielectric increases, the dispersion curve tends to curve 8, which corresponds to a free ferrite film. The transition from the first type of dispersion curves to the second corresponds to the first critical value of the dielectric layer thickness $p_{c1} \vec{k}$ d from the second to the third – to the second critical value $p_{c2} \vec{k}$ $2d$. In the increasing sections of the dispersion curves (before the maximum and after the minimum), the directions of the phase and group velocity vectors coincide, that is, the waves are straight. On the falling section between the maximum and minimum, the directions of these vectors are opposite, that is, the waves are reverse.

Let us now consider the same properties of the dispersion of the SMSW in the FDM structure using isofrequency curves as an example.

First, let us turn to Fig. 3.3, built for the same values p/d as in Fig. 3.1 at a frequency Ω = 0.74 (3626 MHz).

It can be seen from this figure that in the range of values from to and less isofrequency curves have a complex intertwining character due to the simultaneous existence of maxima and minima on the corresponding dispersion curves (Fig. 3.2). This region is shown in

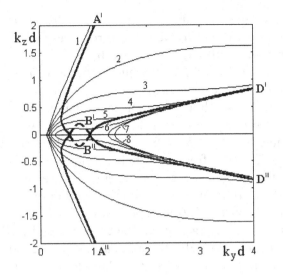

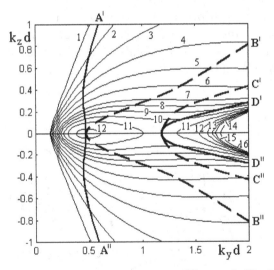

Fig. 3.3. Isofrequency curves under the same conditions as in Fig. 3.1 and frequency $\Omega = 0.74$ (3626 MHz).

more detail in Fig. 3.4, plotted at a slightly higher frequency (3650 MHz), but at values with a smaller step, the same as in Fig. 3.2.

It can be seen that with an increase in the frequency from to (that is, from 3626 MHz to 3650 MHz) by only 0.67%, the nature of the isofrequency curves changed sharply, which becomes clear from the very close relative position of the maxima and minima of the dispersion curves in Fig. 3.2.

As can be seen from the above figures, the isofrequency curves for the FDM structure can be divided into three classes in appearance:
closed, resembling ellipses ('pseudoelliptic'),
open, resembling hyperboles ('pseudohyperbolic'),
open, similar to an ellipse with a gap, the edges of which pass into the branches of the hyperbola ('transitional').

Analysis of the dispersion relation (3.1) shows that in the frequency range from 0.559 (the lower boundary of the SMSW spectrum) to 0.734 (the frequency below which the maximum and minimum on the dispersion curve merge and are replaced by an inflection) all isofrequency curves are pseudohyperbolic. In the frequency range from 0.750 (the upper limit of the SMSW spectrum for a free film) to 1.250 (a similar limit for a metallized film), all curves except 1 are pseudoelliptic, and curve 1 is pseudohyperbolic. In the frequency range from 0.734 to 0.750 at, where the dispersion curve has a minimum, for frequencies above the frequency of the minimum, the isofrequency curves have a transitional type, and below they contain pseudoelliptical and pseudohyperbolic parts. In Fig. 3.3, the curves 1 and 8 are pseudohyperbolic, and the curves 2–7 are transitional. Curve 1 corresponds to (the surface of the ferrite film is metallized), curve 8 - (the surface of the ferrite film is free).

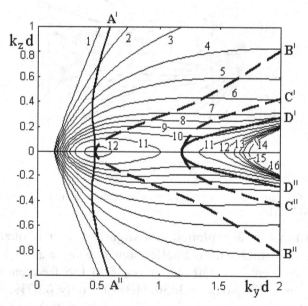

Figure 3.4. Isofrequency curves under the same conditions as in Fig. 3.2 and frequency (3650 MHz).

The rest of the isofrequency curves (2–7) are enclosed between these two, and their shape, like dispersion curves, is determined by the tendency to curve 1 at small values, and to curve 8 at large ones.

As indicated in Section 2.7.1 (Fig. 2.9), each point of the isofrequency curve defines the vectors of the phase and group velocities of the wave, and the phase velocity corresponds to the vector from the origin to this point, and the vector of the group velocity is perpendicular to the tangent to the isofrequency curve at the same point. In this case, if the projection of the group velocity vector onto the phase velocity vector is positive, then the wave is direct, if negative, it is reverse.

In the right half-plane of the plane $Ok_y k_z$, forward waves always propagate in the positive direction of the axis, and the direction of propagation of backward waves relative to the axis can be either positive or negative. On this basis, a number of areas can be distinguished, delimited in Fig. 3.3 by lines $A'A''$, $B'B''$ and $D'D''$. Here lines $A'A''$ and $D'D''$ connect the points corresponding to the tangents to the isofrequency curves constructed from the origin, and the lines $B'B''$ are the points where the isofrequency curves have horizontal tangents (parallel to the axes). Between the isofrequency curve 1 and the line $A'A''$, as well as between the isofrequency curve 8 and the line $D'D''$, the waves are straight.

For further consideration, it is convenient to distinguish two types of direct waves in each of these areas:

straight lines SMSW of the first type – between the isofrequency curve 1 and the line $A'A''$, corresponding to the growing section of the dispersion curves in Fig. 3.1 from zero to the first maximum;

straight lines SMSW of the second type – between the line $D'D''$ and the isofrequency curve 8, corresponding to the growing section of the dispersion curves in Fig. 3.2 from minimum to infinity.

Between the lines $A'A''$ and $D'D''$ SMSWs are reversed. Outside the area bounded by the $B'B''$ the lines, waves propagate in the positive direction of the axis, inside this area - in the negative direction.

Let us now take a closer look at Fig. 3.4, which corresponds to a slightly higher frequency. Lines $A'A''$ *and* $D'D''$, as in Fig. 3.3, connect the points through which the tangents to the isofrequency curves, constructed from the origin, pass. The lines B'B" and C'C" connect points where isofrequency curves have horizontal tangents. As before, outside the area bounded by the lines $A'A''$ and $D'D''$, SMSWs are straight lines, and inside this area, they are reverse. In

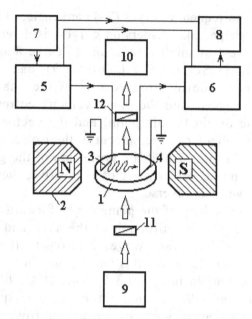

Fig. 3.5. Simplified block diagram of the measuring stand. 1 – ferrite film, 2 – magnet, 3 – emitting transformer, 4 – receiving transducer, 5 – microwave generator, 6 – microwave receiver, 7 – synchronization generator, 8 – oscilloscope, 9 – light source, 10 – microscope, 11 – polarizer, 12 – analyzer.

areas between lines $A'A''$ and $B'B''$, as well as $C'C''$ and $D'D''$, waves propagate in the positive direction of the axis, in the area between lines $B'B''$ and $C'C''$ – in negative. Comparison of Figs. 3.3 and 2.4 b shows that in the second case, in contrast to the first, the region of the negative direction of SMSW propagation is significantly expanded and strongly elongated towards large values of k_y and k_z.

The isofrequency curves 1, 15, 16, plotted with the same parameters, are pseudohyperbolic. Isofrequency curves 1–10 are transitional, and curves 13 and 14 are pseudohyperbolic. The curves 11 and 12 consist of two parts and have both pseudoelliptical and pseudohyperbolic components.

Above, we considered the dispersion and isofrequency curves for the FDM structure depending on the thickness of the dielectric layer. Similar curves can be obtained by changing other parameters of the waveguide structure – magnetization and thickness of the ferrite film, as well as the field. In this case, however, the general picture of the behaviour of forward and backward waves does not change significantly, so we will not dwell on this issue here, and in the future, as necessary, we will only clarify some details.

3.1.2. Experimental study of the dispersion of the SMSW in the structure of the PDM

Let us now consider the results of experimental studies of the propagation of SMSWs in inhomogeneous structures. We will use the results of the works generalized in [292, 297]. First, let us briefly dwell on the experimental technique, after which we proceed directly to the description of the experimental results.

3.1.2.1. Basic experimental technique

Before proceeding to the description of the results of the experimental study of the SMSW, let us consider the main measuring technique used in this work.

3.1.2.1.1. Measuring stand

Most of the experiments described in this work were carried out on a measuring stand, a simplified block diagram of which is shown in Fig. 3.5. Here, 1 is a ferrite film, 2 is a magnet, 3 is an emitting converter, 4 is a receiving converter, 5 is a microwave generator, 6 is a microwave receiver, 7 is a synchronization generator, 8 is an oscilloscope, 9 is a light source, 10 is a microscope, 11 is a polarizer, 12 is an analyzer. In most experiments, the generator 5, receiver 6, and oscilloscope 8, were replaced by a single device – a meter of complex transmission coefficients of the P4-23 type, or a panoramic meter of the P2-52–P2-58 type. If separate devices were used, then G4-79–G4-81 generators were used as a microwave generator 5, depending on the range. As a receiver 6 – measuring receivers P5-3–P5-5. Synchronization signal generator 7 – pulse generator G5-15 or G5-71. Oscilloscope 8 – C1-54 or C1-72. Light source 9 is an illuminator for a microscope OI-9M or an incandescent lamp. Microscope 10 – MBS, MBD or other similar. The polarizer 11 and the analyzer 12 are based on a polaroid film. A permanent or electromagnet was used as magnet 2. The permanent magnet had round poles with a diameter of 20 cm, the gap between which was adjustable from 5 to 25 cm, which provided a field between the centers of the poles from 200 to 700 Oe. The electromagnet had poles

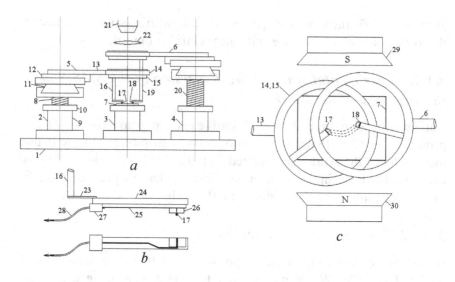

Fig. 3.6. Probe setup diagram.

with a diameter of 12 cm and provided a field from 0 to 1700 Oe in a gap of 9 cm. The emitting and receiving transducers were made in the form of movable antennas from a straight wire 3–4 mm long and 12 μm in diameter, superimposed on the film plane. The mechanical system provided for the movement of the antennas over the entire plane of the film (70 mm in diameter) and rotation around an axis perpendicular to the plane of the film. The geometrical arrangement and orientation of the movable antennas were determined using a measuring microscope 10 equipped with a micrometer object. The observation was carried out in reflected light, for which an additional illumination source in the form of an incandescent lamp, placed on the side of film 1 (not shown in the figure), was used.

The measuring stand works as follows. The microwave signal from the generator 5 is fed to the emitting transducer 3, superimposed on the plane of the ferrite film 1. Here, the electromagnetic signal is converted into an MSW, which propagates along the ferrite film 1, which is in the field of the magnet 2. When the MSW reaches the receiving transducer 4, it is again converted into the electromagnetic signal, which is received, amplified and detected by the receiver 6, and then fed to the oscilloscope 8. The generator 5 can operate in a continuous mode, in a pulse mode set by the generator 7, as well as in a frequency sweep mode. In the continuous mode, the signal propagation, orientation and field dependences, the dispersion law of the MSW are investigated, in the pulsed mode – the time delay of the pulse signal, in the frequency sweep mode – the frequency response, phase characteristics, dispersion of the MSW. The reference signal for measuring the phase characteristics is fed to the receiver 6 from the generator 5 through an additional channel (bypassing the film). When operating in a pulse mode, a synchronization signal is sent from the pulse generator 7 to the oscilloscope 8. The domain structure of the films is observed by the Faraday effect in transmitted polarized light. For this, a light source 9, a polarizer 11, an analyzer 12 and a microscope 10 are used. The mutual independence of the optical and microwave channels allows microwave measurements and observation of the domain structure at the same time, which ensures the required high measurement accuracy.

Practically all the experimental studies described below were carried out using such a stand. At the same time, the structure of the stand, instruments and details of the measurement technique were modified based on the requirements of a particular experiment. More details of such details will be reported later in each relevant section.

3.1.2.1.2. Movable probes

Let us give a brief description of the design of the probe setup, which made it possible to study the propagation of SMSW along the plane of the film. A.V. Vashkovsky, S.V. Gerus, D.G. Shahnazaryan, A.V. Voronenko, E.G. Lokk and other members of the authors' teams of the corresponding works took part in the initial design development.

The study of the propagation of SMSWs in the plane of the film magnetized by an inhomogeneous field was carried out on a setup containing movable transmitting and receiving transducers, which were rectilinear wire antennas 3–5 mm long and 12–15 μm in diameter, superimposed on the film plane. The mechanical design of the installation is illustrated in Fig. 3.6.

Figure 3.6 a shows the general setup in a side view between the magnetic poles, that is, when viewed along the field from one pole to the other.

The installation is based on optical bench 1, on which three optical tables 2, 3, and 4 are installed. Tables 2 and 4 are intended for fastening brackets 5 and 6, carrying movable antennas; sample 7, that is, the ferrite film under study, is attached to Table 3.

As an example, consider the design of the left-hand table 2 according to the figure, the designs of the other two tables 3 and 4 are similar.

The table is made with the ability to move the upper bearing platform in the vertical direction, which is provided by the translational movement of the cylindrical stand 8 inside the cylindrical tube 9. On the surface of the stand, a screw thread is applied, onto which the nut 10 is screwed, resting on the upper edge of the tube 9. Fixing the stand from turning around the vertical axis is provided by a screw (not shown in the figure) screwed into the side wall of the tube and with its end entering a rectilinear vertical slot on the surface of the post, passing through the entire thread covering the post. At the upper end of the rack, a supporting base is fixed, which consists of two dovetail type sliders 11 and 12, which are fastened crosswise, adjustable by longitudinal screws.

Thanks to the described structure, the upper surface of the supporting base has the ability to smoothly move in three mutually perpendicular directions — vertical and two horizontal along the direction of the optical bench and perpendicular to it.

In the upper part of the support base, the actual bracket 5 is fixed, which is a rod 13, at the end of which a circular ring 14 is fixed, into which another ring 15 is coaxially inserted, which can freely rotate relative to the first. The ring 15 is prevented from falling into the ring 14 by a rim protruding in its upper part. The position of the ring 15 relative to the ring 14 is fixed with a screw (not shown in the figure) screwed into the ring 14 from the side and abuts against the side surface of the ring 15.

On the inner side of the ring 15, a holder 16 is fixed in a vertical position, to the lower end of which, bent at a right angle, the actual antenna 17 is connected, the centre of which coincides with the axis of the rings 14 and 15.

Thus, the rotation of the inner ring 15 relative to the ring 14 rotates the antenna about the vertical axis,

The entire system as a whole provides the ability to move the antenna over the entire plane of the film in two mutually perpendicular coordinates, as well as to rotate the antenna relative to the vertical axis perpendicular to the plane of the film at any angle.

The second antenna 18 is attached to the post 4 in the same way using the bracket 6 and the holder 19. At the same time, in order to avoid mutual interference when moving the antennas, the post 20 of the table 4 is pushed up so that its bearing surface with the bracket 6 and the rings fixed to it is above similar rings of the bracket 5. In this case, the holder 19 is made longer than the holder 16 by an amount of about 1-2 cm, sufficient for the unhindered movement of the brackets 5 and 6 with the rings attached to them relative to each other.

The surface of the film 7 with antennas 17 and 18 located on it is observed from above using a microscope 20 equipped with an auxiliary lens 21, which makes it possible to increase its object distance to several centimeters. Registration of the position of the antennas and their orientation were measured using the object-micrometer supplied with the microscope.

Figure 3.6 b shows a diagram of the design of the mobile antenna holder in more detail. In the upper part of the figure there is a side view (similar to that in Fig. 3.6 a), in the lower part – a bottom view from the side of the film.

The vertical part of the holder 16 itself ends with a horizontal platform to which a spring plate 23 is attached with two screws (not shown). The plate is made springy to ensure soft mechanical contact between the antenna 17 and the surface of the film 7 in order to avoid mechanical damage to the film.

A rigid plate 24 is attached to the other end of the spring plate, on the lower surface of which is fixed a microstrip line 25 made on a polycor substrate. A polycor patch 26 is placed at the end of the microstrip line. The antenna itself is made in the form of a conductor 17 located on the surface of this patch perpendicular to the axis of the strip line, one end of the antenna conductor being connected to the central conductor of the strip line, and the other grounded to its substrate. At the other end of the strip line is connector 27 with a coaxial cable 28 connected to it.

Figure 3.6 c shows a schematic diagram of the picture observed through a microscope. Outside, the picture is bordered by rings 14, 15 of both tables fixed on brackets 13 and 6. The displacement of the rings relative to each other is due to the required location of the mobile antennas. Generally speaking, the rings should not fall into the field of view of the microscope, since their internal dimensions should provide unhindered access to the film surface, for which their diameter is chosen large enough (about 7–10 cm). Through the holes of the rings, the surface of the film 7 is visible, over which the upper surfaces of the holders of the movable antennas 17 and 18 are located. The antennas themselves are located on the lower surface of the holders, and only the contours of the edges of the holders are visible from above, according to the coordinates of which, measured by the micrometer object, the location of the antenna conductors is established. The dotted lines between the antennas

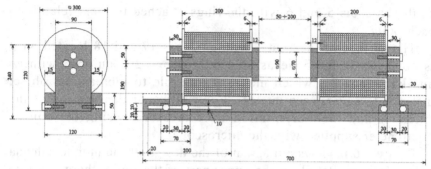

Fig. 3.7. Magnet circuit.

show possible SMSW trajectories, which are not directly observed in the microscope, but can be measured using mobile antennas.

The entire structure is in the field of a magnet, the poles of which 29 and 30 are located so that the field is directed perpendicular to the axis of the installation (axis of the optical bench 1), and in the plane of the film it is sufficiently uniform. That is, the height of the magnet should be set at such a level that the axis passing through the centers of the poles falls on the level of the film plane.

All details of the installation, except for the magnet, must be made of non-magnetic materials, which in this work were used brass, copper or duralumin.

Here are some typical installation sizes. The optical bench – with any convenient profile, about 10–20 cm wide, based on the stability of the installation as a whole on the laboratory bench. The length of the bench should ensure the convenience of placing the tables, that is, be 40–60 cm. The distance between tables 2 and 4 is about 30–40 cm. The height of tables 2 and 4 is about 15–20 cm, table 3 is about 10 cm: vertically – from 10 to 25 cm, along both horizontal axes – up to 10 cm. The diameter of the rings: outer – about 8 cm, inner – about 7 cm. The height of the rings: outer – about 2 cm, inner – about 3 cm. Height holders 16 and 19 – 4 cm and 6 cm, respectively. The distance from the film surface to the microscope objective is up to 12 cm. Lens 22 should have a focal length of about 3–5 cm and be located on a separate holder, fixed relative to the microscope.

3.1.2.1.3. Magnet

Let us briefly describe one of the magnets used in this work, which create a constant field in the film plane.

The dimensions shown in the figure are in millimeters. All dimensions are approximate, that is, when repeating the design, they can be changed depending on the required operating conditions.

The base, poles and cores are made of steel-3 soft iron. For the entire magnetic circuit along its length, it is necessary, if possible, to comply with the condition that the cross-sectional area is in no way less than the cross-section of the coil cores, and even better – that it is larger. That is, the better this condition is met, the more field the magnet can give. In the figure, the dotted line shows the outline of the outer edge of the bed, removed so as not to darken the drawing.

The pole on the right in the figure was fixed on the bed motionlessly. The left one could be moved and fixed motionlessly in the desired position with the help of screws moving in the slots of the bed, thereby setting the required gap between the poles from 50 to 200 mm.

Coil frames with all walls 6 mm thick are made of duralumin. The side cheeks of the coils are fastened to the through cylindrical base by means of M3 screws entering the base from the cheek side along the longitudinal axis of the cylinder. Coil frames are glued with varnished cloth from the inside.

The coils are wound with PEL-0.8 (or PEV-0.8) copper wire, that is, with a 0.8 mm diameter wire in enamel insulation. Winding – in bulk, winding uniformity was established by eye. The winding was carried out using a large lathe (the size of a sofa), set back slowly. The winding time took two days.

Each of the coils contains four sections, so there are a total of eight sections on both coils. The sections are wound on top of each other, that is, one on top of the other. The number of turns was not counted. The number of wires in the sections was adjusted according to the equality of resistance, that is, the innermost section was the thickest, the outermost was the thinnest. The required section thicknesses were calculated in advance, based on the average diameter of the turns, that is, so that the volume of the wire in each section, defined as the product of the section cross-section by the average length of the turn, was kept constant. In accordance with the obtained values of the thicknesses of the sections, holes were drilled in the cheeks through which the ends of the winding of each section were brought out. A layer of varnished cloth was laid between the sections. The ends of the windings were taken out in cambric tubes about 3 mm thick. On each cheek, a textolite board with the layout of the coil leads was fixed outside.

Such a complex sectioned coil arrangement was made for versatility, that is, so that the magnet could be powered from a source with a possibly arbitrary current–voltage ratio. So, when all the sections were connected in series, the total resistance was quite large and for power supply it was possible to use a source of the same type as for lamp circuits, that is, with a voltage of 200–300 V and a current of less than 1 A. When the sections were connected in parallel, the total resistance was rather small and it was possible to use a source of such a type as a battery charger or the batteries themselves, that is, with a voltage of 10–20 V and a current of 5–10 A. In addition, it was possible to supply constant power to most of the sections, and to one or two sections – adjustable, which gave an additional opportunity for variations in the choice of the source, and also provided a finer field adjustment. Due to the independence of the sections from each other, it was possible to feed only some of them, or each section separately from its source.

A magnet without cooling provides a field of up to 3000 Oe in a 5 cm gap, and up to 1500 Oe in a 10 cm gap. In a smaller gap and with some slight heating (ten degrees above room temperature and not for long) it can provide a field of up to 5000 Oe.

One of the working options for the magnet power supplies consists of two LATRs (large), one of which feeds six or seven sections, and the other one or two. Current rectification is carried out using power diodes (about 20 pieces), smoothing – using four electrolytic capacitors 500 µF at 450 V. Measurement of the base and control currents is carried out using two microammeters at 100 µA with shunts.

3.1.2.2. Results of an experimental study of the dispersion properties of SMSW

Let us now proceed directly to the description of the results of an experimental study of the dispersion properties of SMSWs in the FDM structure. Traditional methods for studying the dispersion of SMSWs are reduced to measuring the phase incursion when the frequency or field changes [4, 9]. In this case, rigid fixation of the position of the exciting and receiving antennas creates uncertainty in measuring the SMSW length near the low-frequency edge of the spectrum. The method of mobile antennas used in this work is free from this drawback, which makes it possible to directly measure the SMSW length from the phase change of the output signal by

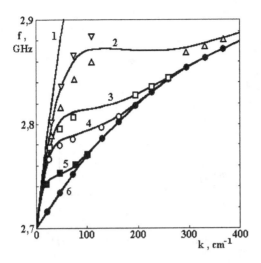

Fig. 3.8. Theoretical (solid lines) and experimental (points) dispersion dependences at different gaps: 1 – 20 μm; 2 – 60 μm; 3 – 100 μm; 4 – 135 μm; 5 – 250 μm; 6 – ∞.

360°. This method, however, is suitable only for a ferrite film (FF), in which the surface is free, and for FM and FDM structures, when the film surface is covered with a metal, it is inapplicable. For this case, the method of measuring the dispersion is used, which consists in changing the distance between the metal strip and the surface of the ferrite film [297]. In this case, the antennas are located outside the band and do not move during its movement. To implement the method, the metal plate is first removed and the dispersion of the SMSW in the free PC is measured by any known method, for example, by moving the receiving antenna. After that, at a fixed frequency, as the metal strip approaches a given distance, an additional phase incursion caused by the metal is measured, which corresponds to a change in the number of SMSW wavelengths in the FDM structure as compared to the free phase transition. Further, from the known width of the metal strip, the wavelength of the SMSW in the FDM structure is determined at a given frequency, which gives a point on the dispersion curve. The full dispersion curve is obtained by repeating the described procedure for the frequency grid in the entire region of existence of the SMSW.

The experiments were carried out on a setup designed to excite and receive SMSWs using mobile antennas, assembled on the basis of a standard meter of complex transmission coefficients. The antenna conductors were 4 mm long and 12 μm thick. The distance between

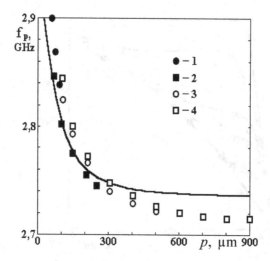

Fig. 3.9. Theoretical (solid line) and experimental (points) dependences of the center frequency of the peak of non-transmission on the gap for different widths of the metal strip: 1 – 400 μm; 2 – 1000 μm; 3 – 2000 μm; 4 – 4000 μm.

the antennas was 6 mm. We used a gallium-doped yttrium–iron garnet film in the form of a disk 30 mm in diameter and 16 μm thick, with a resonance line width of 0.6 Oe and a saturation magnetization of 840 G. A rectangular metal strip 40 mm long and 2 mm wide was superimposed on the top of the film in the middle between the antennas with a gap. The antenna conductors and the axis of the metal strip were parallel to the direction of the field applied in the plane of the ferrite film, which was equal to 632 Oe. In this case, the SMSW excitation was observed in the frequency range from 2704 to 2935 MHz.

The measured dispersion dependences for the FDM structure at various gaps between the metal plate and the ferrite film are shown by dots in Fig. 3.8. In the same figure, solid lines show the curves calculated by formula (3.1).

It can be seen from the figure that the theoretical dependences for gaps larger than 100 μm (3–6) coincide with the experiment within 0.5%. The accuracy of coincidence for gaps less than 100 μm (curves 1, 2) is also high and amounts to a few percent. Some decrease in accuracy, apparently, is due to the difficulty of measuring such small gaps. Since SMSWs are excited and received in those parts of the film where there is no metal, the lower and upper boundaries of the observed frequency spectrum (about 2700 and 2900 MHz) correspond to the boundaries for the free FF (2701 and 2946). For the curves 2 and 3, two regions are well observed, corresponding to direct SMSWs of the first and second types. The discontinuity of these curves in the range of wavenumbers from 100 to 200–300 cm^{-1} is due to an increase in signal attenuation due to the low group velocity of the SMSW.

The indicated discontinuity of the dispersion curves manifests itself in the amplitude–frequency characteristic (AFC) of the SMSW spectrum as a non-transmission peak, which is more pronounced the closer the width of the metal strip is to the distance between the antennas. When the distance between the antennas is 6 mm and the width of the metal strip is 2 mm, the maximum depth of the peak relative to the maximum frequency response approaches 20 dB, with a bandwidth of 4 mm, it exceeds 40 dB.

Figure 3.9 shows the theoretical (solid line) and experimental (dots) dependences of the centre frequency of the non-transmission peak on the gap between the metal strip and the film for different widths of the metal strip.

It can be seen from the figure that the theory describes the experiment the better, the wider the band, which is natural, since the calculation was performed for an infinite FDM structure. With an increase in the gap, the depth of the non-transmission peak decreases and it practically disappears at, since at such a ratio, the disturbance exerted by the band on the SMSW spectrum becomes too small.

3.1.3. On the possibility of experimental observation of backward waves

The dispersion properties of SMSWs in the FDM structure described in section 3.1.1 were obtained theoretically. For direct waves of the first and second types (at those frequencies where their regions of existence do not overlap), these data are well confirmed in the experiment, one of which is described in the previous section 3.1.2. On the other hand, there is no information about the observation of backward waves in the experiment, the reason for which lies, apparently, in the fact that the experiment was usually performed at frequencies below, where the group velocity of backward waves is low and the propagation losses are large. However, as can be seen from Fig. 3.1, at frequencies above, the group velocity of backward waves can be of the same order of magnitude as for forward waves, and a low group velocity below can be useful in designing delay lines. Thus, the problem of experimental detection of backward waves becomes very urgent.

Some recommendations for the experiment can be drawn from this consideration. So, from Figs. 3.1–3.4 it can be seen that the regions of existence of backward waves in frequency and angle are very wide, and increase as the thickness of the dielectric layer decreases and the angle increases. In the experiment, it is necessary to take into account the direction of wave propagation and the orientation of its phase front, and under conditions of low group velocity, the sensitivity of the equipment should be high, and the distance between the emitting and receiving transducers is small. Since, along with the backward wave at the same frequency, there is always a longer forward wave of the first type, the geometry of the exciting transducer must correspond to the length of the backward wave (and not the forward one), for example, it can be made in the form of a grating with a period equal to the length of the backward waves.

It is hoped that, provided the above recommendations are followed, the observation of backward waves will become possible.

3.2. Distribution of SMSW in a two-component environment consists of a free ferrite film and FDM (ferrite–dielectric–metal) structure

As mentioned earlier, the main subject of consideration in this work is the propagation of SMSWs in structures containing ferrite films under conditions of various inhomogeneities. Let us first consider geometrically inhomogeneous structures magnetized by a uniform field. Let the structure be a free ferrite film (FF), over a part of the surface of which, with a certain gap, there is a rectilinear metal strip with parallel edges, which forms a ferrite–dielectric–metal (FDM) structure in this part. SMSWs are excited in the free ferrite film, pass under the metal strip or are reflected from its edge, and are received again in the free ferrite film. The most general case involves the study of refraction and reflection of the SMSW for arbitrary directions of propagation of the SMSW and the orientation of the interface between the media relative to the direction of the field. To get a general idea of the phenomenon, we will consider here in three stages: first, we investigate the propagation of the SMSW in the direction perpendicular to the field, when the interface between the FF and the FDM structure is oriented parallel to the field. Next, we study the refraction of the SMSW in the case when the interface

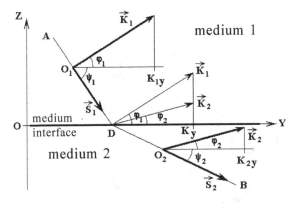

Fig. 3.10. Refraction of SMSW at the interface between two media.

between the media is oriented parallel to the field, and the direction of propagation of the SMSW is arbitrary, after which we consider the refraction of the SMSW for an arbitrary orientation of this boundary. We will mainly use the results of [565, 566, 567, 571, 572, 573, 574, 575]. Some theoretical results on the method of isofrequency curves are generalized in [305].

3.2.1. Analysis of the refraction of the SMSW using the method of isofrequency curves

In this section, we consider the interpretation of the experimentally observed SMSW refraction when passing under a metal strip based on the method of isofrequency curves.

3.2.1.1. Formulation of the problem

Let us first consider the general formulation of the problem of the refraction of an SMSW at the interface between two media. First of all, we note that it is not always convenient to use the concepts of angles of incidence and refraction in a coordinate system associated with the interface between media [545] to describe the refraction of SMSWs, traditionally accepted in classical optics. Indeed, in a magnetic medium, due to its gyrotropic properties set by an external field, the angle of reflection in the general case is not equal to the angle of incidence, and the expression of the angle of refraction in terms of the angle of incidence and refractive indices known in optics is also not fulfilled. Therefore, further, we use a coordinate system traditional for SMSW, in which the plane coincides with the plane

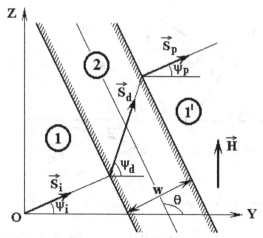

Fig. 3.11. Passage of SMSW through an inhomogheneity as a strip.

of the ferrite film, and the axis is parallel to the external field (Fig. 2.1). Wave propagation is characterized by two angles: – between the wave vector and the axis and – between the group velocity vector and the same axis. The relationship between these angles is determined by the dispersion law for the SMSW in the given medium (for the FF – (2.299), for the FDM – (2.520)). The conditions for refraction and reflection of the SMSW in the case of a rectilinear boundary follow from the same relations between the fields of the incident, reflected, and refracted waves as in optics [545]. Thus, the equality of the exponents of these waves gives the condition for the preservation of the projection of the wave vector onto the interface between the media. From the condition of conservation of the amplitude of the wave field component parallel to the interface between the media, one can find the amplitude coefficients of refraction and reflection of waves, the squares of which give the ratio of the energy densities of the incident, refracted and reflected waves.

The general scheme of refraction of SMSW at the interface between two media is shown in Fig. 3.10.

For simplicity, in the figure, the interface is selected along the axis. The wave propagates from medium 1 to medium 2 along the ADB line. In this case, $\vec{k}_1$ and $\vec{s}_1$ are the wave and ray vectors of the incident wave, $\vec{k}_2$ and $\vec{s}_2$ are similar vectors for the refracted wave. At the interface between the media, the condition is satisfied. Following the tradition adopted in optics [545], below, to designate the parameters of the incident wave, we will use the index, refracted

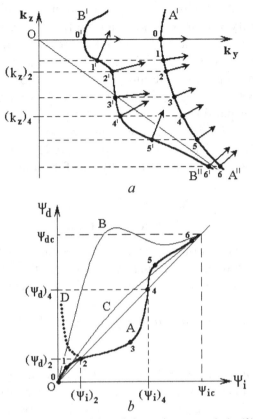

Fig. 3.12. Isofrequency curves (*a*) and dependences $\psi_d(\psi_i)$ (*b*) for the orientation of the metal strip along the field.

– *d* and reflected – *r*. With the known dispersion laws in both media, it is sufficient to use only one pair of angles φ_1 and φ_2 and ψ_1 or ψ_d to analyze the refraction of SMSWs. Based on the convenience of comparison with experiment, further preference is given to the angles ψ_1 and ψ_d, and only in some cases is the angle φ_1 used as a parameter.

In the experiment, the antennas of the exciting and receiving transducers of the SMSW can be installed on the surface of the ferrite film only outside the metal strip; therefore, in order to study the refraction, they must be placed on different sides of it. In this case, the scheme for studying the refraction of the SMSW takes the form shown in Fig. 3.11.

Here, the numbers 1 and 1' denote the regions of a ferrite film with a free surface (FF), and number 2 – a ferrite film with a width equal to a metal strip located above it with a gap equal to it (FDM structure). The angle between the edge of the strip and the axis Oy

is indicated by θ. An incident wave in medium 1 is characterized by a group velocity vector $\vec{s}_i$ making an angle ψ_1 with the axis Oy, refracted in medium 2 by a vector $\vec{s}_d$ and an angle ψ_d and emerging from under a strip in medium 1 – by a vector s_p and an angle ψ_p. In the process of passing under the metal strip, the SMSW experiences two refractive events, in each of which the projection of the wave vector onto a line parallel to the edge of the strip is preserved. In this case, in the media 1 and 1' the group velocity vectors of both waves are parallel, that is $\psi_1 = \psi_p$, and the angle ψ_d is determined through the directly measured values by the formula:

$$\psi_d = \frac{l_y(\operatorname{tg}\alpha - \operatorname{tg}\psi_i)\sin\theta + w\operatorname{tg}\psi_i}{w + l_y(\operatorname{tg}\alpha - \operatorname{tg}\psi_i)\cos\theta}, \qquad (3.2)$$

where l_y is the projection onto the axis of the segment connecting the centres of the exciting and receiving antennas, α is the angle between the axis Oy and this segment, w is the width of the metal strip.

3.2.1.2. Analysis of orientation dependences by the method of isofrequency curves

Let us now consider the refraction of the SMSW using the method of isofrequency curves, which, based on the condition of preserving the projection of the wave vector on the interface between the media, allows us to find the dependence of the angle of the refracted wave on the angle of the incident wave, that is. A qualitative illustration of the construction of such dependences for two different cases of orientation of the metal strip is shown in Figs. 3.12 and 3.13.

3.2.1.3. Strip orientation along the field

Figure 3.12 illustrates the case of SMSW refraction at the boundary between the free FF and the FDM structure oriented parallel to the field ($\theta = 90°$). The case of frequencies sufficiently close to the lower edge of the SMSW spectrum is considered. The thickened lines in Fig. 3.12 a show isofrequency curves for the free phase transition ($A'A''$) and the FDM structure ($B'B''$) (the scale of the curves is distorted to make the figure more clear). The $B'B''$ curve has a bulge near the horizontal axis, which is characteristic only of low frequencies. The thin oblique line corresponds to the cutoff angles ψ_c, the values of which for both structures (at $p \div 0$) coincide ($\psi_{ic} = \psi_{dc}$).

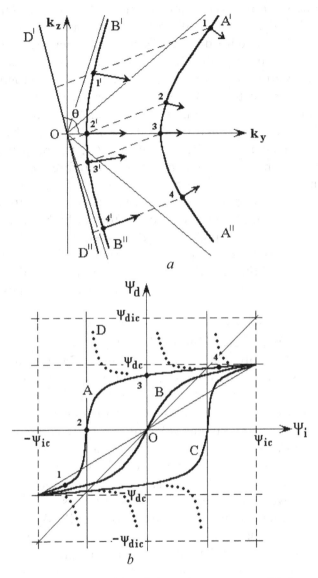

Fig. 3.13. Isofrequency curves (*a*) and dependences $\psi_d = \psi_i(b)$ for an arbitrary orientation of the metal strip.

The arrows show the vectors of the group velocity perpendicular to the tangents at the corresponding points of the isofrequency curves.

The procedure for finding the group velocity vector of the refracted wave is as follows. The direction of the wave vector of the incident wave is set, the corresponding point on the isofrequency curve $A'A''$ is found, from which a straight line is drawn perpendicular to the interface between the media until it intersects with the isofrequency curve $B'B''$, which corresponds to the preservation of the projection of the wave vector onto this boundary. At the obtained intersection point, a tangent to the isofrequency curve and a perpendicular to this tangent are plotted, which gives the direction of the group velocity vector of the refracted wave. Such a procedure in Fig. 3.12a was carried out for points 0–6 on the curve $A'A''$, as a result of which points $0'$–$6''$ were obtained on the curve $B'B''$. Since the interface between the media is parallel to the axis Oz, the projection of the wave vector k onto this axis is preserved, therefore the straight lines connecting the points 0–6 with the points $0'$–$6''$ are horizontal.

The figure shows four pairs of characteristic points in which the directions of the group velocity vectors in both media coincide ($\psi_d = \psi_i$). These are the points 0 and $0'$, where $\psi_d = \psi_i = 0$, 2 and $2'$, where $(\psi_i)_2 = (\psi_d)_2$, 4 and $4'$, where $= (\psi_i)_4 = (\psi_d)_4$, as well 6 and $6'$, remote to infinity, where $\psi_d = \psi_i = \psi_c$. Between points 0–2 and $0'$–$2'$, as well as 4–6 and $4'$–$6'$ – $\psi_d = \psi_i$, between points 2–4 and $2'$–$4'$ $\psi_d < \psi_i$.

This is illustrated in Fig. 3.12 b, which shows the dependences for different frequencies. The thick curve A for low frequencies is plotted from the isofrequency curves in Fig. 3.12a. The thin curves B and C correspond to the middle and high frequencies of the SMSW spectrum and are constructed in a similar way. The oblique line corresponds to $\psi_d = \psi_i$. The points 0–6 correspond to similar points in Fig. 3.12 a. The maximum value of the angle corresponds to the cutoff angle of the SMSW in the free FF ψ_c As noted above, $\psi_{ic} = \psi_{dc}$. It can be seen from the figure that for curve A in the sections between points 0 and 2, as well as 4 and 6, the relation $\psi_d = \psi_i$ is satisfied. Between the points 2 and 4, the inverse relation $\psi_d < \psi_i$. For the frequencies lying in the middle of the SMSW spectrum, the entire curve $\psi_d(\psi_i)(B)$ passes entirely above the straight line $\psi_d = \psi_i$. For frequencies near the upper boundary of the SMSW spectrum, the curve $\psi_d(\psi_i)$ (C) approaches the straight line $\psi_d = \psi_i$.

The curves A, B, C were plotted without taking into account the effect of dispersive splitting of long SMSWs at the exciting antenna (Section 1.3). Since the wavelength in the FDM structure is longer than in the free FF, the splitting phenomena are more noticeable when the SMSW propagates under the metal strip, and are manifested in the tendency of the angle ψ_d of the refracted wave to the value of the cutoff angle ψ_{dc}, the larger the smaller the wavenumber. This character of the dependences is shown in Fig. 3.12 b by the dotted curve D.

3.2.1.4. The orientation of the strip is arbitrary

Let us now consider the refraction of the SMSW at the interface between the free FF and the FDM structure, oriented arbitrarily with respect to the field. For simplicity, we assume that there is no gap between the metal strip and the FF surface ($p = 0$), which corresponds to the FM structure described by the dispersion relation (2.518). This case is illustrated in Fig. 3.13, which shows a construction similar to Fig. 3.12 for the angle between the strip and the axis, at a frequency close to the middle of the SMSW spectrum. The thickened lines in Fig. 3.13a show isofrequency curves for the free phase transition ($A'A''$) and the FM structure ($B'B''$) (the scale of the curves is distorted similarly to Fig. 3.10 a). The direction of the interface is shown by the line $D'D''$. Thin oblique lines correspond to the cutoff angles ψ_c, the values of which for both structures (at $p = 0$) are different ($\psi_{dc} < \psi_{ic}$). The arrows show the vectors of the group velocity. The procedure for finding the group velocity vector of the refracted wave is similar to that described above. In this case, the projection of the wave vector onto the line $D'D''$ is preserved, that is, the straight lines connecting points 1–4 with points $1'$–$4''$ are perpendicular to this line. Two characteristic points can be distinguished in the figure – 3 (for a free FF) and – 2 (for an FM medium), in which the direction of the group velocity vector in one of the media is parallel to the axis Oy. At the first of these points, the incident wave ($\psi_i = 0$) propagates parallel to the axis Oy, and the refracted wave ($\psi_d = 0$) at the second. For $\psi_i < (\psi_i)_2$ (above point 2 along the curve $A'A''$) $\psi_d < 0$. At $\psi_i > (\psi_i)_2$ (below this point). When $\psi_i < (\psi_i)_3 = 0$ $\psi_d > 0$. This is illustrated in Fig. 3.13 b, which shows the dependences $\psi_d(\psi_i)$ for different orientations of the metal plate. Curve A corresponds to the 90° case (it is assumed that $180° - \theta_i < (\psi_{dc})$), curve B – to the case $\theta = 90°$, curve C – to the $\theta < 90°$ case ($\theta > \psi_{dc}$). Curve A is plotted from isofrequency curves in Fig. 3.13 a. The other two curves B and C are constructed in a similar way. Inclined lines connect the origin to the points (ψ_i, ψ_{dc}), as well as (ψ_i, ψ_{dic}), where $\psi_{dic} = \psi_{ic}$. The points 1–4 correspond to similar points in Fig. 3.13 a. The maximum value of the angle corresponds to the cutoff angle of the SMSW in the free FF ψ_{ic}. As noted above, $\psi_{dc} < \psi_{ic}$.

It can be seen from the figure that for curve A in the sections between the points 1 and 2, and the relation $\psi_d < 0$ is fulfilled, and between the points 2 and 4 the inverse relation is $\psi_d > 0$. At point 3 $\psi_d > 0$.

The allowance for the dispersion beam splitting (Section 1.3), which manifests itself during the propagation of the SMSW under the metal strip, gives an increase in those regions where the curves A, B, and C intersect the axis Oy (at small k). In this case, the maximum possible value ψ_d is limited by the cutoff angle in the free FF and is $\psi_{dic} = \psi_{lc}$. The form of the dependences due to dispersion splitting is shown in Fig. 3.13b by dotted curves D.

3.2.1.5. Evaluation of the possibility of manifestation of the effects of dispersive splitting of a wave beam under the conditions of a real experiment

The above theoretical picture of the SMSW propagation is constructed on the basis of the assumption of the infinite extent of the media on both sides of the infinitely long common boundary, and the wavefront is assumed to be infinitely wide, and there is no damping. Nevertheless, in experiments, the dimensions of the ferrite film and the width of the SMSW beam are limited, and the waves are attenuated, which must be taken into account when interpreting the experimental results.

In the experiments, we used films of yttrium–iron garnet (YIG) with a resonance line width of the order of 0.5–1.0 Oe, in the form of a disk 60–70 mm in diameter. The thickness of the antenna conductor was 12 µm, which made it possible to work with SMSWs with a length of several millimeters to 70 µm, which spread over a distance of 3–5 cm. The length of the metal strip was 60–80 mm, and the width was from 100 µm to 10 µm. The size of the gap varied from 5–10 µm to 10–20 µm. The considered phenomena are noticeably affected by the ratio between the lengths of the SMSW and the exciting antenna. Indeed, as indicated in the first chapter (Section 1.3), if the SMSW length approaches the width of the wave beam, which is noticeable near the lower edge of the spectrum, then the beam splits into two, propagating in directions close to the cutoff angles. In this case, if, after refraction, the wave is deflected from the Oy axis more strongly than before it ($\psi_d > \psi_i$), then the width of the refracted beam decreases, and in the opposite case ($\psi_d < \psi_i$) it increases, that is, the splitting conditions change. At the same frequency, the wavelength in the FDM structure, as a rule, is noticeably longer than in the free FF, that is, the splitting is more pronounced.

Let us consider the possibility of manifestation of the effects of dispersive splitting of a wave beam under typical experimental conditions described below. Let the length of the exciting antenna be 4 mm, the saturation magnetization of the ferrite film is 840 G, its thickness is 16 μm, and the constant field is 570 Oe. Suppose that the SMSWs propagate perpendicular to the direction of the field. Under these conditions, the boundaries of the frequency range for the existence of SMSWs for the free phase transition are 2510–2772 MHz. At a frequency of 2530 MHz, the wave number is 25 cm^{-1}, the wavelength is 2512 μm, that is, the initial beam width is less than twice the wavelength. In this case, the splitting is small, but it can be noticeable. At a frequency of 2615 MHz, the wavenumber is 155 cm^{-1}, the wavelength is 405 μm, that is, the initial beam width is ten times the wavelength. In this case, there is no splitting. At a frequency of 2650 MHz, the wavenumber is 230 cm^{-1}, the wavelength is 273 μm, and at the beam width of fifteen wavelengths, splitting is all the more absent. Under the same conditions for the FM structure, the boundaries of the frequency range of existence of SMSWs are 2510–3948 MHz. At a frequency of 2530 MHz, the wave number is 5 cm^{-1}, the wavelength is 12560 μm, that is, the wavelength is three times the beam width. Moreover, the splitting is large. At a frequency of 2650 MHz, the wavenumber is 30 cm^{-1}, the wavelength is 2093 μm, that is, the wavelength is half the width of the beam and splitting can still appear. At 3090 MHz, the wavenumber is 155 cm^{-1}, the wavelength is 405 μm, that is, the wavelength is ten times less than the beam width and there is no splitting.

Thus, if SMSWs are excited in the frequency range 2500–2700 MHz, then in the free phase transition the phenomena of wave beam splitting can manifest themselves only in a small part of this range at frequencies below 2520–2530 MHz, and in the FM structure – over most of the range up to frequencies 2540–2650 MHz.

3.2.2. *Experimental study of the refraction of the SMSW*

Let us now present the results of an experimental study of the refraction of an SMSW at the interface between a free FF and an FDM structure made in the form of a metal strip located above the surface of a ferrite film. The data will be compared with the results

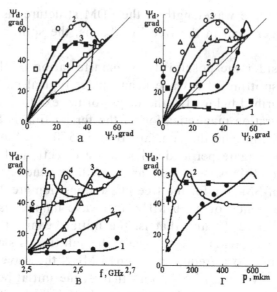

Fig. 3.14. Dependences of the angle of the refracted wave on various parameters of the medium and the incident wave.

a – dependences $\psi_d(\psi_i)$ at $p = 185$ μm and different frequencies: 1 – 2530 MHz; 2 – 2570 MHz; 3 – 2590 MHz; 4 – 2610 MHz.
b – dependences $\psi_d(\psi_i)$ at $f = 2610$ MHz and various gaps: 1 – 0 μm; 2 – 30 μm; 3 – 100 μm; 4 – 140 μm; 5 – 185 μm.
c – dependences $\psi_d(f_i)$ $\varphi_i = -22.5°$ and various gaps: 1 – 0 μm; 2 – 17 μm; 3 – 70 μm; 4 – 140 μm; 5 – 220 μm; 6 – ∞.
d – dependences $\psi_d(p)$ at $\varphi_i = -22.5°$ and different frequencies: 1 – 2530 MHz; 2 – 2570 MHz; 3 – 2650 MHz.

of a machine calculation of the same dependences, performed on the basis of the above theoretical analysis of the refraction of SMSWs.

The experiments were carried out on a ferrite film in the form of a disk 60 mm in diameter made of gallium-doped yttrium iron–garnet with a saturation magnetization of 840 G, a resonance line width of 0.6 Oe, and a thickness of 16 μm. The SMSW was excited and received by movable antennas 3.7 mm long and 12 μm in diameter. The metal strip was 6 mm wide and 40 mm long and was installed above the ferrite film between the antennas parallel to its surface. The film was magnetized by an in-plane field equal to 570 Oe. In this case, the SMSW spectrum for the free FF is limited to frequencies of 2510 and 2772 MHz, and for the FM structure, 2510 and 3948 MHz.

3.2.2.1. Strip orientation along the field

Let us first consider the case of orientation of a metal strip along the direction of the magnetic field. The strip is installed with a gap that can vary. The results obtained are illustrated by figures (3.14a–3.14d, 3.15a–3.15d), where the calculated curves are presented by solid lines, and the experimental data are represented by dots.

Figure 3.14a shows the dependences for SMSWs of various frequencies in the range from 2530 to 2630 MHz (curves 1–6) with a constant gap $p = 185$ μm. The oblique line corresponds to $\psi_d = \psi_i$. It can be seen from the figure that all curves start at the point $\psi_d = \psi_i = 0$ and in the initial section go above the straight line $\psi_d = \psi_i$. Further, the curves increase monotonically, passing both above and below the straight line $\psi_d = \psi_i$, after which they again rush to it, where they end at points coresponding to the cutoff angles $\psi_{dc} = \psi_{ic}$, which are different for different frequencies.

By the nature of the behaviour, three types of curves can be distinguished.

Curves of the first type (1) between the starting and ending points intersect the straight line $\psi_d = \psi_i$ twice.

Curves of the second type (2-4) lie entirely above the straight line $\psi_d = \psi_i$ and depart from it at considerable distances (up to 20°–30°).

Curves of the third type (5.6) also lie above the straight line $\psi_d = \psi_i$, but deviate only little from it (no more than 5°–10°).

These three types of curves correspond to curves A, B, and C shown in Fig. 3.12b. In this case, curves of the first type correspond to the case of low frequencies of the SMSW spectrum (2510 MHz < f < 2540 MHz), where the wavelength under the metal significantly exceeds the gap in the FDM structure, as a result of which the isofrequency curve for this structure has a bulge near the horizontal axis (Fig. 3.12a). The curves of the second type correspond to the case of medium frequencies (2540 MHz < f < 2600 MHz, where the wavelength under the metal is of the same order of magnitude as the gap in the FDM structure, there is no bulge on the isofrequency curve, and the isofrequency curves for the two structures are spaced far apart. The curves of the third type correspond to the case of high frequencies (2600 MHz < f < 2772 MHz), where the wavelength under the metal is much less than the gap in the FDM structure, as a result of which the isofrequency curves for the two structures are close.

The figure shows that the calculated curves coincide with the experimental points, the better, the higher the SMSW frequency. At a frequency of 2630 MHz (curve 6), the coincidence is good over the entire range of angle variation, at a frequency of 2610 MHz (curve 5) – at 20°, at a frequency of 2590 MHz (curve 4) at 30°, at a frequency of 2570 MHz (curve 3) at 40°. At angles smaller than the indicated values, the experimental points for frequencies of 2570, 2590 and 2610 MHz lie above the calculated curves 3–6. This discrepancy is due to the fact that at these angles the SMSW wavelength becomes comparable with the length of the exciting antenna used in the experiments, as well as with the width of the metal plate; therefore, the effects of dispersive splitting, discussed earlier, begin to manifest themselves (curve D in Fig. 3.12b). At frequencies of 2530 MHz and 2550 MHz (curves 1 and 2), the splitting effects are absent only near the cut-off angles (because of the unobservability of the refraction effect in the experiment, measurements were not carried out for these frequencies).

Let us now consider how the dependences described above change when the gap between the surfaces of the ferrite film and the metal plate changes. Figure 3.14b shows the dependences for a SMSW of a fixed frequency of 2610 MHz (curves 1–6) at various gaps p from 0 to ∞. The oblique line corresponds to $\psi_d = \psi_i$. The general course of the curves is close to that shown in Fig. 3.14 a. Here, the same three types of curves can also be distinguished, and the curves of the first type (1,2) are observed at $0 < p < 50$ μm, the second type (3–5) at 50 μm $< p <$ 150 μm, the third type (6) – at 150 μm $< p < \infty$. Similarly to Fig. 3.14 a, at small angles the experimental points lie above the calculated curves, which is also explained by the effects of dispersive splitting.

The dependences shown in Figs 3.14 a and 3.14 b describe the refraction of the SMSW quite completely, however, for practical purposes, it is not always convenient to use them, since in devices the radiating antenna is usually installed in one fixed position, i.e. given an angle φ_i. In this case, the frequency f or gap p may change. Therefore, for practice, it is desirable to know the dependence of the angle ψ_d on the frequency or the gap ($\psi_d(f)$) or $\psi_d(p)$ at $\varphi_i = $ const. These dependences are illustrated in Figs. 3.14 c and 3.14 d.

Figure 3.14 c shows the dependences $\psi_d(f)$ at a fixed value of the angle $\varphi_i = -22.5°$ and various gaps. Curve 1 corresponds to the FM structure, and curve 7 - to free FF, curves 2–6 - to intermediate values of the gap for the FDM structure. As the frequency increases, all curves monotonically increase the stronger the larger the gap, then some fall off and all break off at the frequency for which the specified angle value $\varphi_i = -22.5°$ is equal to the cutoff angle in the free FF. Three types can be distinguished by the character of the increase in the curves. The first (curves 1–3) corresponds to small gaps ($0 < p < 50$ μm) and is characterized by a monotonic increase in the entire frequency spectrum of the SMSW. These curves begin at one point corresponding to the FM structure (curve 1), and each end at its own angle, the larger the larger the gap. Curves of the second type (4-6) correspond to average gaps ($50 < p < 250$ μm) and first increase sharply, pass through a maximum, decrease, and then all tend to curve 7, corresponding to a free FF. These curves start at the same point as the curves of the first type and end at the same point as curve 7. Curves of the third type (7) correspond to large gaps (250 μm $< p < \infty$) and also increase monotonically over the entire frequency spectrum of the SMSW. Unlike curves of the first type, these curves begin at much larger angles and end at one point corresponding to a free FF (7). The described properties of the dependences $\psi_d(f)$ can be explained by constructing isofrequency curves, similarly to Fig. 3.12. So, for example, the beginning of the curves 1–6 at one point can be explained by the fact that near the lower boundary of the SMSW spectrum (2510 MHz), the wavelength is large and within wide limits significantly exceeds the size of the gap, therefore, the properties of the wave do not practically depend on the specific size of the gap (wave spreads as in the FM structure). On the other hand, the maximum on the curves of the second type (4–6) is due to the presence of a flat section on the isofrequency curve for the FDM structure.

Figure 3.14d shows the dependences $\psi_d(p)$ at a fixed value of the angle $\varphi_i = -22.5°$ and different frequencies. Curve 1 corresponds to a frequency of 2530 MHz, which lies near the lower boundary of the SMSW spectrum, and curve 5, to a frequency of 2650 MHz, which is approaching its upper boundary. With an increase in the gap, all curves first increase monotonically the stronger, the higher the frequency, then, after passing through the maximum, they fall off and go to the horizontal section corresponding to the larger values of the angle, the higher the frequency. The maxima on the curves are reached at the larger the gap values, the lower the frequency. It is interesting to note that at the maximum point, the ratio of the gap to the wavelength is almost constant, and the product of the wave number and the gap is close to 0.5. The exit of each curve to the horizontal section at large gaps is due to the excess of the gap

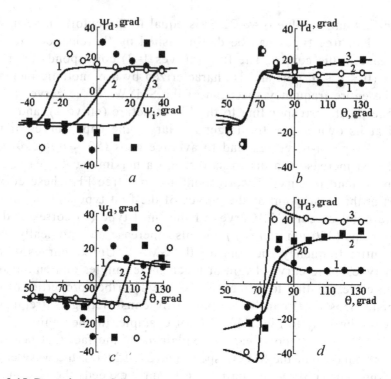

Fig. 3.15. Dependences of the angle ψ_d of the refracted wave on various parameters of the medium and the incident wave.
a - dependences $\psi_d(\psi_i)$ at $f = 2610$ MHz, $p = 0$ and different angles of rotation of the band θ: 1 – 100°, 2 – 90°, 3 – 80°.
b – dependences at $\psi_d(\theta) = -22.5°$, and different frequencies: 1 – 2550 MHz, 2 – 2590 MHz, 3 – 2630 MHz.
c – dependences at $f = 2610$ MHz, and different angles: 1 – 2°, 2 – 14°, 3 – 27°.
d – dependences at $f = 2610$ MHz, $\psi_d(\theta) = -22.5°$ and various gaps: 1 – 5 μm, 2 – 20 μm, 3 – 95 μm.

over the wavelength, as a result of which the SMSWs propagate, as in a free FF.

3.2.2.2. The orientation of the strip is arbitrary

Let us now consider the refraction of the SMSW at an arbitrary orientation of a metal strip mounted close to the film plane (with a zero gap).

Figure 3.15a shows the dependences for the SMSW frequency of 2610 MHz, zero gap $p = 0$, and various angles of rotation of the strip θ. The curves 1–3 correspond to angles of 100°, 90° and 80°. It can be seen that these curves completely correspond to the analogous curves A, B and C shown in Fig. 3.13 b. In this case, $\psi_i = 57.4°$, $\psi_{dc} = 13.6°$. The calculated curves coincide with the experimental points, the better, the more the angle differs from its value corresponding to $\psi_d = 0$. The coincidence can be considered good if the specified difference exceeds 20°–30°. With a smaller difference, the experimental points deviate towards large (in absolute value) values of the angle ψ_d, which is explained by the effects of dispersive splitting and corresponds to curves of type D in Fig. 3.12 b.

The dependences shown in Figs 3.14 a and 3.14 b describe the refraction of the SMSW quite completely, however, for practical purposes, it is not always convenient to use them, since in devices the radiating antenna is usually installed in one fixed position, i.e. given an angle. In this case, the frequency or gap may change. Therefore, for practice, it is desirable to know the dependence of the angle on the frequency or the gap ($\psi_d(f)$ or $\psi_d(p)$) at φ_i = const. These dependences are illustrated in Figures 3.14c and 3.14d.

Figure 3.14 c shows the dependences at a fixed value of the angle $\varphi_i = -22.5°$ and various gaps p. Curve 1 corresponds to the FM structure, and curve 7 to free FF, curves 2–6 to intermediate values of the gap for the FDM structure. As the frequency increases, all curves monotonically increase the stronger the larger the gap, then some fall off and all break off at the frequency for which the specified angle value $\varphi_i = -22.5°$ is equal to the cutoff angle in the free FF. Three types can be distinguished by the character of the increase in the curves. The first (curves 1–3) corresponds to small gaps ($0 < p < 50$ μm) and is characterized by a monotonic increase in the entire frequency spectrum of the SMSW. These curves begin at one point corresponding to the FM structure (curve 1), and each end at its own angle, the larger the larger the gap. Curves of the second type (4–6) correspond to average gaps ($50 < p < 250$ μm) and first increase sharply, pass through a maximum, decrease, and then all tend to curve 7, corresponding to a free FF. These curves start at the same point as the curves of the first type and end at the same point as curve 7. Curves of the third type (7) correspond to large gaps (250 μm $< p < \infty$) and also increase monotonically over the entire frequency spectrum of the SMSW. Unlike curves of the first type, these curves begin at much larger angles and end at one point corresponding to a free FF (7). The described properties of the dependences can be explained by constructing isofrequency curves, similarly to Fig. 3.12. So, for example, the beginning of curves

1-6 at one point can be explained by the fact that near the lower boundary of the SMSW spectrum (2510 MHz), the wavelength is large and within wide limits significantly exceeds the size of the gap, therefore, the properties of the wave do not practically depend on the specific size of the gap (wave spreads as in the FM structure). On the other hand, the maximum on the curves of the second type (4-6) is due to the presence of a flat section on the isofrequency curve for the FDM structure.

Figure 3.14 d shows the dependences at a fixed value of the angle $\varphi_i = -22.5°$ and different frequencies. Curve 1 corresponds to a frequency of 2530 MHz, which lies near the lower boundary of the SMSW spectrum, and curve 5, to a frequency of 2650 MHz, which is approaching its upper boundary. With an increase in the gap, all curves first increase monotonically the stronger, the higher the frequency, then, after passing through the maximum, they fall off and go to the horizontal section corresponding to the larger values of the angle, the higher the frequency. The maxima on the curves are reached at the larger the gap values, the lower the frequency. It is interesting to note that at the maximum point, the ratio of the gap to the wavelength is almost constant, and the product of the wave number and the gap is close to 0.5. The exit of each curve to the horizontal section at large gaps is due to the excess of the gap over the wavelength, as a result of which the SMSWs propagate, as in a free FF.

3.2.2.2. The orientation of the strip is arbitrary

Let us now consider the refraction of the SMSWs at an arbitrary orientation of a metal strip mounted close to the film plane (with a zero gap).

Figure 3.15 a shows the dependences for the SMSW frequency of 2610 MHz, zero gap $p = 0$, and various angles of rotation of the strip θ. Curves 1-3 correspond to angles of 100°, 90° and 80°. It can be seen that these curves completely correspond to the analogous curves A, B and C shown in Fig. 3.13 b. In this case, $\psi_i = 57.4°$, $\psi_d = 13.6°$. The calculated curves coincide with the experimental points, the better, the more the angle differs from its value corresponding to $\psi_d = 0$. The coincidence can be considered good if the specified difference exceeds 20°-30°. With a smaller difference, the experimental points deviate towards large (in absolute value) values of the angle ψ_d, which is explained by the effects of dispersive splitting and corresponds to curves of type D in Fig. 3.12 b.

Figure 3.15b shows the dependences of the angle ψ_d of the refracted wave on the angle of rotation θ of the strip at a fixed angle $\varphi_i = -22.5°$, zero gap $p = 0$, and different frequencies of the SMSW. Curves 1-3 correspond to frequencies of 2550, 2590 and 2630 MHz. The general course of the curves can be explained by plotting isofrequency curves for different frequencies, similarly to Fig. 3.13. All curves cross the horizontal axis near the angle $\theta \sim 67.5°$, which is determined, first of all, by the value of the angle = -22.5o and depends little on the frequency, which is due to the weakness of the frequency dependence of the shape of isofrequency curves for the FM structure in this interval. Similarly to the previous case, the calculated curves coincide with the experimental points, the better the more the angle differs from the value of 67.5o, and the observed difference near this angle is due to the effects of dispersive splitting.

Figure 3.15c shows the dependences of the angle of the refracted wave on the angle of rotation of the strip at the SMSW frequency equal to 2610 MHz, zero gap $p = 0$, and various angles of the incident wave. Curves 1-3 correspond to angles 2°, 14° and 27°, and intersect the horizontal axis near angles equal to 92°, 104° and 117°, respectively. The relationship between theory and experiment is similar to the previous case.

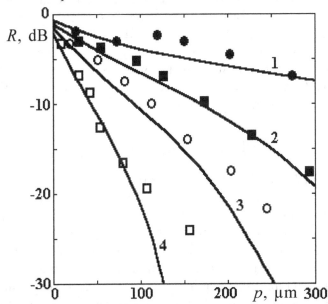

Fig. 3.16. Dependences of the reflection coefficient on the size of the gap between the surfaces of the metal and the PC for SMSW of various frequencies: 1 - 2530 MHz, 2 - 2550 MHz, 3 - 2570 MHz, 4 - 2610 MHz.

Figure 3.15d shows the dependences of the angle φ_d of the refracted wave on the angle of rotation θ of the strip at the SMSW frequency equal to 2610 MHz, a fixed angle $\varphi_i = -22.5°$, and various gaps. Curves 1-3 correspond to gaps of 5, 20 and 95 µm. Increasing the gap leads to a decrease in the angle θ corresponding to the intersection of the curves with the horizontal axis, as well as an increase in angles φ_d away from this intersection. These features become clear if we take into account that, with an increase in the gap, the branches of the isofrequency curves deviate from the vertical the more, the further the points of these branches are removed from the origin. The theoretical curves are related to the experimental points in the same way as described above.

3.2.3. Reflection coefficient of the SMSW from the interface

Let us now consider the reflection coefficient of the SMSW from the interface between the free ferrite film (FF) and the ferrite-dielectric-metal (FDM) structure. Let's use the same coordinate system as before, with the field along the axis (Fig. 2.1). Let the interface between the free FF and the FDM structure pass along the axis ($\theta = 0°$), and let the free correspond to a half-plane, where $z < 0$, and to the FDM structure, a half-plane, where $z > 0$. Assume that the SMSW goes from the lower half-plane to the axis Oy where the interface between the media, after which it is reflected again into the lower half-plane. Moreover, due to the symmetry $\psi_r = -\psi_i$. The condition of conservation of the amplitude of the wave field component parallel to the interface between the media makes it possible to find the reflection coefficient (in terms of energy) in the form:

$$R = \left(|k_{iz}| - |k_{dz}|\right)^2 \cdot \left(|k_{iz}| + |k_{dz}|\right)^{-2}. \tag{3.3}$$

where k_{iz} and k_{dz} are the projections of the wave vectors of the SMSW on the direction of the field in the free FF and in the FDM structure, respectively. This formula is a special case of a more general relation obtained in [296] by the method of the conjugation problem.

In the experiment, we used a gallium-doped YIG film with a saturation magnetization $4\pi M_0 = 840$ G and a thickness of 16 µm. The FDM structure was made in the form of a metal strip 6 mm wide located at a distance of $p = 0 \div 5$ mm above the FF surface. The field was 571 Oe. Excitation and reception of the SMSW was carried out using movable antennas 4 mm long and 12 µm in diameter, placed at

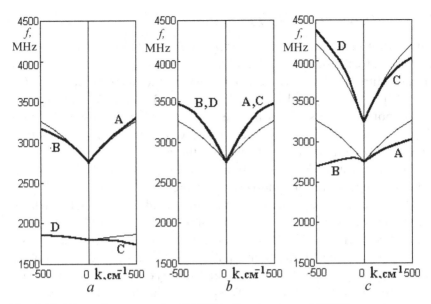

Fig. 3.17. Dispersion curves for SMSW in the structure of MDFF DM at various magnetizations of ferrite layers. $a - 4\pi M_d = 1750$ G, $4\pi M_r = 875$ G; $b - 4\pi M_d = 1750$ G, $4\pi M_r = 1750$ G; $c - 4\pi M_d = 1750$ G, $4\pi M_r = 2625$ G.

a distance of 5–10 mm from the metal strip. The reflection coefficient of the SMSW was determined as the ratio of the powers of the supplied and received signals minus the losses for the transformation and propagation of the SMSW.

The results obtained are illustrated in Fig. 3.16, which shows the experimental (dots) and calculated (solid lines) dependences at $\psi_i = -22.5°$ for different frequencies f: 1 – 2530 MHz, 2 – 2550 MHz, 3 – 2570 MHz, 4 – 2610 MHz.

It can be seen from the figure that with an increase in the frequency of the SMSW and the size of the gap between the PC and the metal strip, the reflection coefficient rapidly decreases, and the theory describes the experimental data quite adequately. The study of the dependence $R(\psi_i)$ at fixed p and f revealed an increase in the reflection coefficient with decreasing angle, which also agrees well with theoretical concepts.

3.3. Dispersional properties of SMSW in structures containing two ferrite layers

Let us now consider the dispersion properties of surface magnetostatic waves in structures containing two ferrite layers. We will mainly use the materials of works [568–570, 576–579,].

3.3.1. Ferrite–ferrite (FF) structure

Let us first consider a two-layer ferrite–ferrite (FF) structure, which is a special case of the MDFDM structure at $p \to \infty$ and $q \to \infty$. Let the structure contain two ferrite layers of the same thickness: $d = r = 7.5$ μm, magnetized by a field of $H = 437.5$ Oe. We choose the same coordinate system as before, with the Oz axis along the field (Fig. 2.1) and restrict ourselves to the case of SMSW propagation along the Oy axis. Let us assume that the magnetization of the layer remains constant and trace the transformation of the SMSW spectrum with a change in the magnetization of the layer r.

The resulting dependences are illustrated in Fig. 3.17, where the thickened lines show the dispersion curves at: $4\pi M_0 = 1750$ G, $4\pi M_r = 875$ G (a), 1750 G (b) and 2625 G (c). Thin lines show dispersion curves at the same parameters for each of the layers taken separately (without the second layer). In this case, the cutoff frequencies of the SMSW spectrum in the isolated layer (at $4\pi M_r = 1750$ Gs) are: $f_b = 2739$ MHz, $f_t = 3675$ MHz. The cutoff freqencies of the SMSW spectrum in the isolated layer at $4\pi M_r = 875$ Gs (a) are $f_b = 2122$ MHz and $f_t = 2450$ MHz, at $4\pi M_r = 1750$ G (b) they are: $f_b = 2739$ MHz and $f_t = 3675$ MHz and at $4\pi M_r = 2625$ G (c) are equal to: $f_b = 3241$ MHz and $f_t = 4900$ MHz. The letters A and B represent the branches belonging to the layer d, and the letters C and D represent the layer r. The distribution of dynamic magnetization over the thickness of the layers is obtained by solving the problem of forced vibrations. An analysis similar to that carried out in [223, 224] shows that branches A and D correspond to the propagation of the SMSW along the outer surfaces of the layers d and r, accordingly, branch B corresponds to the propagation of the SMSW along the inner surface of the layer d, and C – along the inner surface of the layer r. The interaction between the magnetizations of the layers leads to a shift in the frequencies of all branches, and the branches of the internal SMSWs (B and C) are shifted more than the branches of the external ones (A and D). With a strong difference in the magnetizations of the layers (cases a and c), the branches of the external SMSWs (A and D) mutually repulse, and the branches of the internal SMSWs (B and C) are shifted downward in frequency. With a small difference in the magnetizations of the layers (case b), the branches of the external and internal SMSWs shift upward in frequency and merge (A with C and B with D), passing into branches corresponding to a single layer of total thickness. Both outer SMSWs

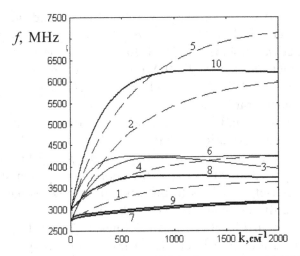

Fig. 3.18. Dispersion curves for SMSW in the structure of MDFDM in the absence (1-6) and presence (7-10) of interaction between the ferrite layers ($f = r = 7.5$ μm). Ferrite layers magnetization: 1–3 – $4\pi M_d = 1750$ G, $4\pi M_r = 0$; 4–6 $4\pi M_d = 0$, $4\pi M_r = 2187.5$ G; 7–10 – $4\pi M_d = 1750$ G, $4\pi M_r = 2187.5$ G. Dielectric layer thicknesses: 1, 4, 7, 8 – $p \to \infty$, $q \to \infty$; 2, 5 – $p = 0$, $q \to \infty$; 3, 6, 9, 10 – $p = 5$ μm, $q \to \infty$.

(A and D) are always straight. Each internal SMSW (B or C) at wavenumbers less than a certain critical value is a forward one, and at large – an inverse. The critical value of the wavenumber is the greater, the greater the magnetization of the layer to which this wave corresponds.

All of the above refers to the propagation of SMSWs along the Oy axis. For an arbitrary direction of propagation in the plane, the general nature of the spectrum is generally preserved, and both internal and external SMSWs propagate within the permissible cutoff angles. With a decrease in the thickness of the dielectric layers, the cutoff angles increase, for external waves up to 90°, and for internal waves – up to values that are closer to 90°, the less the magnetization of the ferrite layers differ from each other.

3.3.2. Metal–dielectric–ferrite–ferrite–dielectric–metal structure (MDFFDM)

Let us now consider the features of the dispersion of the SMSW in the complete structure of the MDFFDM. Let the structure again consist of two ferrite layers of the same thickness: $d = r = 7.5$ μm, magnetized by a field of $H = 437.5$ Oe, and from the side of the layer at a distance of $p = 5$ μm there is a metal layer, and the outer surface of the layer r is left free ($q \to \infty$). Let us assume that the layer has a magnetization $4\pi M_0 = 1750$ G. In this case, the lower limit of the SMSW spectrum is $f_b = 2739$ MHz, and the upper limit for the free surface is $f_b = 3675$ MHz, and for the metallized one $f_{tm} = 6125$ MHz. Of considerable interest is the case of a strong overlap of the SMSW spectra of individual ferrite layers. Based on this, we choose the layer magnetization equal to $4\pi M_r = 2187.5$ G, which gives the boundary frequencies of the spectrum: $f_b = 3001$ MHz, $f_t = 4287$ MHz, $f_{tm} = 7350$ MHz. Using the coordinate system shown in Fig. 2.1, we restrict ourselves to the case of SMSW propagation along the positive direction of the axis. The dispersion curves for this case are shown in Fig. 3.18. Dashed lines 1, 2 and 4, 5 show the curves for separate (non-interacting) layers and, respectively, for the case of free (1, 4) and metallized (2, 5) surfaces. Thin solid lines 3 and 6 show the curves for the same separate (non-interacting) layers in the presence of a metal layer d at a distance of $p = 5$ μm from the layer. The thickened solid lines 7 and 8 show the dispersion curves for bonded layers and in the absence of metal ($p \to \infty$). The thickened solid lines 9 and 10 show similar curves for connected layers d and r in the presence of a metal layer at a distance of $p = 5$ μm from the layer.

Comparing curves 3 and 6 with curves 1 and 4, it can be seen that in the absence of interaction between the ferrite layers, the metal has a much stronger effect on the layer d than on the layer r (curve 3 differs more from curve 1 than curve 6 from curve 4), which is associated with a large distance between the ferrite layer and the metal layer in the second case (12.5 μm, compared to 5 μm). The interaction between the ferrite layers leads to the repulsion of the branches, as a result of which, without metal, the branches for both layers are reduced (curves 7 and 8 compared to curves 1 and 4), similar to curves A and C in Fig. 3.17c. The metal located on the side of the layer changes the dispersion curve for this layer rather weakly (curve 9 is close to curve 7), while for the layer a similar curve changes much more (curve 10 passes much higher than curve 8). Thus, the layer is, as it were, a 'conductor' and 'transfers' the action of the metal to the layer, which is apparently due to the higher magnetization of the layer compared to the layer.

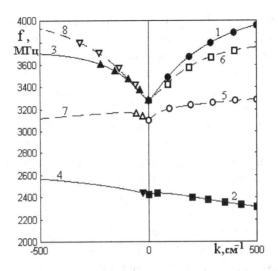

Fig, 3.19. Dispersion curves for SMSW in the structure of MDFFDM at and and various parameters of ferrite layers. 1–4 = 1790 G, $4\pi M_r$ = 735 Gs, d = 15 microns, r = 14 microns; 5–8 – = 1555 G, $4\pi M_r$ = 1830 G, d = 16 microns, r = 6 microns.

3.3.3. Experimental study of the variance of SMSW

The experiments were carried out on in-plane magnetized two-layer films of gallium-doped yttrium iron–garnet grown on gadolinium-gallium garnet (111) substrates. Metallic layers were not considered ($q \to \infty$ and $p \to \infty$). The dispersion curves for SMSWs propagating perpendicular to the direction of the field were measured using the mobile antenna method described above. Two films were investigated: the first had a layer magnetization equal to $4\pi M_d$ = 1790 G, and a layer r magnetization equal to $4\pi M_r$ = 735 G, the layer thickness was respectively d = 15 μm and r = 14 μm. For the second film, the same values were: $4\pi M_d$ = 1555 G, $4\pi M_r$ = 1830 G, d = 16 μm, r = 6 μm. The magnetizing field was 573 Oe. The calculated values of the cutoff frequencies of the SMSW spectrum for the first film in the isolated layer (at $4\pi M_d$ = 1790 Gs) are: f_b = 3258 MHz, f_t = 4110 MHz, in the isolated layer r (at = 735 Gs) are: f_b = 2424 MHz, f_f = 2633 MHz. For the second film, similar frequencies for the layer (at $4\pi M_d$ = 1555 Gs) are equal to f_b = 3092 MHz, f_t = 3781 MHz, for the layer (at $4\pi M_r$ = 1830 Gs) they are equal to: f_b = 3286 MHz, f_t = 4166 MHz.

The results obtained are illustrated in Fig. 3.19, where the solid curves (1–4) are calculated for the first film, the dashed curves (5–8) for the second. The dots show the experimental values.

It can be seen from the figure that the dispersion curves for all four branches of the SMSW spectrum were measured for the first time in the experiment, and, along with the straight lines, backward waves were also observed (the right side of curve 2). The agreement between the theoretical curves and the experimental data is no worse than 2%.

Along with the above-described case of relatively thick films (10–20 μm), spin-wave resonances (SWR) caused by pinning of surface spins, manifested in discontinuities of dispersion curves, were observed on two-layer films less than 5–7 μm thick. Discontinuities were observed in all branches of the spectrum, and their frequencies satisfactorily (20–30%) coincided with those calculated only for the standing SWR modes with numbers from 3 to 7, and in some cases with numbers 13–17. Some SWR frequencies could be described only on the basis of the model of incomplete pinning of spins and inhomogeneity of the magnetization of the film over its thickness, which indicates a high complexity of the observed phenomenon.

Chapter 3. Conclusions

The main issues discussed in this chapter are as follows.

1. Dispersion and isofrequency curves are plotted for surface magnetostatic waves (SSSW) in the FDM structure at different thicknesses of the dielectric layer. Two critical values of this thickness were revealed, the first of which is close to the thickness of the ferrite film, and the second to the doubled thickness of the same film. It is shown that for a dielectric thickness less than the first critical value, the dispersion curve has a single extremum – a maximum, for a dielectric thickness between the first and second critical values – two extrema – a maximum and a minimum, with a dielectric thickness greater than the second critical value - the dispersion curve has no extrema. The corresponding isofrequency curves can be of three types: pseudoelliptic, transitional and pseudohyperbolic. On the plane of isofrequency curves, the boundaries of the regions of existence of forward and backward SMSWs are plotted. Two types of direct waves are identified, differing in large and small values of the wave number. It is shown that in the case of the simultaneous existence of forward waves of both types, between them lies a region of backward waves, the wave number in which takes intermediate values. The possibilities of experimental observation of backward waves are discussed, recommendations are given on the choice of the

geometry of the experiment and the implementation of the receiving transducer SMSW.

2. The propagation of the SMSW in spatially inhomogeneous structures magnetized by a uniform field is considered, in particular, the laws of refraction and reflection of the SMSW at the interface between a ferrite film with a free surface and a FDM structure are studied. In solving these problems, the high efficiency of the isofrequency curves method was demonstrated, which allows, based on the condition of maintaining the projection of the wave vector on the interface between the media, to find the dependence of the orientation of the group velocity vector of the refracted wave on the orientation of the group velocity vector of the incident wave. For the case of the orientation of the interface between the free film and the FDM structure along the field, it is shown that, depending on the frequency of the SMSW, the angle of refraction can be either greater or less than the angle of incidence, and the maximum difference is observed at intermediate values of the angle of incidence between zero and the cutoff angle. For the case of an arbitrary orientation of the interface between the free film and the FM structure with respect to the direction of the field, it is shown that, along with the difference in the absolute values of the angles of refraction and incidence, only when the interface is oriented perpendicular to the direction of the field, the signs of the angles of refraction and incidence always coincide, and for all others orientations, there are regions of angles of incidence where these signs are different. The features of refraction of the SMSW are considered under the conditions of a real experiment, where the finite length of the emitting converter of the SMSW leads to the phenomena of dispersive splitting. It is shown that splitting manifests itself in the tendency of the direction of the group velocity vector of the refracted wave to the cutoff angle, which is most noticeable for the FM structure in the low-frequency part of the SMSW range.

3. An experimental investigation was made of the dispersion of the SMSW in the FDM structure at various thicknesses of the dielectric layer. The regions of existence of direct waves of the first and second types are revealed in the SMSW spectrum. In the frequency domain corresponding to the reverse SMSW, a strong signal attenuation is noted due to the smallness of the group velocity of the SMSW. An experimental study of the orientational dependences of the refraction of the SMSW at the interface between the free film and the structure of the FDM, as well as the FM, has been carried out. Good (within

10%) agreement between the experimental results and theoretical conclusions was noted in all cases where the effects of dispersive splitting are absent. A noticeable manifestation of splitting effects was noted only in the lower part of the SMSW spectrum, which for a free ferrite film is no more than 10% of the total spectrum width, for an FDM structure – from 10% to 20%, and for an FM structure - up to 40–50%. The observed anomalies in the orientation dependences of the angle of refraction on the angle of incidence in these regions are in complete agreement with theoretical predictions. The reflection coefficient of the SMSW from the interface between the free ferrite film and the FDM structure is considered. It is shown that with an increase in the thickness of the dielectric layer, the reflection coefficient decreases the more, the higher the frequency of the SMSW. Experimental verification confirms the conclusions of the theory with an accuracy of 10–20%.

4. The dispersion properties of SMSW in structures containing two ferrite layers are considered. It is shown that the FF structure has four SMSW branches, two of which are external and two are internal, and their frequencies mutually repulse with a strong difference in the magnetizations of the layers. External SMSWs are always straight. Each internal SMSW at wavenumbers less than a certain critical value is direct, and at large, it is reverse. Similar dependences are considered for the structure of MDFFDM. It is shown that an additional ferrite layer located between the main ferrite layer and the metal significantly enhances the effect of the metal on the SMSW spectrum of the main layer. The dispersion of SMSW in two-layer ferrite films was studied experimentally. For the first time, the presence of all four branches predicted by theory was revealed in the SMSW spectrum; the repulsion of individual branches was observed, and on one of the internal branches, a section corresponding to backward waves. The theoretical curves coincided with the experimental ones within no worse than 2%. Dispersion curves of thin two-layer films (5–7 μm and less) in all branches showed discontinuities due to spin-wave resonances, the frequencies of most of which satisfactorily (up to 20–30%) coincided with those calculated by the model of pinned surface spins.

4

Methods of research and analysis of the propagation of SMSW under conditions of magnetization by a longitudinal inhomogeneous field

This chapter presents the results of a theoretical and experimental study of the propagation of SMSWs in ferrite films and structures based on them, magnetized by a longitudinally inhomogeneous field. The main types of field inhomogeneities and the spatial configuration of the SMSW propagation regions are considered. Two main methods for calculating the SMSW trajectories are presented: isofrequency curves and Hamilton–Auld curves, their advantages and disadvantages are discussed. The method of isofrequency curves is used to analyze the possible types of trajectories of forward and backward SMSWs in various structures. The trajectories, wavenumber, phase and group velocities of SMSWs in ferrite structures magnetized by a linear field, as well as fields of the 'valley' and 'shaft' type are calculated by the Hamilton–Auld method. The phase incursion and the group delay time of the SMSW are investigated. The main experimental results on the observation of the propagation of SMSWs under conditions of longitudinally inhomogeneous magnetization are described.

In the sections of the chapter devoted to the propagation of SMSWs in a free film, papers [580–592] were used, papers [593–599] were devoted to forward waves in FDM structures, and backward waves were discussed in papers [600–607], the phase incursion and delay time were considered in papers [608–614]. The rest of the necessary links are given in the text of the chapter.

4.1. Basic types of inhomogeneities of a magnetizing field

Real configurations of an inhomogeneous magnetizing field are very diverse. In this case, a necessary condition is the satisfaction of the equations of magnetostatics:

$$\operatorname{rot}\vec{H}=0 \;,\; \operatorname{div}\vec{B}=0 \qquad (4.1)$$

In the coordinate system shown in Fig. 2.1, the inhomogeneous field in the plane coinciding with the plane of the ferrite film changes in magnitude and direction. The arbitrary dependence of the field on y and z at the origin of the coordinates can be approximated with a certain accuracy using the Taylor series expansion to terms of the second order.

An important special case is a 'longitudinally inhomogeneous' field, the direction of which along the axis Oz does not change, and the intensity depends only on the coordinate:

$$H_z(z) = H_0\left(a_0 + a_1 z + a_2 z^2\right) \qquad (4.2)$$

From (4.1) we obtain: $H_y = 0$, $H_x \sim x$. In this case, if the surface of the ferrite film corresponds to the coordinate $x = 0$, then on it $H_x = 0$. The case $a_0 = 1$, $a_1 = a_2 = 0$, corresponds to a homogeneous field. The longitudinally inhomogeneous field can be of three types:

 1) linear $a_0 = 1$, $a_1 \neq 0$, $a_2 = 0$:,, - increases linearly ($a_1 > 0$) or decreases ($a_1 < 0$) along the coordinate z;
 2) of the 'valley' type $a_0 = 0$, $a_1 = 0$, $a_2 > 0$ – increases quadratically along the coordinate z to both sides of the Oy axis;
 3) of the 'shaft' type: $a_1 = 0$, $a_2 = 0$, $a_2 < 0$ – quadratically d`ecreases along the coordinate z to both sides of the Oy axis.

If the field gradient is positive, we will call such a field 'increasing', if negative – 'decreasing'.

Another important case is a 'transversely inhomogeneous' field, the direction of which along the Oz axis also does not change, and the intensity depends only on the coordinate:

$$H_z(y) = H_0\left(b_0 + b_1 y + b_2 y^2\right) \qquad (4.3)$$

the same basic configurations are possible – linear ($b_0 = 1$, $b_1 \neq 0$, $b_2 = 0$), 'valley' ($b_0 = 1$. $b_1 = 0$, $b_2 > 0$) and 'shaft' ($b_0 = 1$, $b_1 = 0$, $b_2 < 0$). At $x = 0$, we get $H_x = 0$, however, $H_y \sim (b_1 + 2b2y$ that is $H_y = 0$, only at $z = 0$..

An important special case of a field of a changing direction is a 'radially symmetric' field oriented along the polar radius vector from a point $y = z = 0$, with an intensity determined by the distance from the coordinates origin:

$$H_r(r) = H_0\left(c_0 + c_1 r + c_2 r^2 + c_3 r^3 + ...\right), H_\varphi = 0. \quad (4.4)$$

where $r = \sqrt{y^2 + z^2}$. The symmetry implies that $c_0 = 0$. When $x = 0$ we get:,,.

$$H_y = H_0 y\left(c_1 + c_2\sqrt{y^2 + z^2} + ...\right), H_z = H_0 z\left(c_1 + c_2\sqrt{y^2 + z^2} + ...\right) \quad (4.5)$$

The lines corresponding to a constant field strength are concentric rings centred at the origin. Ring configurations are possible: an annular linear field, an annular 'valley', an annular 'shaft'.

The listed cases allow us to consider with sufficient accuracy the features of the SMSW propagation for any field configurations in real devices. Later in this chapter, a longitudinally inhomogeneous field will be considered, and in the following, some cases of other configurations. Chapter 7, on applied results, describes the fields produced by real magnetic systems used in devices.

4.2. Spatial configuration of the areas if distribution of the SMSW

Setting a fixed frequency value determines the field interval in which SMSWs can exist. In an inhomogeneous field, this interval gives the spatial configuration of the regions in which the SMSW propagation is possible. So, for example, for a ferrite film with a free surface, the field boundaries of the regions of existence of SMSW have the form:
;

$$H_1 = \sqrt{\left(\frac{\omega}{\gamma}\right)^2 + (2\pi M_0)^2} - 2\pi M_0 \quad (4.6)$$

$$H_2 = \frac{\omega}{\gamma} - 2\pi M_0. \tag{4.7}$$

For a longitudinally inhomogeneous linear field with, we find that SMSWs can propagate within a strip, the boundaries of which are parallel to the axis and intersect the axis at the points:

$$z_1 = \frac{\sqrt{\left(\frac{\omega}{\gamma}\right)^2 + (2\pi M_0)^2} - (2\pi M_0 + H_0)}{a_1 H_0} \tag{4.8}$$

$$\ldots z_2 = \frac{\frac{\omega}{\gamma} - (2\pi M_0 + H_0)}{a_1 H_0} \tag{4.9}$$

It's clear that $z_1 > z_2$. Expressions (4.8) and (4.9) do not impose any restrictions on the value H_0; however, it is obvious that the saturation field of the film H_S must exceed H_0.

In the case of a field of the 'valley' type at $a_2 > 0$ we get:

$$z_1 = \pm\sqrt{\frac{\sqrt{\left(\frac{\omega}{\gamma}\right)^2 + (2\pi M_0)^2} - (2\pi M_0 + H_0)}{a_2 H_0}} \tag{4.10}$$

$$z_2 = \pm\sqrt{\frac{\frac{\omega}{\gamma} - (2\pi M_0 + H_0)}{a_2 H_0}} \tag{4.11}$$

The two signs in the obtained expressions reflect the symmetry of the field about the axis Oy. It is obvious that $|z_1| > |z_2|$. From the field interval of existence of SMSWs (4.6) and (4.7), we find that at $H_S < H_0 < H_2$ when SMSWs propagate within two strips parallel to the Oy axis, for which the coordinate z satisfies the inequalities: $-z_1 < z < -z_2$. or $z_1 < z < z_2$. When $H_2 < H_0 < H_1$ SMSW is distributed within only one band, for which z satisfies the inequalities: $-z_1 < z < -z_1$. With $H_1 < H_0 < +\infty$ the SMSWs cannot spread.

For a field of the 'shaft' type $z_{1,2}$ are also determined by formulas (4.10) and (4.11), with the difference that $a_2 < 0$. Moreover, if $H_2 < H_0 < H_1$, then SMSWs cannot be distributed. If $H_2 < H_0 < H_1$, then SMSWs are distributed within the same band for which

$-z_2 < z < -z_1$. If $H_2 < H_0 < +\infty$, then SMSWs spread within two bands at $-z_2 < z < -z_1$ or $z_1 < z < z_2$.

In the case of the FM structure, the picture is completely analogous, except for the replacement in the expressions for z_2 and H_2 the value $2\pi M_0$ by $4\pi M_0$ and, as a result of which the bands expand. In the FDM structure, z_2 and H_2 take intermediate values determined by the thickness of the dielectric layer. The boundaries of the SMSW propagation areas for other types of inhomogeneous fields are determined in a similar way.

4.3. Methods for analysis of SMSW propation under the conditions of inhomogeneous binding (frequency curves and Hamilton–Auld)

The propagation of SMSWs under the conditions of inhomogeneous magnetization is further investigated by two mutually complementary methods: isofrequency curves and the Hamilton–Auld method. Let's consider briefly each of them separately.

4.3.1. Isofrequency curve method

The method of isofrequency curves consists in dividing the plane of an inhomogeneously magnetized ferrite film into sections, within each of which the field change is so small that it can be considered uniform, after which it is assumed that the propagation trajectory of the SMSW within each such section remains rectilinear, and at the boundaries between sections is refracted provided that the projection of the wave vector on the interface is preserved. The relationship between the angles of incidence and refraction is determined using isofrequency curves as described in Section 3.2.1.

An example of constructing a trajectory for the case of a longitudinally inhomogeneous field is schematically illustrated in Fig. 4.1 a, b.

In Fig.4.1 a, the *Oyz* plane of the coordinate system coincides with the plane of the film, and the *Oz* axis is directed along the field, which monotonically increases along this axis and does not depend on the coordinate *y*. The plane of the film is divided by horizontal lines into six strips (1–6), within each of which the fields can be considered constant, and $H_1 > H_2 > H_3 > H_4 > H_5 > H_6$. Figure 4.1 b shows isofrequency curves 1–6 plotted for a given frequency of the SMSWs, corresponding to the same fields H_{1-6}. Let the SMSW be

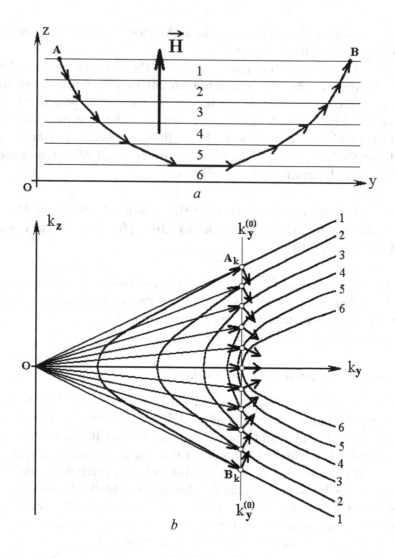

Fig. 4.1. Scheme for constructing the MSW trajectory by the method of isofrequency curves. a – FF (ferrite film) plane, 1-6 – areas of constant field value; b – plane of isofrequency curves, 1–6 – isofrequency curves corresponding to fields in regions

excited at a point A (Fig.4.1a), and the direction of the wave vector is specified by the excitation conditions, and the value is determined by the dispersion law. On the plane Ok_yk_z (Fig.4.1b), the end of the wave vector corresponds to a point A_k. The perpendicular to the tangent to the isofrequency curve 1 at this point determines the direction of the group velocity vector in the field region H_1, which gives a trajectory segment within the first strip. To construct the further course of the trajectory, we will take into account the condition of preserving the projection of the wave vector onto the boundaries of the strips. In Fig.4.1a, these boundaries are parallel to the axis Oy; therefore, the component of the initial wave vector is preserved, which we denote by $k_y^{(0)}$. Figure 4.1b shows that the ends of the wave vectors for bands 2–6 coincide with the intersection points of the vertical straight line drawn through the point A_k with the isofrequency curves 2–6. The perpendiculars to the tangents to the isofrequency curves at the intersection points give the directions of the group velocity vectors in the same strips, shown by short arrows. Thus, passing from strip to strip, it is possible to reduce the procedure for constructing the wave trajectory to sequentially finding the directions of the group velocity vectors in separate strips. Figure 4.1 a shows that the trajectory first in the bands 1–5 goes towards lower field values, then in band 6 it experiences a turn towards higher field values, and passes the same bands in the opposite order. The rotation of the trajectory in band 6 is due to the transition of the end of the wave vector from the upper branches of the isofrequency curves to the lower ones.

The described algorithm for constructing SMSW trajectories is suitable for fields of any configuration. The requirement that the direction of the field remains unchanged in the plane of the film is not strict. So, if the direction of the field along the film changes, then the orientation of the coordinate systems $Ok_y k_z$ used to construct the isofrequency curves on different sides of the interface between the regions will be different. Preservation of the projection of the wave vector onto the interface remains a necessary condition for refraction. The construction of the trajectory can be continued until the field at the point of wave propagation exceeds the permissible limits determined by its frequency and the saturation magnetization of the film. The smaller the size of the split areas, the higher the construction accuracy. Along with the trajectory, one can study the change in the wave number and the orientation of the vectors of the phase and group velocities of the wave. The algorithm can be easily implemented as a machine program.

4.3.2. The Hamilton–Auld method

The Hamilton–Auld method, based on an analogy between an MSW and a particle moving in an inhomogeneous potential, was first described in [272] and used to calculate the trajectories of SMSWs under conditions of inhomogeneous magnetization in [273]. In the coordinate system, the plane of which coincides with the plane of the ferrite film, the Hamilton–Auld system of equations describing the propagation of the SMSWs has the form [273] (2.649)–(2.651):

$$\frac{dk}{dy} = k\left(\frac{\partial k}{\partial y}\cos\phi + \frac{\partial k}{\partial z}\sin\phi\right)\cdot\left(k\cos\phi + \frac{\partial k}{\partial\phi}\sin\phi\right)^{-1} \quad (4.12)$$

$$\frac{d\phi}{dy} = -\left(\frac{\partial k}{\partial y}\sin\phi - \frac{\partial k}{\partial z}\cos\phi\right)\cdot\left(k\cos\phi + \frac{\partial k}{\partial\phi}\sin\phi\right)^{-1} \quad (4.13)$$

$$\frac{dz}{dy} = \left(k\sin\phi - \frac{\partial k}{\partial\phi}\cos\phi\right)\cdot\left(k\cos\phi + \frac{\partial k}{\partial\phi}\sin\phi\right)^{-1} \quad (4.14)$$

where: k is the length of the SMSW wave vector, φ is the angle between the SMSW wave vector and the axis, and the partial derivatives included in the right-hand side of the above equations are determined from the dispersion relation $f(\omega, k, H(y,z)) = 0$. A detailed derivation of these equations is given in Section 2.8 (formulas (2.649)–(2.651)).

The system of equations (1.28)–(1.30) is integrated numerically using the traditional algorithm for the numerical solution of a system of three first-order differential equations, for example, by the Euler or Runge–Kutta method [482]. In this case, the first equation gives the length of the SMSW wave vector, the second – the direction of this vector through the angle $\varphi(y)$, and the third – the direction of the group velocity vector and the SMSW trajectory, on the plane Oyz the form $z(y)$. Specific calculation examples are given below.

The described method for constructing SMSW trajectories is suitable for fields of any configuration. The permissible field values are determined from the cutoff frequencies of the SMSW spectrum. Thus, the lower frequency boundary of the SMSW spectrum gives $\omega_b = \gamma\sqrt{H(H+4\pi M)}$ for the upper boundary of the field:

$$H_t = -2\pi M + \sqrt{(2\pi M)^2 + (\omega/\gamma)^2} \quad (4.15)$$

The upper frequency boundaries of the spectrum for free $\omega_c = \gamma(H+2\pi M)$ and metallized films $\omega_{tm} = \gamma(H+2\pi M)$ give the lower boundaries of the field for the same cases:

$$H_z = H_0(1+a_1 z) \text{ and } H_{hm} = -4\pi M_0 + \omega/\pi = \gamma \quad (4.16)$$

The smaller the integration step of the system of equations (4.12)–(4.14), the more accurate trajectory is constructed. The wavenumber and orientations of the phase and group velocity vectors of the wave are obtained simultaneously with the wave trajectory. The machine implementation of the algorithm is very convenient.

4.3.3. Comparison of methods for analyzing SMSW trajectories

Both of the described methods equally allow one to study the properties of the trajectories of SMSWs propagating under conditions of inhomogeneous magnetization.

An important advantage of the isofrequency curve method is its clarity. However, when the direction of the field is changed along the plane of the film, the geometric constructions become rather cumbersome, and the machine calculation becomes complicated. At the points where the trajectory passes perpendicular to the field (region 6 in Fig.4.1 a), the machine algorithm must be supplemented with a rule that ensures the transition from one branch of the isofrequency curve to another. The study of the wavenumber, phase and group velocities requires additional actions.

The advantage of the Hamilton–Auld method is the simplicity of its implementation on a computer. In this case, the change in the orientation of the field and the rotation of the trajectories are taken into account automatically, without causing additional difficulties, which leads to a decrease in the computer time of the calculation, in comparison with the method of isofrequency curves, while maintaining the same accuracy. The disadvantage of this method is the lack of a convenient geometric interpretation and low clarity.

In accordance with the foregoing, further investigation of the propagation of the SMSWs under conditions of inhomogeneous magnetization will be carried out mainly by the Hamilton–Auld method, and the isofrequency curves method will be used as needed for a qualitative interpretation of the results.

4.4. Distribution of SMSWs in ferrite films with free surfaces

Let us first consider the simplest case of a ferrite film with free surfaces. To reveal the most detailed picture of the SMSW propagation, we use both methods: isofrequency curves and Hamilton-Auld.

4.4.1. Analysis of SMSW trajectories by the method of isofrequency curves

Let us first investigate the qualitative picture of the SMSW propagation, for which we use the method of isofrequency curves. Let us consider the main configurations of a longitudinally inhomogeneous field: linear, of the 'valley' and 'shaft' type.

4.4.1.1. Linearly inhomogeneous field

Let us first consider a linearly inhomogeneous increasing field $H_z = H_0(1+a_1 z)$, where $a_1 > 0$. The trajectory construction for a fixed frequency is illustrated in Fig. 4.2. Figure 4.2 a shows the plane of the film. The field H is directed along the Oz axis and increases linearly with increasing z. SMSWs can propagate in a strip bounded by horizontal dashed lines passing through z_1 and z_2, defined by formulas (4.8) and (4.9). Figure 4.2b shows isofrequency curves 1–5 for different field values: a larger curve number corresponds to a larger field. Straight lines 6 correspond to cut-off lines at H_z and H_1 (4.7).

Let us assume that the SMSWs are excited on the film plane at a point A (Fig. 4.2 a). The angle between the wave vector and the axis Oy determines the nature of the trajectory. Three different cases are possible:

1) $\varphi < 0$ – trajectory 'minus-type'. At the point of excitation the wave vector $\vec{k}_1$ corresponds to the group velocity vector $\vec{s}_1$ directed in the direction of increasing the field. When the SMSW propagates, the end of the wave vector in the plane of isofrequency curves $Ok_1 k_2$ (Fig. 4.2 b) moves downward up to the cutoff line. The complete trajectory shown by the curve L_1 in Fig.4.2 a, after excitation, monotonically tends upward up to the boundary of the region of existence of the SMSW, determined by the coordinate z_1.

2) $\varphi = 0$ trajectory 'zero-type'. At the point of excitation the wave vector $\vec{k}_2$ corresponds to the group velocity vector $\vec{s}_2$ directed parallel to the axis Oy, as a result of which the trajectory cannot deviate in the direction of a larger or smaller field and is all a straight line parallel to the axis. This trajectory is unstable and is not realized in the experiment.

3) φ > 0 'plus-type' trajectory. At the point of excitation, the wave vector $\vec{k}_3$ corresponds to the group velocity vector $\vec{s}_3$ directed towards the decrlease in the field. When the SMSW propagates, the end of the wave vector on the plane $Ok_y k_z$ (Fig. 4.2b) moves downward, passing from the upper branches of the isofrequency curves through the axis Ok_y to the lower ones up to the cutoff line. The trajectory of the SMSW L_3 (Fig. 4.2a), from the point A goes down to the point P from where it turns up and rushes to the upper boundary of the region of existence of the SMSW, determined by the coordinate z_1.

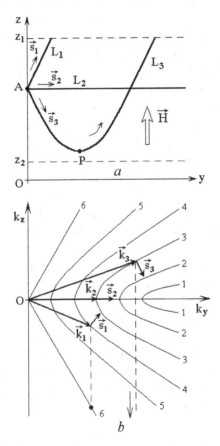

Fig. 4.2. Scheme for constructing the trajectories of the SMSW for a linearly inhomogeneous field. a – the plane of the film and the trajectory of the SMSW; b – isofrequency curves 1–6 for different values of the field, increasing with a constant step.

The considered picture corresponds to an increasing field ($a_1 > 0$). In a decreasing field ($a_1 < 0$), the same three types of trajectories are observed, mirrored about the axis Oy.

These three types of trajectories are basic and make it possible to explain the features of the behaviour of SMSWs in longitudinally inhomogeneous fields of any other configurations.

4.4.1.2. Valley-type field

Let us now consider the distribution of SMSWs in a field of the 'valley' type $H_z = H_0(1 + a_2 z^2)$, where $a_2 > 0$. The trajectory formation scheme is illustrated in Fig. 4.3a. The type of trajectories is determined by the value of the 'valley bottom' field H_0. At $H_S < H_z < H_2$ the SMSWs propagate within two bands: $z_2 < z < z_1$ or $-z_1 < z < -z_2$ (here H_S is the saturation field,, H_S, z_1, z_2 are given by the formulas (4.7), (4.10) and (4.11)). In the first strip, the field is increasing, in the second it is decreasing, so the trajectories are similar to the case of a linear field: L_1 and L_1' - minus-type, L_2 and L_2' - zero-type, and L_3 and L_3' plus-type. With an increase in the field H_0, the inner edges of the bands (at $\pm z_2$), tend to each other and at $H_0 = H_2$ converge on the axis Oy, and the region of existence of the SMSWs merges into one strip with boundaries. In this case, the plus-type trajectory that has reached the 'bottom of the valley' (axis Oy) turns into a zero-type trajectory that propagates exactly along this 'bottom' (at $z = 0$). The resulting trajectory $L_4(L_4')$ is shown by a dashed line. As in the case of a linear field, this trajectory is unstable.

In the interval of fields on the plane, the family of isofrequency curves is limited from the inside by the 'limiting' isofrequency curve corresponding to the field of the 'bottom of the valley'. For plus-type trajectories, excited at, the end of the wave vector on the plane, when moving down, reaches the limiting isofrequency curve, after which it goes up to the cut-off line. In this case, the trajectory of the plus-type of the species, having reached the 'bottom of the valley', passes through it into region c, 'merges' into the corresponding trajectory of the plus-type of the species, and, rushing to the lower edge of the 'valley', ends at $z = -z_1$ The resulting path is shown with a dashed line. Excitation of the SMSW at $z < 0$ ives a similar trajectory. With an increase in the field, the band of existence of SMSWs narrows and, with their propagation, becomes impossible.

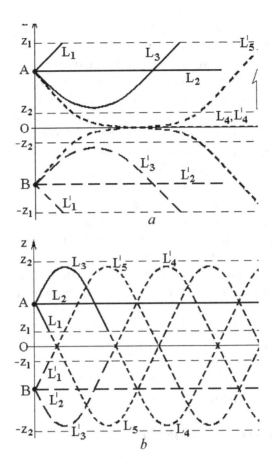

Fig. 4.3. Scheme for constructing the trajectories of the SMSW for a quadratically inhomogeneous field. *a* – field of the 'valley' type; *b* – field of the 'shaft' type.

Thus, due to the presence of the limiting isofrequency curve, which is absent for a linear field, in a 'valley'-type field, there can be two new types of trajectories, which are a combination of the studied ones: trajectories of the plus-type, turning into zero-type and trajectories of the plus-type of an increasing field, passing into the plus-type of a decreasing field (and vice versa). We will call them 'plus-zero-type' and 'plus-plus-type' trajectories.

4.4.1.3. Shaft-type field

Let us now consider the propagation of SMSWs in a field of the 'shaft' type $H_z = H_0(1 + a_2 z^2)$, where $a_2 < 0$. The trajectory formation scheme is illustrated in Fig. 4.3 b. Similarly to the previous case, the nature of the trajectories is determined by the field of the 'shaft top'. With SMSWs, they spread within two bands: $-z_2 < z < -z_1$ (increasing field) or $z_1 < z < z_2$ (decreasing field). In each band, trajectories typical for a linear field are possible (L_1 and L'_1) minus-type (L_2 and L'_2) zero-type and plus-type (L_3 and L'_3). When $H_2 < H_0 < H_1$ SMSW is distributed within one band, for which $-z_2 < z < -z_1$. On the plane $Ok_y k_z$, the family of the isofrequency curves is bounded outside by the 'limiting' isofrequency curve corresponding to the field of the 'shaft top' H_0. In this case, the minus-type trajectory of the decreasing field () excited at $z > 0$, passing through the 'top of the shaft' (axis Oy), continues further as a plus-type trajectory of the increasing field (similarly to L'_3), that is, it first moves away from the 'top of the shaft', and then turns again the top. This process is repeated periodically, resulting in a 'pseudo-sinusoidal' trajectory. Excited at a minus-type trajectory, gives a pseudosine-wave trajectory L'_4. Plus-type trajectories L_3 and L'_3 give pseudo-sinusoidal trajectories L_5 and L'_5, which are identical with L_4 and L'_4 up to phase shift. The trajectories of the L_2 and L'_2 zero type are as before unstable. With $H_0 < H_2$ SMSW cannot spread.

Thus, in a field of the 'shaft' type, there is another new kind of trajectories – periodic pseudosine wave.

4.4.2. Analysis of SMSW trajectories by the Hamilton–Auld method

Let us now consider the quantitative characteristics of the SMSW propagation for various parameters of the field and the wave itself, for which we use the Hamilton–Auld method. We will pay most attention to the case of a linear field as the main one, and then consider the features of the trajectories in fields of other configurations. In all considered cases, the saturation magnetization of the film $4\pi M_0$ was taken equal to 1750 G, thickness $d = 15$ μm.

4.4.2.1. Linearly inhomogeneous field

The linear field corresponds to Figs. 4.4–4.7. This field is described by formula (4.2). For Figs. 4.4 and 4.5, the values included in it are equal: $H = 437.5$ Oe, $a_0 = 1$, $a_1 = 1/16$ cm^{-1}. The variable parameter is the coordinate y measured along the axis Oy perpendicular to the direction of the field (axis Oz). Each figure contains four parts, designated by letters a–d, and part a corresponds to the SMSW trajectories, b – to the wavenumber dependences k on y, c – to the dependences of the angle φ between the wave vector and the axis Oy, and d – to the dependences of the angle φ between the group velocity vector and the axis Oy on y.

Figure 4.4 shows the above dependences for the same initial angle φ = 30° for different frequencies. The SMSW trajectories shown in Fig. 4.4a are 'plus-type' trajectories, that is, they first go towards a smaller field, then turn to a larger field and break off at points whose coordinates are the greater, the higher the frequency, in accordance with the formula (4.8). The trajectory minima are also the deeper and the further they are from the axis, the higher the frequency. In this case, deeper minima correspond to the greater length of the trajectories along the axis Oy. Figure 4.4 b shows the dependences $k(y)$ for the same frequencies. It can be seen that high frequencies correspond to large wavenumbers, and the length of these curves along the Oy axis is obviously equal to the length of the curves for the same frequency in Fig. 4.4 a. The wavenumber is minimum at values corresponding to the minimum on the trajectories. Figures 4.4c and 4.4d show the dependences $\varphi(y)$ and $\psi(y)$ for the same frequencies. At the starting point of the trajectory, the angle φ for all frequencies is positive and equal to 30°, and the angle ψ is negative, and its absolute value increases with increasing frequency. As the trajectories develop, the absolute value of both angles first decreases, then passes through zero at the coordinate corresponding to the minimum of the trajectory in Fig. 4.4a, as a result of which the angles change sign, and then increase, reaching a value equal to the cutoff angle. The monotonic character of all the given frequency dependences is slightly violated only for curve 6, corresponding to 3300 MHz. Indeed, at y > 3 cm, this curve intersects curve 5, corresponding to a lower frequency of 3200 MHz. This behaviour of this curve is due to the proximity of its frequency to the upper boundary of the SMSW spectrum.

Figure 4.5 shows similar dependences for a fixed frequency of 3000 MHz and vasrious starting angles φ. Curves 1 and 2 correspond to the trajectories of the minus type, curve 3 to the zero type, and curves 4 and 5 to the plus type. The minimum on the trajectories, as well as on the dependences, is observed only for curves 4 and 5. All trajectories break off at a value $z = z_1$ determined by formula (4.8) at a frequency of 3000 MHz. The angle φ, the value of which at y = 0 is determined by the initial conditions, as y increases, always tends

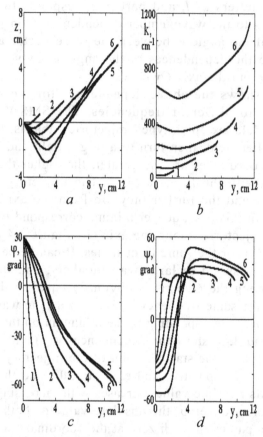

Fig.4.5. SMSW trajectories (*a*) and the corresponding dependences (*b*), (*c*), and (*d*) in a linear field at different initial angles: 1 – –40°, 2 – –20°, 3 – 0°, 4 – 20°, 5 – 40°.

towards negative values, and the angle ψ, mainly towards positive values. The maximum values of the φ and ψ angles correspond to the cutoff angles for a given frequency. Figure 4.6 shows similar dependences for a fixed frequency of 3000 MHz and a varying field gradient. In formula (4.2), the parameters H_0, a_0 and a_2 remain the same, and the parameter a_1 takes on varying values. It is seen that a change in the field gradient does not change the general character of the curves, but only leads to a change in the scale along the coordinate axes. So, when the gradient decreases, the trajectories are stretched along both Oy and Oz axes and the dependences $k(y)$, $\varphi(y)$ and $\psi(y)$, remaining within the same limits along the Oz axis, are stretched along the axis Oy. This behaviour of these curves is due to a change in the boundaries of the SMSW propagation bands, in accordance with the formulas (4.8) and (4.9).

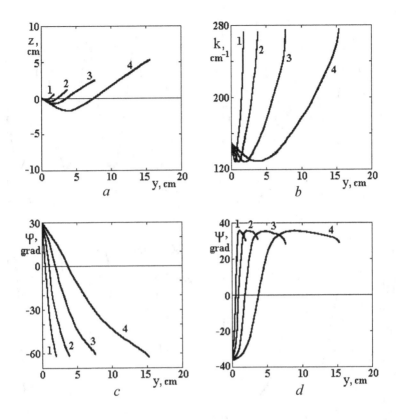

Fig. 4.6. SMSW trajectories (*a*) and the corresponding dependences (*b*), (*c*), and (*d*) in a linear field at its various gradients. Parameter: 1 – 1/4 cm^{-1}, 2 – 1/8 cm^{-1}, 3 – 1/16 cm^{-1}, 4 – 1/32 cm^{-1}.

Let us now consider what happens to the wave when it reaches the point where the field is determined by formula (4.8). In this case, the wavenumber remains finite and takes on values of the order of cm-1. This value is not small enough for the SMSW to be converted into an electromagnetic wave by radiation (for this, k should be no more than 10 cm^{-1}). At the same time, it is not large enough for the SMSW to be transformed into a short exchange spin wave and then into heat (for this it must be at least 10^4 cm^{-1}). Therefore, it can be assumed that at this point the SMSW is reflected from the line of constant field value according to the law of specular reflection. Such reflection will certainly take place if the edge of the film is placed on the line corresponding to the coordinate determined by formula (4.8). The SMSW trajectories, as well as dependences, and for this

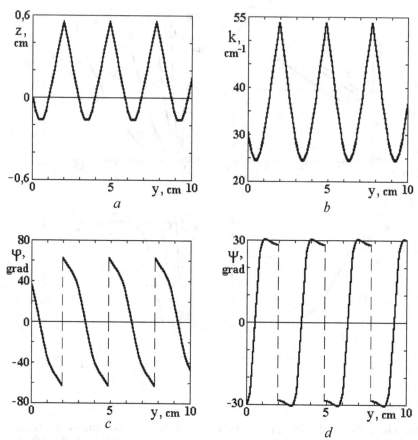

Fig. 4.7. SMSW trajectories (*a*) and the corresponding dependences (*b*), (*c*), and (*d*) in a linear field for a channel of the first type.

case are shown in Fig. 4.7. The curves are plotted for a frequency of 2800 MHz at an initial angle of $\varphi = 30°$ and a field parameter of $a_1 = 1/16$ cm^{-1}. It can be seen that all dependences are periodic with a period of about 3.4 cm. Thus, the SMSW propagates along the Oy axis as if in a channel, the upper boundary of which is determined by the formula (4.8), and the lower boundary is determined by the coordinate at which the trajectory undergoes a rotation, that is $\varphi = \psi = 0$. The channel and paths are asymmetrical about the Oy axis. Let's call such a channel the 'first type channel'.

Comment. The assumption about the reflection of the SMSW from the line of constant field value was made here purely intuitively without sufficient reason. The reality of this assumption should be verified experimentally; however, the authors of this monograph are currently not aware of such experiments. At the same time, reflection from the edge of the film is unconditional and has been experimentally confirmed, for example, in [139, 303].

4.4.2.2. Valley-type field

As the next example, consider the propagation of SMSW in a film magnetized by a 'valley'-type field determined by formula (4.2) with the parameters: $H_0 = 437.5$ Oe, $a_0 = 1$, $a_1 = 0$, $a_2 = 1/16$ cm^{-2}. Here, as before, $4\pi M_0 = 1750$ G, $d = 15$ μm. The SMSW (y) trajectories, as well as the dependences, and for this case at different frequencies are shown in Fig.4.8. In this case, in accordance with the accepted terminology, the curves 1–4 in Fig.4.8 a are plus-type trajectories, curve 5 is a plus-zero-type trajectory, and the curves 6–9 are plus-plus-type trajectories. From above and below, these trajectories terminate at coordinates $z = \pm z_1$ where z_1 is determined by formula (4.8). The critical frequency corresponding to the infinite plus-zero trajectory (curve 5) with an accuracy of 1 Hz is equal to 3126.664425 MHz. A deviation from this frequency by 5 Hz leads to a deviation of the trajectory up or down at $y = 8$ cm (the curves 4 and 6). The wave number here (Fig. 4.8b), as for a linear field, is the larger, the higher the frequency. The angle φ (Fig. 4.8c) at the initial part of the trajectory for all frequencies decreases and for frequencies below the critical one it continues to change in the same direction up to the cutoff angle (curves 1-4). For the critical frequency (curve 5) at $y > 4$ cm, this angle becomes equal to zero, for frequencies above the critical (curves 6–9), passing through a minimum near the zero value, it increases up to the cutoff angle. The angle ψ (Fig. 4.8d) behaves in a similar way up to a sign.

In a 'valley'-type field, it is also possible for the SMSW to propagate in the channel, the formation of which for a frequency of 3129 MHz is illustrated in Fig. 4.9. Here, the SMSW reflection occurs at two coordinates $z = z_1$ and $z = -z$, where z_1 is determined by formula (4.8). The trajectories are periodic, symmetric about the Oy axis, which is due to the symmetry of the field. Let's call such a channel the 'channel of the second type'. Note that here the SMSW reflection from the lines corresponding to the channel boundaries should be treated in the same way as in the case of the first type channel (section 4.4.2.1)

4.4.2.3. Shaft-type field

Let us now consider the propagation of SMSWs in a film magnetized

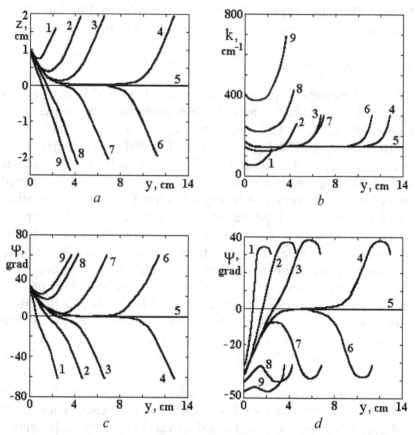

Fig. 4.8. SMSW trajectories (a) and the corresponding dependences (b), (c), and (d) in a 'valley'-type field for different frequencies: Curves 1 correspond to a frequency of 3000 MHz, 2 – 3100 MHz, 3 – 3124 MHz, 4 – 3126.664420 MHz, 5 – 3126.664425 MHz, 6 – 3126.664430 MHz, 7 – 3129 MHz, 8 – 3200 MHz, 9 – 3300 MHz.

by a field of the 'shaft' type, determined by formula (4.2) with the parameters: $H_0 = 437.5$ Oe, $a0 = 1$, $a1 = 0$, $a2 = 1/16$ cm−2. Here also $4\pi M_0 = 1750$ G, $d = 15$ μm. The SMSW trajectories $z(y)$, as well as the dependences $k(y)$, $\varphi(y)$ and $\psi(y)$ for this case at different frequencies are shown in Fig. 4.8. At any frequency, SMSWs come out of the origin with an initial angle $\varphi = 30°$. In accordance with the accepted terminology, all trajectories (1–3) in Fig. 4.10a are periodic pseudosine and lie within the limits, where it is determined by formula (4.9). The range of the pseudo-sinusoidal curves along the axis and the value of the period along the axis are the greater, the higher the frequency of the SMSW. Due to the periodic nature

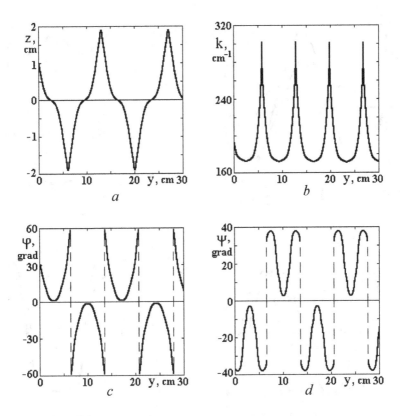

Fig. 4.9. SMSW trajectories (*a*) and the corresponding dependences $k(y)$ (*b*), $\varphi(y)$ (*c*) and $\psi(y)$ (*d*) in a field of the 'valley' type for a channel of the second type. The frequencies are the same as in Fig. 4.8.

of the trajectories, the dependences are also periodic. The wave number here (Fig. 4.10 b), as before, the greater, the higher the frequency. The angle φ (Figure 4.10 c) at the initial section of the trajectory for all frequencies decreases and, having passed through zero, continues to change in the same direction up to a value equal to the initial one (30°). Thus, the amplitude of all dependences is the same and equal to 30°. The dependences (Fig.4.10 d) resemble the previous ones with the opposite sign, but here the amplitude is the higher, the higher the frequency.

Obviously, due to the pseudo-sinusoidal character of the trajectories, the 'shaft'-type field forms a channel for the SMSW propagation. Such a channel is caused not by reflection, but by the reversal of the direction of the SMSW trajectories relative to the

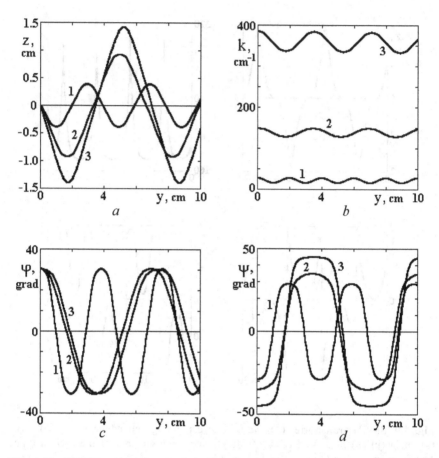

Fig. 4.10. SMSW trajectories (a) and the corresponding dependences $k(y)$ (b), $\varphi(y)$ (c), and $\psi(y)$ (d) in a 'shaft'-type field for different frequencies: 1 – 2800 MHz, 2 – 3000 MHz, 3 – 3200 MHz.

Oz axis in regions of a sufficiently strong field. By analogy with the previous ones, it can be called a 'third type channel'. The main properties of this channel are illustrated in Fig. 4.11 a–d.

Figure 4.11 a shows the change in the trajectories of the SMSW at a fixed frequency of 3000 MHz depending on the field gradient. The starting point of the trajectories is the origin, the starting angle is 30°. It can be seen that both the amplitude and the period of the trajectories increase with decreasing field gradient (from curve 1 to curve 3). Figure 4.11 b shows the change in the SMSW trajectories outgoing from the origin of coordinates of the same frequency (3000 MHz) depending on the initial angle at a constant field gradient ($a_2 = -1/16$ cm^{-2}). Here, the amplitude and period of the trajectories

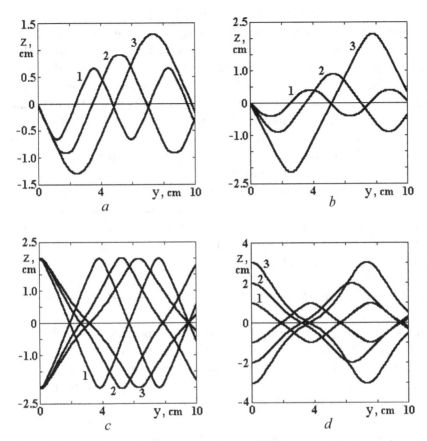

Fig. 4.11. SMSW trajectories in a 'shaft'-type field when changing various parameters. a – the field gradient parameter a_2 changes: 1 – $-1/8$ cm^{-2}, 2 – $-1/16$ cm^{-2}, 3 – $-1/32$ cm^{-2}; b – the initial angle φ changes: 1 – 15°, 2 – 30°, 3 – 45°; c – frequency changes: 1 – 2800 MHz, 2 – 3000 MHz, 3 – 3200 MHz; d – the coordinate z changes: 1 – ± 1 cm, 2 – ± 2 cm, 3 – ± 3 cm.

grow with increasing angle. Figure 4.11c shows the trajectories of different frequencies emerging from a point with coordinates $y = 2$ and $z = \pm 2$ cm parallel to the axis Oy ($\varphi = \psi = 0$). It can be seen that the amplitudes of all trajectories are equal to the initial value of the coordinate z (2 cm), and the higher the frequency, the longer the period. Figure 4.11d shows the trajectories of a fixed frequency of 2800 MHz, extending parallel to the Oy axis (at the beginning $\varphi = \psi = 0$) from points with different coordinate values z. In this case, the coordinate y is always zero. It can be seen that the amplitude of each trajectory is equal to the initial value of the coordinate z, and the period is the greater, the further away its initial point is from the origin of coordinates. In Figs 4.11 c and 4.11 d, for some values of the coordinate y, a strong narrowing of the beam of trajectories is observed, resembling focusing of the beam. This phenomenon, inherent in trajectories, the initial sections of which are parallel to the Oy axis, will be called 'pseudofocusing'. There are several (2–3 or more) pseudofocusing regions along the axis, the weaker the more pronounced, the further they are from the origin (due to the 'dephasing' of the periodic curves). So in Fig. 4.11 c, two pseudofocusing regions are visible, and in the first, at $y \approx 2.5$ cm, the trajectory span is ± 0.5 cm, and in the second at $y \approx 9$ cm, the span is ± 0.8 cm. Figure 4.11 d also shows two pseudofocusing regions: the first is from $y \approx 2.5$ cm to $y \approx 4.5$ cm and the second, which starts at $y \approx 10$ cm. The continuation of the trajectories of Fig. 4.11 d (at $y > 10$ cm) shows that here, as the number of the pseudofocusing region increases, the dephasing and the range of trajectories increases.

4.5. Distribution of SMSW in the ferrite–metal structure

Let us now consider the features of the propagation of SMSW in the ferrite–metal (FM) structure. In this case, the SMSW spectrum is much wider than for a free film, and its upper boundary in the field is described by formula (4.6), and the lower one, by formula (4.7) with replacement of $2\pi M_0$ by $4\pi M_0$. The shape of the isofrequency curves in this case is similar to the analogous curves for a free film, but the point of their intersection with the Ok_y axis is located much closer to the origin, and the opening of the branches is much larger. The main qualitative difference between this case and a film with a free surface is the absence of a wave cutoff at the lower frequency (upper field) boundary of the SMSW spectrum. In this case, the cutoff angle is formally equal to 90°, that is, any directions of SMSW propagation are permissible. In the rest of the SMSW spectrum, the cutoff angle is less than 90°, but always greater than the cutoff angle for the same frequency (or field) in a free film. The qualitative construction of the SMSW trajectories in this case is similar to a free film; therefore, we will immediately consider the quantitative aspect using the Hamilton–Auld method.

4.5.1. Linearly inhomogeneous field

Consider first the case of a linear field described by formula (4.2) with parameters: $H_0 = 437.5$ Oe, $a_0 = 1$, $a_1 = 1/4$ cm^{-1}, $a_2 = 0$. For this case, Fig. 4.12 a shows the trajectories of the SMSW $z(y)$, Fig. 4.12b dependences $k(y)$, in Fig.4.12 c dependences $\varphi(y)$ and in Fig.4.12 d $\psi(y)$ for different frequencies. Curves 1 correspond to a frequency of 3000 MHz, 2 to 3500 MHz, 3 to 4000 MHz, 4 to 4500 MHz, 5 to 5000 MHz. The starting angle φ is everywhere 30°. Film parameters $4\pi M_0$ = 1750 G, thickness 15 μm. Comparison of Fig. 4.12 with Fig. 4.4 (taking into account the difference in frequencies and field gradients for similar dependences) shows that the nature of the curves for the FM structure is similar in form to the nature of similar curves for a film with a free surface (for example, all trajectories are of the plus type and their range increases with frequency). The common difference between the curves for the FM structure in comparison with the free film is the significantly greater (one and a half to two times) length of all curves along the axis. The

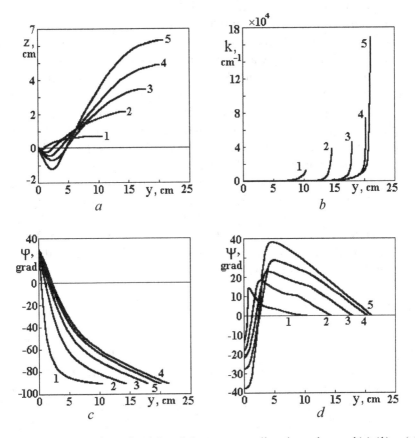

Fig. 4.12. SMSW trajectories (*a*) and the corresponding dependences $k(y)$ (*b*), $\varphi(y)$ (*c*), and $\psi(y)$ (*d*) for the FM structure in a linear field at various frequencies: 1 – 3000 MHz, 2 – 3500 MHz, 3 – 4000 MHz, 4 – 4500 MHz, 5 – 5000 MHz.

range of trajectories along the Oz axis for the FM structure is much smaller than for a free film. Both of these features are explained by the above-mentioned differences in the shape of the isofrequency curves. The trajectories of a given frequency in a free film (Fig. 4.4 a) approach the horizontal line corresponding to the upper boundary value of the field (4.6) at an acute angle corresponding to the cutoff, after which they break off. In the FM structure (Fig. 4.12 a), due to the absence of cutoff at the upper boundary value of the field, the trajectories only asymptotically approach this line without crossing it, which corresponds to the angles $\varphi = 90°$ and $\psi = 0°$. Therefore, the trajectories in such a structure do not break off and theoretically can continue to infinity, turning into horizontal straight lines. To overcome this circumstance, which undoubtedly cannot be realized in the experiment, the trajectories shown in Fig. 4.12 a are cut off artificially at the angle value $\varphi = 89.9°$.

The behaviour of the dependences in the FM structure (Fig. 4.12 b) has a noticeable difference from the analogous dependences in Fig.4.4 b, that here the wavenumber increases sharply as the angle approaches 90° (to 0°), and the stronger the higher the frequency of the SMSW. For a frequency of 5000 MHz at $\varphi = 89.9°$ it reaches $1.7 \cdot 10^5 \, cm^{-1}$ (curve 5). At such large values of the wave number, the SMSW acquires an exchange character, and consideration based on an exchangeless dispersion relation of the type (2.14) loses its meaning. Nevertheless, this feature of shortening the SMSW in the FM structure can be considered as a potential option for exciting exchange MSWs for use in devices. The dependences $\varphi(y)$ (Fig. 4.12 c) and $\psi(y)$ (Fig. 4.12 d) for the FM structure are similar to the same dependences for the free film (Figure 4.4c, d), with the difference that here at the end of the trajectories for all frequencies the angle φ tends to 90°, and the angle ψ tends to 0°.

4.5.2. Valley-type field

Let us now consider the propagation of SMSWs in an FM structure magnetized by fields of other configurations. For simplicity, we will restrict ourselves only to trajectories, since the nature of the remaining dependences can be easily understood by analogy with the case of a linear field. The main cases are illustrated in Fig. 4.13a-d.

Figure 4.13 a shows the trajectories of SMSWs of different frequencies in the FM structure magnetized by a valley-type field with the parameters: $H_0 = 437.5$ Oe, $a_0 = 1$, $a_1 = 0 \, cm^{-1}$, $a_2 = 1/4$

cm−2. Film parameters are the same as before. Here, the frequency 5154.44996 MHz (curve 5), corresponding to the plus-zero-type trajectory, is critical: below it, the trajectories have a plus-type, above it, a plus-plus-type. Comparing these curves with those shown in Fig. 4.8 a, one can see that a serious difference consists only in the bending of the ends of the trajectories for the FM structure to horizontal straight lines, which is completely analogous to the case of a linear field and is explained by the same reasons.

4.5.3. Shaft-type field

Figure 4.13 b shows the trajectories of SMSWs of different frequencies in the FM structure magnetized by a field like a shaft with the parameters: $H_0 = 437.5$ Oe, $a_0 = 1$, $a_1 = 0$ cm^{-1}, $a_2 = -1/4$ cm^{-2}. Film parameters are the same. Similar to the case of a free film, the 'shaft'-type field here also forms a channel of the third type for the propagation of SMSW. The difference from similar trajectories for a free film (Fig. 4.10) is only quantitative. For convenience of comparison, the same Fig. 4.13 b shows curve 4 corresponding to a free film at a frequency of 3000 MHz. It can be seen that curve 1, plotted at the same frequency for the FM structure, has a smaller amplitude and a longer period than curve 4. Both of these features are a consequence of the proximity to the origin and a large opening of the branches of the isofrequency curves for the FM structure compared to the same curves for free film.

4.5.4. Channels of the first and second type

In the FM structure in linear and valley-type fields, as in the case of a free film, it is also possible to create channels of the first and second types. However, here, due to the fact that near the upper boundary in the field, the SMSW length sharply decreases, and the trajectory asymptotically tends to a horizontal straight line, reflection from this boundary is obviously impossible. However, if the edge of the film is placed before this boundary, then the SMSW can be reflected from it, thus providing a channel. For this case, the form of trajectories in a linear field (channel of the first type) and in a field of the 'valley'-type (channel of the second type) is shown, respectively, in Fig. 4.13 c ($a_1 = 1/4$ cm^{-1}, $a_2 = 0$ cm^{-2}, frequency 4000 MHz) and 3.13 d ($a_1 = 0$ cm^{-1}, $a_2 = 1/4$ cm^{-2}, frequency 5170 MHz). The horizontal dashed lines indicate the locations of the edges of the film that provide reflection of the SMSW. Comparison with Figs. 4.7 a and 4.9 a shows that in the FM structure, the form of trajectories in channels of both types is similar to the same trajectories in a free film. The noticeably large value of the SMSW period for the FM structure is due to the same reasons as for non-periodic trajectories (Fig. 4.12 a, 4.13 a, b). A smoother rounding of trajectories in the vicinity of the upper boundary along the field is caused by the same reasons as the bending of the ends of the trajectories in Figures 4.12 a and 4.13 a.

4.6. Distribution of SMSWs in the structure of ferrite–dielectric metal

Let us now consider the features of the propagation of SMSws in the

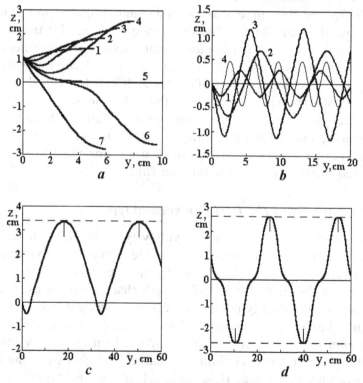

Fig. 4.13. SMSW trajectories for the FM structure in different fields. a – field of 'valley' type, the frequency changes: 1 – 3500 MHz, 2 – 4000 MHz, 3 – 4500 MHz, 4 – 5000 MHz, 5 – 5154.44996 MHz, 6 – 5170 MHz, 7 – 5500 MHz; b – 'shaft' type field, frequency changes: 1 – 3000 MHz, 2 – 4000 MHz, 3 – 5000 MHz, 4 – auxiliary curve; c – linear field, channel of the first type, frequency – 4000 MHz; d – field of the 'valley' type, channel of the second type, frequency – 5170 MHz.

ferrite–dielectric metal (FDM) structure, which is a generalization of those considered above. As in the above-mentioned structures, forward waves of both types can propagate here, as well as backward ones. Isofrequency curves are pseudoelliptic, pseudohyperbolic, or transient. As before, we first use the isofrequency curves method and then turn to the Hamilton–Auld method.

4.6.1. Analysis of SMSw trajectories by the method of isofrequency curves

Let us consider the propagation of SMSws in a FDM structure magnetized by fields of various configurations: linear, of the 'valley' and 'shaft' type.

4.6.1.1. Linearly inhomogeneous field

Consider a linearly inhomogeneous field, where $H_z = H_0(1+a_1 z) =$ 437.5 Oe and $a_1 > 0$. We set the magnetization of the film equal to $4\pi M_0 = 1750$ G. The dispersion curves at $\varphi = 0$, depending on the ratio of the film thickness p to the dielectric thickness d, can be of three types: with one extremum – maximum (like 4–7 in Figure A3.1), with two extrema – maximum and minimum (8–14 in Fig. A3.2) and without extrema (15–16 in Fig. A3.2). Consider here the most difficult case of two extrema (the other two cases are reduced to this one). For this we take $d = 15$ μm, $p = 25.5$ μm. Wherein $p/d = 1.7$. This case is illustrated in Fig. 4.14, in the upper part of which is shown the isofrequency curves (in normalized units) on the plane, plotted for a fixed frequency $\Omega = 0.745$ (3650 MHz) and different fields. The nature of curves 1–4 is pseudoelliptic, 6–10 is transitional, 11–19 is pseudohyperbolic. One of the branches of curve 5 is pseudoelliptic, the other is pseudohyperbolic. The family of isofrequency curves is bounded outside the curve corresponding to the upper field boundary of the SMSW spectrum (3.7), equal to 0.398 (695 Oe) (this curve is outside the figure).

In the lower part of Fig. 4.14, trajectories on a plane are schematically shown. Here, everywhere A is the initial point specified by the excitation conditions, B is the final point determined by the upper field boundary of the SMSW spectrum. Arrows under the trajectories show the direction of propagation of the SMSW relative to the Oy axis. The numbers below the trajectories indicate their form, which is explained below. At the starting point $\varphi > 0$

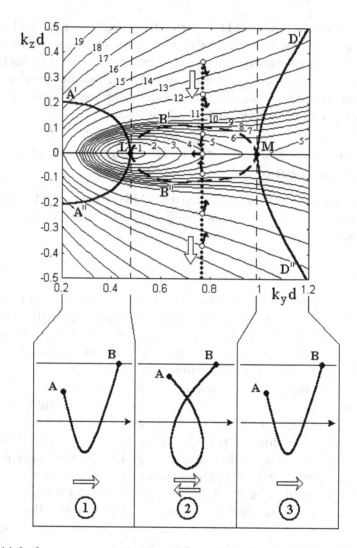

Fig. 4.14. Isofrequency curves (1–19) at different fields (top) and SMSW trajectories (1–3) at different initial conditions (bottom) for an FDM structure in a linear field. All curves are plotted for a fixed frequency $\Omega = 0.745$ (3650 MHz), but different fields. Curve 1 corresponds to $\Omega_H = 0.249$, 2 – 0.250, 3 – 0.251, 4 – 0.252, 5 – 0.253, 6 – 0.254, 7 – 0.255, 8 – 0.256, 9 – 0.257, 10 – 0.258, 11 – 0.260, 12 – 0.270, 13 – 0.280, 14 – 0.290, 15 – 0.300, 16 – 0.320, 17 – 0.340, 18 – 0.360, 19 – 0.380.

As in Figs 3.3, 3.4, the lines $A'A''$ and $D'D''$ connect points corresponding to tangents to isofrequency curves constructed from the origin, and the lines $B'B''$ are points where the isofrequency curves have horizontal tangents. Points L and M correspond to the intersection of these curves with the axis Ok_y. Dashed lines are drawn through these points parallel to the axis Ok_z. In the area lying to the left of the curve $A'A''$ SMSW are straight lines of the first type, to the right of the curve $D'D''$ the straight lines of the second type, between these curves – inverse. Outside the area bounded by the $B'B''$ lines, the waves propagate in the positive direction of the axis, inside – in the negative direction. The field is always directed along the axis Oz, therefore, during the propagation of the SMSW y-component of the wave vector, it is preserved, that is, the end of

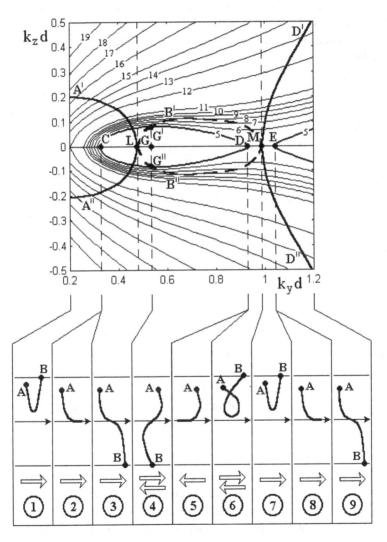

Fig. 4.15. Isofrequency curves (1–19) at different fields (top) and SMSW trajectories (1–9) at different initial conditions (bottom) for a FDM structure in a 'valley'-type field.

the wave vector on the plane is always on a vertical line parallel to the axis Ok_z (similar to Fig. 4.1). For example, a series of successive positions of the end of the wave vector for the backward wave region is shown in Fig. 4.14 by open points connected by a dotted line. The arrows show the successive directions of the group velocity vector. The resulting trajectory view is shown in the same figure below (2). Such an analysis shows that in this case the trajectories can be of three types:

1) V-shaped without rotation of the first type, when the y-component of the wave vector lies to the left of the point L.

2) V-shaped with rotation when the y-component of the wave vector lies between points L and M.

3) V-shaped without rotation of the second type, when the y-component of the wave vector lies to the right of the point M.

Here 'rotation' is called the change in the direction of propagation of the SMSW relative to the Oy axis – from positive to negative and vice versa. For the trajectories of types 1 and 3, starting outside the areas bounded by the lines $A'A''$ and $D'D''$, SMSWs are reverse, inside these areas are straight lines. Thus, at the boundary of the indicated areas, the backward waves are converted into forward waves and vice versa. SMSW, the trajectories of which begin between the dashed lines passing through the points L and M are always reverse.

4.6.1.2. Valley-type field

Let us now consider a field of the 'valley' type, where $H_z = H_0(1+a_2z^2) = 437.5$ Oe and $a_2 > 0$. The corresponding isofrequency curves and trajectories are shown in Fig. 4.15. The parameters of the figure are the same as above.

In contrast to the previous case, the 'valley'-type field is bounded from below by the 'valley bottom' (437.5 Oe), which corresponds to the isofrequency curve 5. There are no isofrequency curves inside it. The end of the wave vector when moving along a straight line parallel to the axis Ok_z, having reached curve 5, turns back, which corresponds to the passage of the SMSW trajectory through the 'bottom of the valley'. There are new characteristic points on the plane: C, D and E, corresponding to the intersection of curve 5 with the axis, as well as G' and G'', corresponding to the intersection of this curve with the line $B'B''$. The point of intersection of the line $G'G''$ with the axis is denoted by G.

Let the trajectories begin at point A at $z > 0$ and $\varphi > 0$. Case $z < 0$ and $\varphi < 0$ is a mirror reflection of the considered one relative to the bottom of the valley. The trajectories end on the positive $(z > 0)$ or negative $(z < 0)$ valley slopes at point B, defined by the boundary of the SMSW spectrum, or go to infinity along the valley bottom. There are nine characteristic types of trajectories:

1) V-shaped without rotation of the first type. The end of the y-component of the wave vector is located to the left of point C. It goes all the time in the positive direction of the Oy axis and is reflected, not reaching the bottom of the valley. Ends on the positive slope of the valley.

2) L-shaped without rotation of the first type. The end of the y-component of the wave vector corresponds to point C. It goes in the positive direction of the Oy axis and, after reaching it, goes to plus-infinity.

3) S-shaped without rotation of the first type. The end of the y-component of the wave vector is between points C and G. Goes in the positive direction of the Oy axis and passes through the bottom of the valley. Ends on the negative slope of the valley.

4) S-shaped with a turn. The end of the y-component of the wave vector is between the points G and D. If the starting point on the plane lies above the $B'B''$ line, then the trajectory first goes in the positive direction of the axis Oy, then turns to negative and, after passing through the bottom of the valley, returns to the positive. If lower, then at first – in negative, and then similarly. Ends on the negative slope of the valley.

5) L-shaped with a turn. The end of the y-component of the wave vector corresponds to point D. It goes in the negative direction of the axis Oy and, having reached it, goes to minus-infinity.

6) V-shaped with a turn. The end of the y-component of the wave vector is between the points D and M. The direction of the initial section is determined similarly to trajectory 4, after which it goes in the negative direction of the Oy axis and, having made a loop, goes over to the positive one. Ends on the positive slope of the valley.

7) V-shaped without turning of the second type. The end of the y-component of the wave vector is between the points M and E. It ends on the positive slope of the valley.

8) L-shaped without turning of the second type. The end of the y-component of the wave vector corresponds to point E. It goes to plus-infinity.

9) S-shaped without turning of the second type. The end of the y-component of the wave vector is to the right of point E. It ends on the negative slope of the valley.

The trajectories 1–3 and 7–9 are similar to each other and correspond to three main types of trajectories in a valley-type field for the FM structure and free film. In this case, the wave numbers for the trajectories 7–9 are larger than the wave numbers for the trajectories 1–3 by about an order of magnitude. The trajectories 4–6 also resemble the same three types, up to 'mirroring' about the Oz axis. This is due to the fact that in the area to the left of point G and to the right of point M, the branches of all isofrequency curves diverge, and between these points (at least for the curves 5–10, close to the bottom of the valley) they converge, that is, with an accuracy of 'reflection' behave in a similar way. The forward or backward character of the waves is analyzed similarly to the case of a linear field. So, the trajectories of the form 1, 2, 3 and 11 can begin both straight (inside the curves $A'A''$ and $D'D''$) and backward (outside these curves) waves, which then turn into straight lines, and always end reverse. The trajectories of the type 4–10 always correspond to backward waves.

4.6.1.3. Shaft-type field

Let us now consider a field of the 'shaft' type, where $H_z = H_0(1 + a_z z^2) = 437.5$ Oe and $a_2 < 0$ The corresponding isofrequency curves and trajectories are shown in Fig. 4.16. The parameters of the picture coincide with those adopted above. The field is bounded from above by the 'top of the shaft' (437.5 Oe), to which the isofrequency curve 5 corresponds. Outside of it, there are no isofrequency curves. The end of the wave vector when moving along a straight line parallel to the axis Ok_z, reaching the positive part of the isofrequency curve 5, turns to its negative part, then back to the positive, and so on, as a result of which the trajectory along the Oy axis has a periodic character and, as a rule, goes to infinity. On the plane $Ok_y k_z$ between L and G there is a 'critical' point H, which is not distinguished in any way on the isofrequency curves and is distinguishable only by the resulting nature of the trajectories. Since the isofrequency curves exist only within the regions bounded by curve 5, then when the y-component of the wave vector is to the left of point C or between D and E, the propagation of the SMSWs is impossible. Outside these areas, there are six types of trajectories:

1) Wavy without rotation of the plus-type, when the end of the y-component of the wave vector is between points C and L. Goes in the positive direction of the Oy axis.

2) Wavy with a plus-type rotation, when the end of the y-component of the wave vector is located between the points L and H. First, it goes in the positive direction of the Oy axis, then, before reaching the top of the shaft, turns to negative, passing the top of the shaft, turns to the positive, and so on. The general view of the trajectory resembles periodically repeating two Greek letters Ω, composed of bases to each other and shifted by half a period. The length of the trajectory sections traversed in the positive direction of the Oy axis exceeds the length of the same sections traversed in the negative direction, as a result of which it generally goes in the positive direction of the Oy axis. If the end of the y-component of the wave vector is close to the point H, the trajectory may experience self-intersection.

3) Eight-shaped (like a figure of eight), for which the end of the y-component of the wave vector exactly corresponds to the critical point H. It is transitional from type 2 to type 4, that is, the length of the sections traversed in the positive direction of the Oy axis is equal to the length of the same sections traversed in the negative direction. Thanks to such equality, it cannot go to infinity.

4) Wavy with a minus-type turn, when the end of the y-component of the wave vector is between the points H and G. It is similar to the previous one, with the difference that the length of the sections traversed in the positive direction of the Oy axis is less than the length of the same sections traversed in the negative direction, resulting in a generally negative axis direction. Self-intersection is also possible here.

5) Wavy without rotation of the minus-y type, when the end of the y-component of the wave vector is between the points G and D. Goes in the negative direction of the Oy axis.

6) Wavelike without rotation of the plus-y type, when the end of the y-component of the wave vector is to the right of the point E. Goes in the positive direction of the Oy axis.

The trajectories 1 and 6 are similar to each other and are similar to the pseudo-sinusoidal trajectory for the FM structure and a free film in a shaft-type field. The wavenumbers for trajectory 6 are larger than the wavenumbers for trajectory 1 by about an order of magnitude. Trajectory 5 also resembles a pseudo-sinusoidal one up to a 'mirror' reflection about the axis. The reason for this is the same as for a valley-type field. The forward or reverse nature of the waves is analyzed similarly to the previous cases. Thus, the SMSWs corresponding to trajectories 1 and 6 are always straight lines, except for a small section between the $A'A''$, the vertical straight line passing through the point L and the isofrequency curve 5, where they are inverse. The SMSWs corresponding to the trajectories 2–5 are always reverse. In the considered case, the isofrequency curve 5, corresponding to the top of the shaft, consists of two parts – pseudoelliptic and pseudohyperbolic, that is, the existence of an SMSW between the points D and E is prohibited. With a different choice of the parameters, this curve can become a transitional one (Section 2.1.3), that is, the SMSWs between these points can exist and three more trajectories, similar to 2–4, will be added to the studied types of trajectories between 5 and 6 with an accuracy of 'mirror' reflection relative to the Oz axes, with the result that the total number of trajectories can increase to nine.

4.6.1.4. General comment

The above consideration refers to the case of two extrema on the dispersion curve at $\varphi = 0$, which is determined by the choice of the frequency and thickness of the dielectric layer (Section 2.1.3). With other parameters, if the dispersion curve has one extremum – a maximum, only the trajectories 1 and 2 are realized in a linear field, 1–6 in a valley-type field, 1–4 in a shaft-type field. In the case of a dispersion curve without extrema, the trajectories 1 are realized in a linear field, 1–3 in a valley-type field, and 1 in a shaft-type field.

4.6.2. Analysis of SMSW trajectories by the Hamilton–Auld method

Let us now carry out a quantitative analysis of the propagation of SMSWs in an inhomogeneously magnetized FDM structure by the Hamilton–Auld method. Let's pay attention only to those features that distinguish this case from the previous ones. As before, the parameters of the film are: $4\pi M_0 = 1750$ G, $d = 15$ μm. The thickness of the dielectric layer is everywhere taken equal to $p = 25.5$ μm ($p/d = 1.7$), which provides two extrema on the dispersion curve. The SMSW frequency is 3650 MHz.

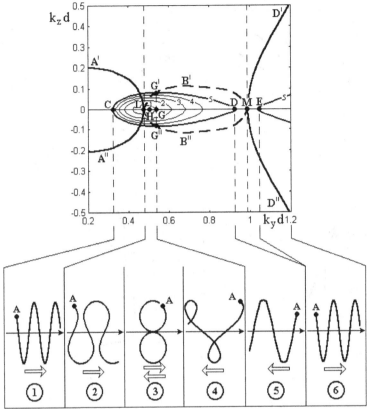

Fig. 4.16. Isofrequency curves (1–5) at different fields (top) and SMSW trajectories (1–6) at different initial conditions (bottom) for an FDM structure in a 'shaft– type field.

4.6.2.1. Linearly inhomogeneous field

Figure 4.17 corresponds to a linear field. This field is described by formula (3.2), where H_0 = 437.5 Oe, a_0 = 1, a_1 = 1/4 cm^{-1},. In the figure, the letter a denotes dependences $z(y)$, $b - k(y)$, $c - \varphi(y)$, $d - \psi(y)$. The starting point of the trajectory is y = 0 cm, z = 1.5 cm, where the field is 601.6 Oe (in normalized units 0.3438). The numbers on the curves correspond to three possible types of trajectories (Fig. 4.14), realized at different initial values of the angle φ: curve 1 – φ = 48.7° (k = 112.7 cm^{-1}) – V-shaped trajectory without rotation of the first type (type 1 in Fig. 4.14), 2 – φ = 41.2° (k = 601.7 cm^{-1}) – V-shaped trajectory with a turn (type 2 in Fig. 4.14), 3 – φ = 38.63° (k = 1868.0 cm^{-1}) – V-shaped trajectory without turning the second type (type 3 in Fig. 4.14). Setting the frequency limits

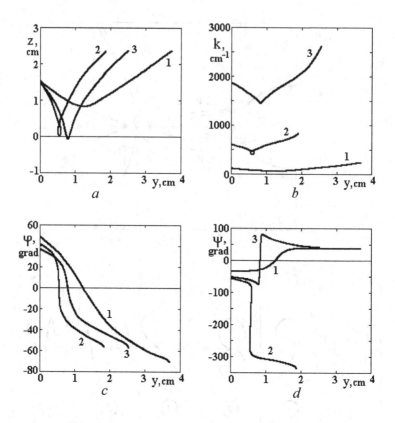

Fig. 4.17. SMSW trajectories (*a*) and the corresponding dependences $k(y)$ (*b*), $\varphi(y)$ (*c*), and $\psi(y)$ (*d*) for the FDM structure in a linear field at a frequency of 3650 MHz and various initial angles: 1 – 48.7°, 2 – 41.2°, 3 – 38.6°.

the trajectories at the coordinate $z = 2.35$ cm, which corresponds to the upper boundary of the SMSW spectrum in the field (3.6), (3.8) – 695.0 Oe (0.3971). Trajectory 1 (Fig. 4.17 a), in comparison with trajectory 3, continues along the axis approximately twice as far, and its range along the Oz axis is one and a half to two times less. At the same time, its wavenumber (Fig. 4.17b) is less than an order of magnitude, the absolute value of the angles φ (Fig. 4.17c) and ψ (Fig. 4.17d) on most of the trajectory is also less than for th

The explanation of the listed features is that curves 1 correspond to relatively long straight waves of the first type, for which the dispersion law is close to the dispersion law in the FM structure, while curves 3 correspond to shorter forward waves of the second type, for which the dispersion law is close to a free film (Fig. 2.3a,

2.4a). The jump on curve 3 of the dependence (Fig. 4.17d) reflects a sharp turn of trajectory 3 at the minimum. Curves 2 occupy an intermediate position between 1 and 3, as they correspond to the falling portions of the dispersion curves (Fig. 3.2). On trajectory 2, a loop-shaped section is clearly visible, which was predicted earlier on the basis of an analysis of isofrequency curves (Figure 4.14, curve 2). The sharp jump in the dependence for curve 2 reflects the loop-like character of the corresponding trajectory.

4.6.2.2. Valley-type field

Let us now consider the propagation of SMSWs in an FDM structure magnetized by a field of the 'valley' type (3.2) with the parameters: $H_0 = 437.5$ Oe, $a_0 = 1$. $a_2 = 1/16$ cm^{-2}. Let us choose the initial angles φ in such a way as to cover all possible types of trajectories, which are fundamentally different from the cases considered earlier. We start the trajectories from the point $y = 0$ cm, $z = 1.5$ cm, where the field is 546.9 Oe (0.3125). The resulting trajectories of the SMSW, as well as the dependences $k(y)$, $\varphi(y)$, $\psi(y)$ are shown in Fig. 4.18 (a–d). The curves 1–5 correspond to different initial angle values φ.

In accordance with Fig. 4.15, curve 1 is S-shaped with a turn (type 4 in Fig/ 4.15), 2 is L-shaped with a turn (type 5 in Figure 4.15), 3 is V-shaped with a turn (type 6 on Fig/ 4.15), 4 – V-shaped without rotation of the first type (type 1 in Fig. 4.15), 5 – S-shaped without rotation of the second type (type 9 in Fig. 4.15). The curves 4 and 5 are similar to those considered above in a valley-type field for a free film (5) and an FM structure (6) and are presented here for comparison. As before, the trajectories break off at coordinates $z = \pm z_1$, where $z_1 = 3.07$ cm, which corresponds to a field of 695.0 Oe (0.3971).. The distance along the axis Oy, traversed by V- and S-shaped trajectories that have a turn (curves 1,3 in Fig. 4.18), is two to three times less than the same distance for similar trajectories without a turn (curves 4,5 in Fig. 4.18), which is due to the passage of some of these trajectories in the negative direction of the Oy axis. The L-type trajectory (2), corresponding to the critical ratio between the frequency (3650 MHz) and the starting angle of 33.8212°), goes to infinity in the negative direction of the Oy axis. The wave numbers of the trajectories 1–3 (Fig.4.18 b) are the larger, the smaller their initial angles. The relative proximity of the values to each other (the difference is no more than 30%) is due to the proximity of the mutual arrangement of the extrema of the dispersion curve under the

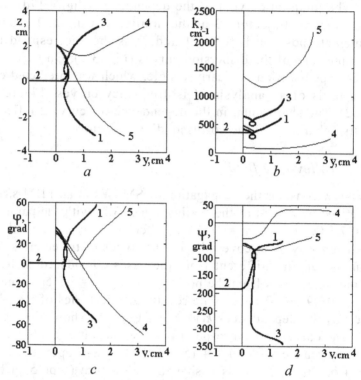

Fig. 4.18. SMSW trajectories (a) and the corresponding dependences $k(y)$ (b), $\varphi(y)$ (c), and $\psi(y)$ (d) for the PDM structure in a 'valley'-type field at a frequency of 3650 MHz and various initial angles: 34.5° ($k = 451.5$ cm^{-1}), 2 – 33.8212° ($k = 490.39$ cm^{-1}), 3 – 32.3° (= 610.02 cm^{-1}), 4 – 36.0° ($k = 138.0$ cm^{-1}), 5 – 30.9° ($k = 1381.1$ cm^{-1}).

conditions under consideration (Fig. 3.2). Differences in the course of the dependences $\varphi(y)$ (Fig. 4.18 c) and $\psi(y)$ (Fig. 4.18 d) for the curves 1–3 from the course of similar curves 4 and 5 are due to the presence of a rotation on these curves. The loop-like nature of the curves reflects the opposite direction of wave propagation in the turning region. Note that in the presence of conditions for wave reflection from a line parallel to the axis Oy here (as well as in a linear field), the formation of channels of the first or second type is possible.

4.6.2.3. Shaft-type field

Let us now consider the propagation of SMSW in the FDM structure magnetized by a field of the 'shaft' type (3.2) with the parameters: $H_0 = 437.5$ Oe, $a_0 = 1$, $a_1 = 0$, $a_2 = -1/32$ cm^{-2}. As before, we will choose the initial angles in such a way as to cover all possible types of trajectories, which are fundamentally different from the cases considered earlier. We start the trajectories from the point $y = 0$ cm, $z = 0$ cm, where the field is 437.5 Oe (0.250). The region of existence of the SMSW (3.7) and (3.11) is limited by the coordinates $z = \pm z_1$, where $z_1 = 0.808$ cm, corresponding to the field 428.6 Oe (0.245).

The resulting SMSW $z(y)$ trajectories, as well as the dependences $k(y)$, $\varphi(y)$ and $\psi(y)$, and corresponding to different values of the initial angle φ, are shown in Fig. 4.19 (*a–d*). The curves 1 and 3, shown only partially, have periods of 0.390 and 0.842 cm, respectively. In accordance with Fig. 4.16, curve 1 is wavy without turning the plus-type (type 1 in Fig. 4.16), 2 – wavy without turning the minus-type (type 5 in Fig. 4.16), 3 – wavy without turning the plus-type (type 6 in Fig. 4.16), 4 – wavy with a turn of the minus

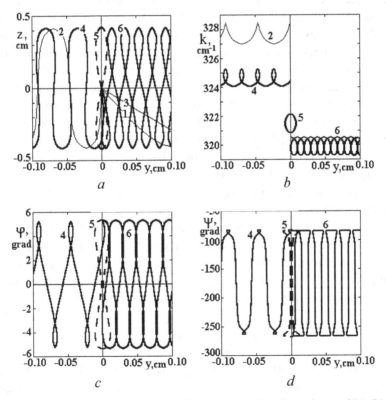

Fig. 4.19. SMSW trajectories (*a*) and the corresponding dependences $k(y)$ (b), $\varphi(y)$ (*c*), and $\psi(y)$ (*d*) for the PDM structure in a 'shaft'-type field at a frequency of 3650 MHz and various initial angles: $1 - 5.0°$ ($k = 295.9$ cm^{-1}), $2 - 5.0°$, ($k = 328.7$ cm^{-1}), $3 - 5.0°$ ($k = 1226.4$ cm^{-1}), $4 - 5.05°$ ($k = 325.1$ cm^{-1}), $5 - 5.086°$ ($k = 321.9$ cm^{-1}), $6 - 5.10°$ ($k = 320.4$ cm^{-1}).

type (type 4 in Fig. 4.16 without self-intersection), 5 – eight-shaped (type 3 in Fig. 4.16), 6 – wavy with a turn of the plus-type (type 2 in Fig. 4.16 with self-intersection). The curves 1 and 3 are similar to those considered above in a shaft-type field for a free film (3) and an FM structure (1) and are presented here for comparison. Curve 2 is intermediate (in wavenumber) between the curves 1 and 3 and corresponds to the falling part of the dispersion curve, that is, backward waves, as a result of which the trajectory propagates in the negative direction of the Oy axis. Otherwise, its properties are similar to the curves 1 and 3. There is no rotation, as a change in the direction of propagation along the Oy axis, for curves 1–3. The curves 4–6 are fundamentally different from curves the 1–3 by the presence of a turn, and the positive (6), negative (4) or neutral (5) direction of wave propagation as a whole, as well as the phenomenon of self-intersection are determined by the reasons explained in the analysis of isofrequency curves. Like the L-type trajectories in the case of the valley, the eight-shaped curve 5 is observed only at the critical ratio between the frequency (3650 MHz) and the starting angle (5.086°). The wave numbers of the trajectories 4–6 (Fig.4.18b) are the larger, the smaller their initial angles. The reason for the extreme closeness of the values k to each other (the difference is no more than 2%) is the same as for the 'valley' type field. The course of the $\varphi(y)$ (Fig. 4.18 c) and $\psi(y)$ (Fig. 4.18 d) dependences for the curves 4–6 is also fully explained by the presence of rotation. The loop-like nature of the curves reflects the opposite direction of wave propagation in the turning region. An important difference is that the period of all curves having a rotation (4–6) is significantly (sometimes by an order of magnitude or more) less than the period of similar curves without rotation (the curves 1–3). This circumstance becomes clear from considering the isofrequency curves (Fig. 4.16). Indeed, in the range of values of the y-component of the wave number corresponding to waves with a turn (between the points L and G in Fig. 4.16), the course of the isofrequency curve 5 corresponding to the top of the shaft is close to horizontal, as a result of which the -component of the group velocity is small and the trajectory along the Oy axis little is displaced. Everywhere outside this region, the curve 5 is steeper and the period of the trajectories becomes larger.

[1]

4.7. Phase rise and delay time

Let us now consider the most important integral characteristics of a propagating wave, determined by its phase and group velocities – phase incursion and delay time. In a non-uniformly magnetized film, they are calculated using formulas (2.589) and (2.594). Let us compare the features of the behaviour of these characteristics for a free film and a ferrite–metal structure, magnetized by fields: linear, such as 'valley' and 'shaft'.

4.7.1. Linearly inhomogeneous field

The linear field described by formula (4.2), where $H_0 = 437.5$ Oe, $a_0 = 0$, $a_1 = 1/2$ cm^{-1}, $a_2 = 0$ corresponds to Fig.4.20, where the letter a denotes the SMSW trajectories $z(y)$, b – similar dependences for the wavenumber $k(y)$, c – the phase incursion $\Phi(y)$ and d – the delay time $T(y)$. The starting point of the trajectory: $y = 0$ cm, $z = 0$ cm (field 437.5 Oe), at this point the angle between the wave vector and the axis Oy: $= 30°$. Film parameters: $4\pi M_0 = 1750$ G, $d = 15$ μm. The different curves correspond to two types of waveguide structures and different frequencies. The curves 1 and 2 are plotted for a film with a free surface, 3–5 for the FM structure at different frequencies. In Fig.4.20 b–d, the scale of the curves 1 and 2 corresponds to the left scale, the curves 3–5 – to the right.

Figure 4.20 a shows that all trajectories are V-shaped. Their nature completely coincides with the previously considered similar trajectories for a free film (Fig. 4.4) and an FM structure (Fig. 4.12). The same applies to the dependences for the wavenumber $k(y)$ (Fig.4.20 b). Both in the free film and in the FM structure, the phase incursion (Fig.4.20 c) and the delay time (Fig.4.20d) are the longer, the higher the SMSW frequency, which is explained by the monotonic growth of the dispersion law. Due to additivity, both of these quantities always increase monotonically as the wave propagates, and the stronger the further the trajectory goes along the Oy axis. The law of increasing phase incursion and delay time in a free film is close to linear throughout the SMSW trajectory, except for the place corresponding to the minimum of the dependence, where the growth of these quantities slows down somewhat (by a factor of 2–3), which is caused by the minimum of the wavenumber and slope of the dispersion curve at this point. In the FM structure, the phase incursion and the delay time at the beginning of the

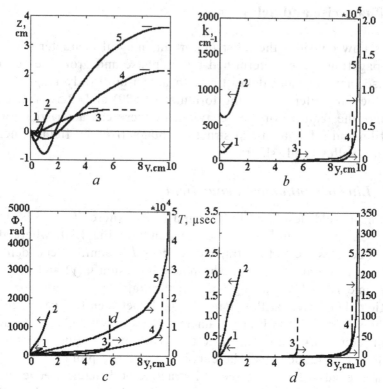

Fig. 4.20. SMSW trajectories (*a*) and the corresponding dependences $k(y)$(*b*), $\Phi(y)$ (*c*), and $T(y)$ (*d*) in a linear field for various types of waveguide structures and frequencies. Structures: 1, 2 free FT, 3–5 – FM structure. Frequencies: 1 – 3000 MHz, 2 and 3 – 3300 MHz, 4 – 4300 MHz, 5 – 5300 MHz.

trajectory also increase almost linearly, as a rule, even more slowly than in a free film, and toward the end of the trajectory, their growth sharply increases and the corresponding curves tend to infinity. The reason for this behaviour is that in a free film the trajectories of the SMSW 1 and 2 intersect the horizontal line corresponding to the maximum value of the field (according to formula (4.8)) at an acute angle and break off at the point of intersection, whereas in the FM structure the trajectories of SMSW 3, 4, 5 this line does not cross, but, asymptotically approaching it, go to infinity. Similar behaviour of trajectories for the structures under consideration was discussed earlier in Sections 4.4 and 4.5. Here, as in Section 4.5, the construction of trajectories for the FM structure (3–5) ended when the angle reached 89.9°. From a comparison of Fig.4.20 a with Fig.4.20 b, where the dependences $k(y)$ are shown for the same conditions, it can be seen that as the trajectories tend to a larger field, the wavenumber always increases. For a free film (curves 1, 2), this growth is limited due to the limitation of the trajectories (for curve 2, at the level of 1150 cm^{-1}). For the FM structure (curves 3–5), there is no limitation, the wavenumber tends to infinity, and at $\varphi = 89.9°$, for example, for curve 5 it reaches the value $2 \cdot 10^5$ cm^{-1}. The phase incursion for any wave, other things being equal, is the greater, the larger the wavenumber. In this case, as can be seen from Fig.4.20 c, for a free film, the growth of the phase incursion as it increases is limited (for curve 2 at a level of 1500 rad). For the FM structure, there is no restriction, the phase incursion $\Phi(y)$ tends to infinity and at $\varphi = 89.9°$ reaches $5 \cdot 10^4$ rad (curve 5). The delay time $T(y)$ behaves similarly, the dependences for which are shown in Fig.4.20 d. For a free film, the delay time $T(y)$ is limited (for curve 2 at a level of 2.1 μsec), and for the FM structure tends to infinity and at $\varphi = 89.9°$ reaches 300 μsec (curve 5).

Thus, in an FM structure magnetized by a linear field, the phase incursion and delay time can exceed those in a free film by one or two orders of magnitude or more.

4.7.2. Valley-type field

As the next example, let us consider the phase incursion and the delay time of the SMSW in a 'valley'-type field of the form (3.2) with the parameters: $H_0 = 437.5$ Oe, $a_0 = 1$, $a_1 = 0$, $a_2 = 1/8$ cm^{-2}. Figures 4.21 and 3.22 correspond to this case. In Fig. 4.21, letters b denote the trajectories for a free film (a) and an FM structure (b), letters c and d denote the dependences for the same cases. In Fig. 4.22, letters a and b refer to the phase incursion for a free film (a) and an FM structure (b), letters c and d – to the delay time for the same cases. The starting point of the trajectory: $y = 0$ cm, $z = 1$ cm (field 492.2 Oe). At this point, the angle between the wave vector and the Oy axis is $\varphi = 30°$. Film parameters $4\pi M_0 = 1750$ G, $d = 15$ μm.

Different curves correspond to two types of waveguide structures and different frequencies. In Section 4.4.2.2, it is shown that the trajectories with frequencies below the critical have a V-shaped form, and above the critical one – S-shaped; therefore, to reveal a more complete picture, the trajectory frequencies for each structure were chosen both near the edges of the SMSW spectrum and along both sides are near the critical frequency, which is 3377.4 MHz for a free film and 4289.2 MHz for an FM structure. Thus, trajectories

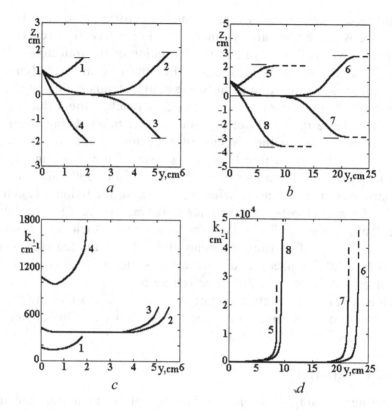

Fig. 4.21. SMSW trajectories $z(y)$ (a, b) and the corresponding dependences $k(y)$ (c, d) in a 'valley'-type field for a free FT (a, c) and an FM structure (b, d). The figures on the curves correspond to the frequencies: 1 - 3200 MHz, 2 - 3377 MHz, 3 - 3378 MHz, 4 - 3500 MHz, 5 - 3600 MHz, 6 - 4289 MHz, 7 - 4290 MHz, 8 - 5000 MHz.

1, 2, 5, 6 are V-shaped, and trajectories 3, 4, 7, 8 are S-shaped, and trajectories 2, 3 and 6, 7, due to the proximity to the critical frequency, pass the distance along the axis significantly greater than trajectories 1, 4, 5, 8. Note first of all that all dependences, and go higher, the higher the frequency of the SMSW, which is completely analogous to the case of a linear field. Further, as in the case of a linear field, the trajectories of both types in a free film (1–4), having reached the limiting value of the field, break off, and the trajectories in the FM structure (5–8) asymptotically tend to horizontal straight lines, going to infinity (at this, as before, the calculation was carried out only up to the value $\varphi = 89.9°$). For this reason, the dependences of the wavenumber on the coordinate y for a free film (Fig. 4.21 c) also break off at the level of 300–1700 cm^{-1}, and for the FM structure (Fig. 4.21 c) they reach (2÷5) cm^{-1} and more. Similarly, the phase incursion in a free film (Fig. 4.22) is limited to 500–3900 rad, and in the FM structure (Fig. 4.22 b) it reaches 4000–14900 rad and more. The delay time in a free film (Fig. 4.22c) is also limited at the level of 1–6 µs, and in the FM structure (Fig. 4.22 d) it reaches 20–46 µs and more.

Thus, in the case of a valley-type field, the phase incursion and the delay time for both structures increase with frequency, and for the FM structure these values exceed those for a free film by one or two orders of magnitude or more.

4.7.3. Shaft-type field

Let us now consider the phase incursion and the delay time of the SMSW in a 'shaft'-type field having the form (3.2) with the parameters: $H_0 = 437.5$ Oe, $a_0 = 1$, $a_1 = 0$, $a_2 = -1$ cm^{-2}. Figures 4.23 and 4.24 correspond to this case. In Fig. 4.23, the letters a and

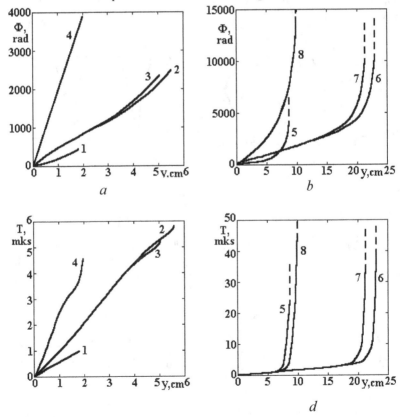

Fig. 4.22. Dependences $\Phi(y)$ (a, b) and $T(y)$ (c, d) in a 'valley'-type field for the same cases as in Fig. 4.21. The frequencies are the same as in Fig. 4.21.

b denote the trajectories *z*(*y*) for a free film (*a*) and an FM structure (*b*), letters *c* and *d* denote dependences for the same cases. In Fig. 4.24, letters *a* and *b* refer to the phase incursion $\Phi(y)$ for a free film (*a*) and an FM structure (*b*), the letters *c* and *d* refer to the delay time *T*(*y*) for the same cases. The starting point of the trajectory: $y = 0$ cm, $z = 0.5$ cm (field 328.1 Oe). At this point, the angle between the wave vector and the axis Oy: $\varphi = 30°$. Film parameters: $4\pi M_0 = 1750$ G, $d = 15$ μm.

The fundamental difference between the field of the 'shaft' type and the above-considered configurations of fields – linear and of the 'valley' type, where at a sufficient distance from the axis the field can be arbitrarily large, consists in the limitation of the field from above by the value corresponding to the top of the 'shaft' (here 437.5

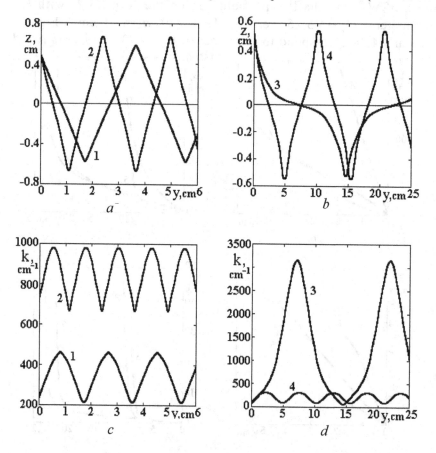

Fig. 4.23. SMSW trajectories (*a*, *b*) and the corresponding dependences *z*(*y*) (*c*, *d*) in a 'shaft'-type field for a free FT (*a*, *c*) and an FM structure (*b*, *d*). The numbers on the curves correspond to the frequencies: 1, 3 – 2740 MHz, 2, 4 – 3000 MHz.

Oe). In this case, if the frequency of the SMSW is such that its entire field interval of existence (4.6)–(4.7) lies below the field of the top of the 'swell', then on each slope of the 'swell' the trajectories are not periodic, but behave similarly to the case of a linear field, that is, in a free film they break off, and in the FM structure they go to infinity along the axis Oy. If the SMSW frequency is such that for it the upper field boundary of the wave existence (3.6) exceeds the field of the top of the 'shaft', then the trajectory becomes periodic - pseudosinusoidal (Section 3.4.2), and the possible phase incursion and delay time are the greater, the smaller the upper field the border differs from the field of the 'shaft' top. Based on this requirement, we further consider the frequency 2740 MHz, for which (at $4\pi M_0 =$ 1750 G) the upper field boundary of the SMSW is 437.72 Oe, which exceeds the field of the top of the 'shaft' by 0.22 Oe. The frequency 3000 MHz was chosen as the control one. Thus, the curves 1, 3 in Figs. 4.23 and 3.24 correspond to a frequency of 2740 MHz, the curves 2 and 4 to 3000 MHz.

From the given figures it can be seen that all trajectories (Fig.4.23 a and Fig.4.24 b) have a periodic pseudosine-wave character, considered earlier (in Sections 3.4, 3.5). In the FM structure, on trajectory 3 of the frequency of 2740 MHz, at its intersection with the Oy axis, wide shallow sections are observed, due to the proximity of the upper field boundary of the SMSW spectrum for a given frequency (437.72 Oe) to the field on this axis (437.5 Oe). On trajectory 4 of the 3000 MHz frequency, such sections are practically absent, since the upper field boundary of the SMSW spectrum for this frequency is quite high (508.32 Oe). The dependences $k(y)$ shown in Figs. 4.23 c and 4.23 d for the free film (1, 2) and the FM structure (3, 4) are also similar to those considered earlier (Sections 3.4, 3.5). An important feature is a sharp (by more than an order of magnitude) increase in he wavenumber for a frequency of 2740 MHz in the FM structure, corresponding to the gentle sections of trajectory 3 in Fig. 4.23 b. The phase incursion in a free film (Fig. 4.24a) at both frequencies monotonically increases according to a law close to linear. Curve 2 lies above curve 1, since it corresponds to a higher frequency. In the FM structure (Fig. 4.24), a similar close to linear phase incursion is observed only at a frequency of 3000 MHz (curve 4). Curve 3 for a frequency of 2740 MHz has a number of sharply increasing sections, which also correspond to flat sections of trajectory 3 in Fig. 4.23 b. Such a rapid growth of the phase incursion in this case is due to the large values of the wave

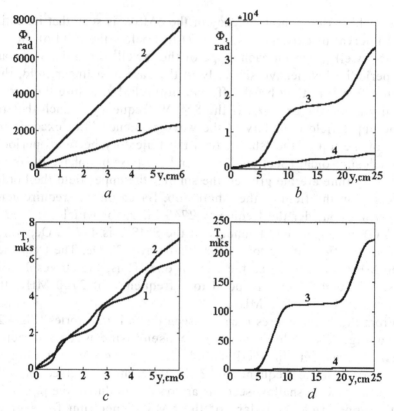

Fig.4.24. Dependence $\Phi(y)$ (*a*, *b*) and $T(y)$ (*c*, *d*) in the shaft=type field for the same cases as the Fig. 4.23.

number (Fig. 4.23 d). At the same time, the total phase incursion at a given distance is due not only to the regions where the wavenumber is large (about 3000 cm^{-1}), but also to those lying between them, where it is relatively small (less than 100 cm^{-1}). Therefore, the total phase incursion in the FM structure at a frequency near the lower boundary of the SMSW spectrum turns out to be comparable to the phase incursion in a free film at frequencies lying in the upper part of the SMSW spectrum. So the total phase incursion in the FM structure at $y = 25$ cm, at a frequency of 2740 MHz is about 32000 rad, while in a free film under the same conditions it is 10000 rad. On the other hand, in a free film at a frequency of 3000 MHz at $y = 25$ cm, the phase incursion reaches 29000 rad, which approaches a similar value in the FM structure.

The delay time $T(y)$ in a free film (Fig. 4.24c) at both frequencies increases monotonically, and at a frequency of 3000 MHz (curve

2) according to a law close to linear, and at a frequency of 2740 MHz (curve 1) it has a number of sharply increasing sections corresponding to the intersection axis trajectory (curve 1 in Figure 4.23 a), on which the slope of the dependence increases by 7–9 times. In the FM structure (Fig. 4.24 d), a close to linear increase in the delay time is observed only at a frequency of 3000 MHz (curve 4), and at a frequency of 2740 MHz (curve 3), the same features are observed as for a free film, expressed more sharply: the slope dependence on increasing areas increases here more than 100 times. As a result, curve 3 goes much higher than curve 4, and after the second trajectory crossing the horizontal axis ($y = 25$ cm), the delay time for curve 3 is 220 μs, while for curve 4 it barely reaches 10 μs. Under similar conditions, the delay time in a free film is close to 30 μsec.

Thus, the maximum values of the phase incursion for any frequencies in both structures differ by no more than 10–20%, while the maximum attainable delay time in the FM structure can exceed the analogous value in a free film by at least an order of magnitude and is maximum for frequencies close to the lower boundary of the SMSW spectrum.

4.8. Experimental study of SMSW trajectories

To test the theoretical provisions developed above, an experimental study of the propagation of wave beams of SMSW in ferrite films magnetized by longitudinally inhomogeneous fields of various configurations: linear, of the 'valley'- and 'shaft'-type was carried out.

4.8.1. The main parameters of the experiment

The experiments were carried out on a measuring stand, the scheme of which is shown in Section 3.1.2.1. (Fig. 3.5). We used a film of yttrium–iron garnet (YIG), with magnetization $4\pi M_0 = 1810$ G, thickness $d = 14.5$ μm and line width FMR $2\Delta H = 0.6$ Oe, grown on a round (76 mm in diameter, 0.5 mm thick) substrate of gadolinium-gallium garnet, oriented perpendicular to the crystallographic axis of the [111] type. The excitation and reception of the SMSW was carried out using movable antennas superimposed on the film plane, made of gold-plated tungsten wire with a diameter of 12 μm and a length of 4 mm. The film was placed in the centre of symmetry of the round poles of a permanent magnet with a diameter of 20 cm with a distance of 10–20 cm between them.The field varied from 300 to 700 Oe, the uniformity within the film plane was at least 3–5%, and in the working area at least 2 %. The coordinate system $Oxyz$ was chosen in such a way that the plane coincided with the film plane, and the Ox axis was perpendicular to it. The Oz axis was oriented along the direction of the field of the permanent magnet in the centre between its poles. The field inhomogeneity was created using a cylindrical soft iron shunt 10–25 mm in diameter and 20–40 cm long, placed near the film plane along the Oy axis. In the experiments, the exciting antenna was stationary, and the receiving antenna moved along the film plane. The course of the trajectories of the SMSW wave beams was recorded using the maximum signal at the receiving antenna. The maximum measurable length of trajectories, determined by the sensitivity of the equipment and the attenuation of the SMSW, in the case of a uniform field was 3–4 cm. The results of the study of the SMSW trajectories are shown in Fig. 4.25–4.27. Each figure under the letter 'a' shows the dependence of the field on the coordinate, and the points correspond to the actually measured values of the field, the solid line – to the approximate law used in the calculations of trajectories. Under the letter 'b' the trajectories of the SMSW are shown, the points are the experiment, the lines are the theoretical curves calculated by the Hamilton–Auld method.

4.8.2. Linearly inhomogeneous field

Figure 4.25 corresponds to a linear field. To describe the dependence of the field H on the coordinate z, the expression was used $H = 532+50\ z$: where H is measured in Oe, z in cm.The starting point of all trajectories: $y = 0$ cm, $z = 3.25$ cm.At this point $H= 694.5$ Oe. The initial angle between the Oy axis and the wave vector $\varphi = 32°$. The boundaries of the SMSW spectrum at the excitation point: 3693 and 4479 MHz. The curves correspond to several frequencies ranging from 3790 MHz to 4070 MHz.

From an examination of Fig. 4.25 b, it can be seen that the best agreement of the calculated curves with the experiment (about 5% with respect to the full range of the trajectories along the Oz axis) is observed closer to the middle of the SMSW spectrum – for the curves 2 and 3, somewhat worse (10%) – closer to the boundaries of the SMSW spectrum are the curves 1 and 4. In this case, in the low-frequency region of the spectrum (the curves 1 and 2), the calculated

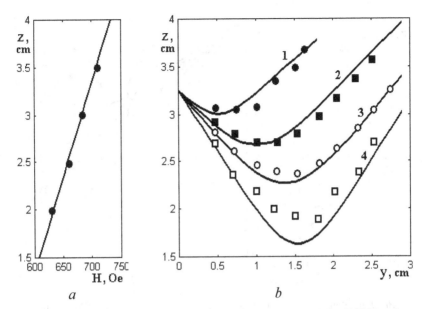

Fig. 4.25. Dependence of the field H on the coordinate z (a) and the SMSW trajectory $z(y)$ (b) in this field. Lines – theory, dots – experiment. Trajectory frequencies: 1 – 3790 MHz, 2 – 3890 MHz, 3 – 3980 MHz, 4 – 4070 MHz.

results exceed the experimental ones, and in the high-frequency region (the curves 3 and 4) they lie below them.

4.8.3. Valley-type field

The trajectories in the field of the 'valley' type are illustrated in Fig. 4.26. An interpolation polynomial of the form was used to describe the dependence of the field on the coordinate:

$$H_z(z) = 554 - 4U_1 - 16U_2 + 8U_3 - 4U_4 + 3.2U_5 + 2.32U_6 - 5.78U_7$$

$U_1 = z + 10$, $U_n = \left[z - 0.5(n-1)\right] \cdot U_{n-1}$, $n = 2,...4$, where H – Oe, z in cm).

This polynomial corresponds to the actually measured values of the field, however, for the best agreement of theory with experiment, when calculating the trajectories, the field values were used lower by 19 Oe (3.4%). A possible reason for the need to introduce such a correction may be the anisotropy of the YIG film that was not taken into account in the calculations. The horizontal dashed line drawn at $z = 2.4$ cm corresponds to the 'bottom of the shaft' (506 Oe). The starting point of all trajectories: $y = 0$ cm, $z = 0$ cm. At

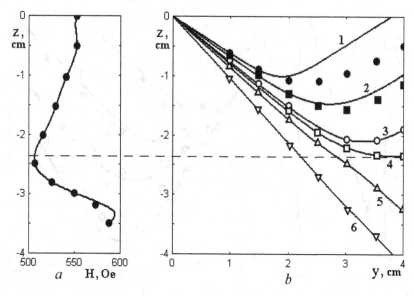

Fig. 4.26. Dependence of the field H on the coordinate z (*a*) and the SMSW trajectory $z(y)$ (*b*) in this field. Lines – theory, dots – experiment. Trajectory frequencies: 1 – 3216 MHz, 2 – 3276 MHz, 3 – 3356 MHz, 4 – 3380 MHz, 5 – 3420 MHz, 6 – 3500 MHz.

this point $H = 554$ Oe. The initial angle between the Oy axis and the wave vector $\varphi = 37°$. The boundaries of the SMSW spectrum at the excitation point are 3204 and 4085 MHz.

Figure 4.26 b shows that curve 4 (3380 MHz) corresponds most of all to the critical frequency corresponding to the L-shaped trajectory. The trajectories 1–3 have frequencies below the critical and are reflected, without reaching the 'bottom of the valley', the trajectories 5 and 6, whose frequencies are higher than the critical, pass through the 'bottom of the valley'. The calculated trajectories for frequencies above 3356 MHz (curves 3–6) coincide with the experiment with an accuracy of no worse than 1%. The trajectories of lower frequencies before reflection also coincide well; however, after reflection, the calculated curves pass slightly below the experimental points and for a frequency of 3276 MHz (curve 2) the discrepancy is 5%, and for a frequency of 3216 MHz (curve 1) it reaches 20%. The reason for this discrepancy may be the finite width of the SMSW wave beam, which is not taken into account in the calculations.

4.8.4. Shaft-type field

The trajectories in the 'shaft'-type field are illustrated in Fig. 4.27.

An interpolation polynomial of the form was used to describe the dependence of the field on the coordinate:

$$H_z(z) = 295 + 54U_1 - 18U_2 - 38.66U_3 + 11.32U_4$$

$U_1 = z + 10$, $U_n = [z+1.0-0.5(n-1)] \cdot U_{n-1}$. $n = 2,...4$, where H – Oe, z in cm).

This polynomial also corresponds to the actually measured field, and in the calculations of the trajectories, the same correction was introduced – 19 Oe (3.4%). The horizontal dashed line drawn at $z = 0$ cm corresponds to the 'top of the shaft' (340 Oe). The starting point of all trajectories: $y = 0$ cm, $z = 0$ cm. At this point $H = 340$ Oe. The initial angle between the Oy axis and the wave vector $\varphi = 19.5°$. The boundaries of the SMSW spectrum at the excitation point: 2394 and 3486 MHz.

Figure 4.27 b shows that the calculated trajectories for all frequencies within the first half-period coincide with the experiment with an accuracy of no worse than 1%. Further, the coincidence begins to deteriorate gradually, the stronger the lower the frequency. So, at the end of the second half-period for a frequency of 3150 MHz (curve 4), the discrepancy does not exceed 2%, while for a frequency of 2570 MHz (curve 1) it reaches 20–25%. The reason for the gradual increase in the discrepancy with the growth of the

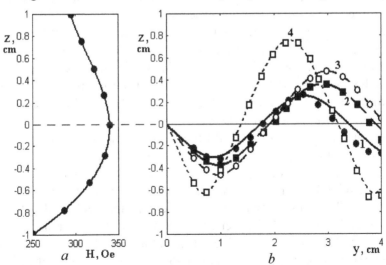

Fig. 4.27. Dependence of the field H on the coordinate z (*a*) and the SMSW trajectory $z(y)$ (*b*) in this field. Lines – theory, dots – experiment. Trajectory frequencies: 1 – 2570 MHz, 2 – 2710 MHz, 3 – 2870 MHz, 4 – 3150 MHz.

coordinate is the high sensitivity of the shape of pseudo-sinusoidal trajectories in a 'shaft'-type field to the initial conditions of wave excitation, as well as to the local configuration of the field, due to their large length. In this case, with decreasing frequency, the 'sensitivity' of the trajectories to these factors increases, since in an inhomogeneous field the trajectories are curved the more, the lower their frequency.

4.8.5. Change of various parameters of the experiment

Along with the dependence on the frequency, the behaviour of the SMSW trajectories was investigated with a change in other parameters of the experiment. Thus, measurements of trajectories in the range of initial angles from $\pm 5°$ to $\pm 45°$ also showed good agreement between theory and experiment. It was found that a much more critical characteristic is the gradient of the inhomogeneous field. Thus, measurements of trajectories in fields with a gradient from 20 to 300 Oe/cm showed that the discrepancy between theory and experiment in the range of 20÷200 Oe/cm does not exceed that described above, while at field gradients above 200 Oe/cm the discrepancy increases and at 300 Oe/cm reaches 10–20% even for high frequencies. Especially critical is the field gradient in the area of the excitation transducer. Thus, the discrepancy noticeably increases (up to 5–10% or more) starting from 50 Oe/cm, which is due to the spreading of the wave beam and even splitting it into two, especially when the exciting transducer is located at the 'bottom of the valley'. In the experiment, this manifests itself as a sharp (up to 90 dB) attenuation of the SMSW at distances less than 1 cm from the exciting transducer. The reason for the spreading is that the different ends of the transducer are in different fields and emit wave beams in different directions. In this case, the spreading of the beam from frequency within the SMSW spectrum changes little (no more than 5%), but it increases with an increase in the initial angle of the wave vector. So, if the ends of the transducer are at a distance of 0.4 cm from each other along the axis, the field in the middle of the transducer is 500 Oe, the (linear) field gradient is 20 Oe/cm, and the SMSW frequency is 3200 MHz, then at a distance of 1 cm from the exciting transducer the width beam (along the extreme trajectories) at $\varphi = 35°\pm5°$ is equal to 0.43 cm (exceeds the initial one by 7.5%). If the field gradient is 100 Oe/cm, then the same value at $\varphi = 30°$ is 0.73 cm (1.8 times more than the initial one), and at $\varphi = 40°$ it reaches 1.07 cm (2.7 times more than the initial one). For this reason, in the experiments to reduce the spreading of the beam, the exciting transducer was placed in the region with the lowest possible field gradient. Beam spreading is especially pronounced in a 'valley'-type field for frequencies close to critical. In a 'roll' type field, on the contrary, spreading is practically absent at any initial angles and frequencies of the SMSW. Such a field has strong channeling properties and, as it were, 'suppresses' the dispersive beam divergence (Section 1.3), as a result of which the SMSW attenuation in such a channel becomes record low. Since the dispersion divergence is strong near the lower boundary of the SMSW spectrum, the decrease in attenuation is especially large at low frequencies. So at a frequency of 3150 MHz, the attenuation in a uniform field is 8.0 dB cm^{-1}, and in a 'shaft'-type field (Fig. 4.27a) it is 7.5 dB cm^{-1}, while at a frequency of 2570 MHz the same values are respectively equal to 5.0 dB cm^{-1} and 1.0 dB cm^{-1}.

Conclusions for chapter 4

The main issues discussed in this chapter are as follows.

1. For the purpose of classification, the main types of magnetizing field inhomogeneities are considered. It is shown that in the plane of the ferrite structure, it is convenient to approximate the real field based on the expansion in the Taylor series. Important special cases of longitudinally and transversely inhomogeneous as well as radially symmetric fields are considered. It is shown that, up to terms of the second order, three main field configurations are possible: linear, of the 'valley' and 'shaft' type. Based on the analysis of the dispersion relation, the field boundaries of the SMSW spectrum were obtained, from which the spatial boundaries of the SMSW propagation regions were found for the cases of fields: linear, of the 'valley' and 'shaft' type.

2. Two main methods for calculating the trajectories of SMSWs in nonuniformly magnetized ferrite structures are considered: isofrequency curves and Hamilton–Auld curves. A comparison of the methods is carried out, their advantages and disadvantages are revealed. It is shown that the method of isofrequency curves, combining high simplicity and clarity for the case of a field of constant direction, experiences certain difficulties in calculating the trajectories of SMSWs in fields of other configurations, as well as in studying the wavenumber, phase and group velocities. The

Hamilton–Auld method, losing out to the isofrequency curves method in simplicity and clarity, is much more universal, suitable for fields of any configuration, makes it easy to calculate the wavenumber, phase and group velocities, and is also very convenient for numerical implementation on a computer.

3. For the case of a ferrite film with a free surface magnetized by a longitudinally inhomogeneous field, the trajectories of SMSW in fields: linear, of the 'valley' and 'shaft' type are investigated. It is shown by the method of isofrequency curves that three types of trajectories can exist in a linear field: 'minus-type', having the form of weakly curved C-shaped curves, monotonically tending towards a higher field up to the boundary of the region of existence of the SMSW; 'Plus-type', having the form of V-shaped curves, first going in the direction of decreasing the field, and then turning in the direction of a larger field up to the boundary of the region of existence of the SMSW; 'Zero-type' in the form of straight lines perpendicular to the direction of the field. With respect to small changes in the initial conditions, the plus and minus-type trajectories are stable, and the zero-type trajectories are unstable, easily pass into the plus or minus-type trajectories and are not realized in the experiment. In a field of the 'valley' type on each of its slopes, similarly to the case of a linear field, there are trajectories of 'plus', 'minus' and 'zero-type' that do not reach the bottom of the 'valley'. An important difference from the case of a linear field is the possibility of the existence of unstable 'zero-type' trajectories in the form of straight lines going exactly along the bottom of the 'valley'. As a result of this circumstance, two new types of trajectories arise: the first is a 'plus-zero-type', consisting of a 'plus-type' trajectory, which turns into a 'zero-type', having the form of L-shaped curves, and the second – plus-type', consisting of two' plus-type 'trajectories passing into each other, located on different slopes of the' valley', having the form of S-shaped curves. Plus-zero-type trajectories are unstable, plus-plus-type trajectories are stable. In a field of the 'shaft' type, on each of its slopes, similarly to the case of a linear field, there are trajectories of 'plus', 'minus' and 'zero-type' that do not reach the top of the 'shaft'. The most important difference from the previous cases is the possibility of the existence of periodic trajectories of the 'pseudo-sinusoidal' type, formed by alternating trajectories 'plus' and 'minus-type', having the form of wavy curves similar to sinusoids. These trajectories are stable and continue indefinitely along the normal to the direction of

the field, while maintaining a constant vibration amplitude relative to this normal, that is, the 'shaft' -type field along the specified normal forms a channel for the propagation of the SMSW.

4. For the case of a ferrite film with a free surface magnetized by a longitudinally inhomogeneous field, the Hamilton-Auld method analyzed the quantitative characteristics of all the analyzed types of trajectories. For the case of a linear field, it is shown that the swing of the V-shaped 'plus-type' trajectories is the greater, the higher the frequency of the SMSW, the greater the initial deviation of the phase velocity vector from the normal to the direction of the field, and also the smaller the field gradient. Under the same conditions, analogous dependences for the wavenumber and angles, which make up the vectors of the phase and group velocities with the normal to the direction of the field, acquire a wider range. All other trajectories inherent to fields of the 'valley' and 'shaft' type have similar properties. For a field of the 'valley' type, a critical frequency is revealed, below which the trajectories have the form of V-shaped curves of the 'plus-type', and above them – S-shaped curves of the 'plus-plus-type'. The trajectory corresponding to the critical frequency has the form of an L-shaped curve and is unstable. For the case of a linear field, provided that the SMSW is reflected from the line corresponding to the upper field boundary of the SMSW spectrum, the possibility of the existence of a channel (of the first type) for the propagation of the SMSW, the axis of which is oriented perpendicular to the direction of the field, is predicted. The SMSW trajectories in such a channel are formed by adjacent V-shaped 'plus-type' trajectories. For the case of a field of the 'valley' type, the possibility of the existence of a similar channel (of the second type), formed by adjacent S-shaped trajectories of the 'plus-plus-type', is predicted. The quantitative characteristics of the channel for the SMSW formed by the field of the 'shaft' type are investigated. It is shown that the range of trajectories in the channel is completely determined by the wave parameters at the initial point of the trajectory, and the total channel width is the larger, the smaller the field gradient. For trajectories, the initial sections of which are perpendicular to the direction of the field, the phenomenon of 'pseudofocusing' was discovered, consisting in a periodically repeating and gradually decreasing narrowing of the beam of trajectories emerging from different points of the emitting converter of the SMSW.

5. The trajectories of the SMSW in the FM structure magnetized by longitudinally inhomogeneous fields: linear, of the 'valley' and 'shaft' type are considered. It is shown that in this structure the main dispersion properties of SMSWs are close to the same properties for a free ferrite film, but the quantitative difference is a significantly wider spectrum width, a larger opening of isofrequency curves, and their relative proximity to the origin of coordinates. The main qualitative difference is the absence of a wave cutoff at the lower frequency boundary of the SMSW spectrum. In linear fields, such as 'valley' and 'shaft', the main types of trajectories for this case are the same as for a ferrite film with a free surface, however, the range of trajectories along both axes significantly (two to three times or more) exceeds the range of trajectories for free film. The most important difference from the case of a free film is the fact that, due to the absence of cutoff, when the trajectories approach the line corresponding to the upper field boundary of the SMSW spectrum, they do not break off, but only asymptotically approach it, passing into straight lines going to infinity perpendicular to the direction fields. The tendency of the trajectory to infinity is accompanied by an unlimited increase in the wave number, which in a real experiment should lead to a strong wave attenuation. In a field of the 'valley' type, the same course of trajectories is observed on both its slopes, where the field reaches the value of the upper field boundary of the SMSW spectrum. In a 'shaft' type field, this feature is possible only if the field of the 'shaft' top exceeds the upper field boundary of the SMSW spectrum, in other cases, the SMSW trajectories have a periodic 'pseudosinusoidal' character, similar to the case of a free film. In linear fields, such as 'valley' and 'shaft', channels are possible, similar to the case of a free film.

6. The trajectories of the SMSW in the FDM structure magnetized by longitudinally inhomogeneous fields: linear, of the 'valley' and 'shaft' type are considered. It is shown that, owing to a more complex dispersion law than for a free film and an FM structure, the nature of the trajectories in the FDM structure becomes much more complicated. So, if the dispersion curve has two extrema, then in the case of a linear field, three types of trajectories are possible: V-shaped without rotation of the first type corresponding to small values of the wave number; V-shaped without rotation of the second type, corresponding to large values of the wavenumber; V-shaped with a turn, corresponding to intermediate values of the wavenumber, where turn means a change in the direction of wave propagation

relative to the normal to the direction of the field to the opposite. For a field of 'valley' type, nine types of trajectories are possible: V-, L- and S-shaped without turning of the first type; V-, L- and S-shaped with a turn; V-, L- and S-shaped without turning of the second type. For a field of the 'shaft' type, when a wave propagates along the 'shaft' axis, six types of trajectories are possible: wavy without turning the first and second plus-type; undulating minus-type without turning; undulating with a turn of the plus and minus type; eight-shaped. For all three cases of the field, on the plane of the isofrequency curves, the boundaries of the regions in which the SMSWs have a direct or reverse character are determined. It is shown that in the process of propagation, waves can cross the boundaries of these regions, which corresponds to a change in the nature of the wave from forward to reverse and vice versa. Along with the dispersion curve, which has two extrema, the cases are considered when the dispersion curve has only one extremum or none. It is shown that in this case, of the full set of possible trajectories, only the part corresponding to the corresponding parts of the dispersion curve is realized. For all three types of fields, the Hamilton-Auld method calculated the quantitative characteristics of the trajectories, as well as the coordinate dependences of the wave number and angles that determine the directions of the phase and group velocity vectors of the SMSW. The intervals of the initial orientations of the phase velocity vector, which are necessary for the realization of all the considered types of trajectories, are revealed. For a field of the 'valley' type, the critical ratios of the initial wave parameters corresponding to L-shaped trajectories are determined. For a field of the 'shaft' type, the critical ratios of the initial parameters of the wave corresponding to an eight-shaped trajectory are determined.

7. For the cases of a ferrite film with a free surface and a FM structure magnetized by a longitudinally inhomogeneous field, the phase incursion and delay time in fields: linear, of the 'valley' and 'shaft' type are considered by the Hamilton-Auld method. It is shown that, both in the free film and in the FM structure, the phase incursion and the delay time with the propagation of the wave always monotonically increase the more the higher the SMSW frequency. The growth rate of the phase incursion and the delay time in the free film throughout the entire trajectory remains close to its initial value. A similar velocity in the FM structure, at first being significantly (more than an order of magnitude) less than the same velocity in a free film, as the trajectory approaches the region of

the film where the field is close to the upper field boundary of the SMSW spectrum, sharply (by orders of magnitude) increases and tends to infinity. In the case of a linear field, this leads to a gigantic increase in the phase incursion (104 rad and more) and the delay time (300-400 µsec and more). In a 'valley' type field, such an increase in the phase incursion and delay time is observed on both its slopes, and in the case of a 'shaft' type field - near its top. The reason for such a gigantic increase in these parameters in the FM structure is the absence of a cutoff near the lower frequency boundary of the SMSW spectrum.

8. An experimental study of the propagation of SMSW wave beams in ferrite films magnetized by a longitudinally inhomogeneous field: linear, of the 'valley' and 'shaft' type has been carried out. V-shaped trajectories are investigated in a linear field. Good agreement of the calculated curves with the experimental ones was revealed. It is shown that for frequencies close to the middle of the SMSW spectrum, the coincidence accuracy is 5%, and near the edges of the spectrum, 10%. In the field of 'valley' type, trajectories of V and S-shaped types are investigated. Found the critical frequency corresponding to the L-shaped trajectory. A good agreement between the calculated data and experiment was revealed: so the calculated value of the critical frequency coincides with the experimental one with an accuracy of 3%, the calculated trajectories for frequencies above the critical one coincide with the experimental ones with an accuracy of 1%, below the critical one − 5%. Somewhat worse agreement was noted for the frequency close to the lower edge of the SMSW spectrum, where it is 20%. In a field of the 'shaft' type, the trajectories of the pseudosine-wave type are investigated. The agreement between the calculated data and experiment remains good here too. Thus, within the first half-period, the calculated trajectories coincide with the experimental ones within the entire SMSW spectrum with an accuracy of 1%. After passing the second half-period, the calculated trajectories for the frequencies of the middle and upper part of the SMSW spectrum coincide with the experiment with an accuracy of 2%. Somewhat worse agreement is also observed here at frequencies close to the lower edge of the SMSW spectrum, where it is 20%. Changes in the observed SMSW trajectories are investigated for various experimental parameters. It is shown that a very critical characteristic is the gradient of an inhomogeneous field, especially at the location of the emitting SMSW converter, since in this case, different parts of the converter,

being in a different field, emit SMSW at different angles and the overall wave beam loses its unity. In a field of the 'valley' type, a strong spreading and even splitting of the wave beam into two, caused by the same reason, was revealed in the case when the emitting transducer is located at the bottom of the 'valley' and the SMSW frequency is close to critical. In a 'shaft'-type field, due to its channeling properties, beam spreading due to field inhomogeneity, as well as splitting due to dispersive divergence, are fully compensated for, which leads to a record low (up to 1.0 dB cm^{-1}) attenuation of the SMSW, especially for frequencies close to to the lower end of the spectrum.

5

Propagation of wave beams of finite width in inhomogeneous magnetized ferrite films

In the previous chapters (3, 4), the apparatus of SMSW trajectories was developed, which makes it possible to determine the magnetic disturbance created by a wave with given initial parameters, emanating from one definite point of the film. Real 'pathogens' of SMSW have a finite length, that is, they excite a wave not at one point, but on a certain segment of the film, the field at different points of which may be different. A full wave beam is a set of a large number of unit trajectories of SMSWs, which begin under different conditions and propagate along their own paths, and it can be approximately described using a unit trajectory only if it is narrow enough or the field inhomogeneity is so small that the field changes little over the beam width. It is this situation that is realized in the experiments described in Sections 3.8 and 4.1.2. In all other cases, it is necessary to take into account the difference in the initial conditions for the trajectories emanating from different points of the exciting transducer. Let us consider the features of propagation of wide wave beams in nonuniformly magnetized films using several examples.

In this chapter, basically the results of works [615–626] are given. The rest of the necessary links are given in the text of the chapter.

5.1. Spatial transformation of wide beams of SMSW propagating in inhohogensouly magnetized films

A wide wave beam in a nonuniformly magnetized film is transformed into space: its width and transverse profile change, and splitting may occur. Let us consider the most important cases of fields: linear, such as 'valleys' and 'shafts'. We will use the Hamilton–Auld method. As before (chapters 3, 4), we will assume $4\pi M_0 = 1750$ G, $a = 15$ µm.

5.1.1. Linearly inhomogeneous field

Let the linear field be described by formula (3.2) with the same parameters as in Section 3.4.2: $H_0 = 437.5$ Oe, $a_0 = 1$. $a_1 = 1/16$ cm^{-1}. Figures 5.1 and 5.2 correspond to this case. The converter that excites the SMSW has the form of a straight line segment of a conductor, the ends of which are designated by the letters S' and S''. The figures show the trajectories emanating from various equally spaced points of the transducer. The edges of the wave beams defined by the trajectories outgoing from the end points of the transducer S' and S'' are shown by thickened lines. In Fig. 5.1, the coordinates of the middle of the transducer: $y = 0$ cm, $z = 0$ cm, points $S' = -0.36$ cm, $z = 1.00$ cm, points S'': $= 0.36$ cm, $z = -1.00$ cm. The angle between the conductor of the transducer and the Oz axis is 20°, which sets the initial angle for SMSW $\varphi = 20°$. Two wave beams were constructed for different frequencies: 1 – 2900 MHz and 2 – 3200 MHz. The first frequency lies near the beginning of the SMSW spectrum (at = 0 cm), the second - near its middle.

Figure 5.1 shows that all trajectories are V-shaped and end at a field corresponding to a given frequency. Beam 1, corresponding to

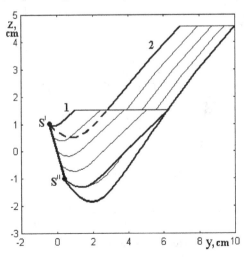

Fig. 5.1. Transformation of wide SMSW wave beams in a linearly inhomogeneous field for different frequencies: 1 – 2900 MHz, 2 – 3200 MHz.

a lower frequency, is strongly deformed. If its initial (at = 0 cm) width along the coordinate is 2 cm, then already at = 1 cm it is about 3 cm, at = 2 cm – more than 3.5 cm, and then sharply narrows and at $y = 4$ cm it is about 1.5 cm , and at $y = 6$ cm the beam ends. The beam intensity, determined by the number of trajectories, at the coordinate $y = 1.5$ cm remains equal to the initial one, but then sharply decreases and at = 3.5 cm is half of the initial one, and at $y = 6$ cm it drops to zero. Beam 2, whose frequency is higher than 1, is retained much better. So, after a slight expansion near $y = 2$ cm (up to 3 cm) up to $y = 7$ cm, its width does not exceed 2.5 cm, and only after that a sharp drop to zero occurs at $y = 10$ cm.

Figure 5.2 corresponds to two different positions of the transducer: 1 – at an angle to the Oz axis 10° and at 2 30°. Accordingly, for a wave beam 1 – φ = 10°, 2 – 30°. In the first position, the coordinates of the middle of the transducer: $y = 0$ cm, $z = 0$ cm, points $S' = -0.18$ cm, $z = 1.00$ cm, points $S'' = 0.18$ cm, $y = -1.00$ cm, in the second - the middle of the transducer: $y = 0$ cm, $z = 0$ cm, points $S' = -0.58$ cm, $z = 1.00$ cm, points $S'' = 0.58$ cm, $z = -1.00$ cm. The frequency of both beams is the same and amounts to 3000 MHz.

From Fig. 5.2 it can be seen that the first wave beam is practically preserved up to $y = 3$ cm, while the second is significantly deformed already at: $y = 1.5$ cm.

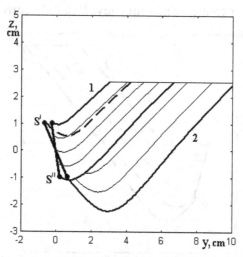

Fig. 5.2. Transformation of wide wave beams of SMSW with a frequency of 3000 MHz in a linearly inhomogeneous field at different angles between the conductor of the transducer and the Oz axis: 1 – 10°, 2 – 30°.

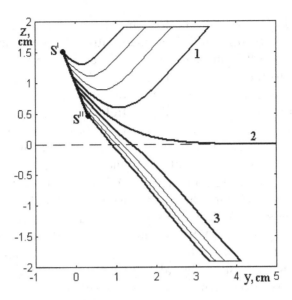

Fig. 5.3. Transformation of a wide SMSW wave beam of frequency 3126.664420 MHz in a 'valley'-type field. The numbering of the beam parts is explained in the text.

Thus, during propagation over the plane of a film magnetized by a linear field, the wave beam is retained the better, the higher its frequency and the smaller the initial angle.

5.1.2. Valley-type field

As the next example, consider the propagation of SMSWs in a film magnetized by a 'valley'-type field, which is described by formula (3.2) with the same parameters as in Section 3.4.2: H_0 = 437.5 Oe Oe, a_0 = 1, a_1 = 0 cm^{-1}, a_2 = 1/16 cm^{-2}. This case corresponds to Fig.5.3. In this case, the coordinates of the middle of the transducer: y = 0 cm, z = 0 cm and its ends: points S' y = –0.58 cm, z = 1.00 cm, points S'' = 0.58 cm, z = –1.00 cm. The angle between the conductor of the transducer and the Oz axis is –30°, that is, for SMSW = 30°. The SMSW frequency is 3126.664420 MHz, which corresponds to the critical frequency for the centre point of the transducer.

Figure 5.3 shows that as it approaches the 'bottom of the valley', the beam splits into three parts, the trajectories in which are V-shaped (1), L-shaped (2) and S-shaped (3). In this case, the trajectories outgoing from that part of the transducer have a V-shape. which is more distant from the bottom of the 'valley', S-shaped – from the one that is closer to the bottom of the 'valley', and L-shaped is the only trajectory outgoing from the centre of the transducer. Part of beam 1 at $y > 0.5$ cm takes the form of a separate beam and deforms similarly to the case of a linear field (Figs. 5.1, 5.2), moreover, to the upper edge of the beam, the trajectories are condensed, which is due to the quadratic law of the field increase with distance from the bottom of the 'valley'. Part of the beam 3 $y > 0.5$ cm is also separated into a separate beam, which is deformed much less than 1: its initial (at $y = 0$ cm) width along the coordinate is 0.5 cm, and at = 3.5 cm it increases to no more than 0.7 cm. thicken towards the lower edge of the bundle for the same reason as above. Part of the beam 2 has no shaped boundaries along the coordinate, and as an increase, the intensity of the SMSW in this place (on the axis) decreases. In fact, trajectory 2 is the boundary between beams 1 and 3.

Thus, when propagating along the plane of a film magnetized by a 'valley'-type field, the wave beam is split into two clearly defined parts corresponding to different parts of the transducer, and the part corresponding to the part of the transducer closest to the 'valley bottom' is better preserved.

5.1.3. Shaft-type field

Let us now consider the propagation of SMSW in a film magnetized by a 'shaft'-type field defined by formula (3.2) with the same parameters as in Section 3.4.2: $H_0 = 437.5$ Oe, $a_0 = 1$, $a_1 = 0$ cm^{-2}. $a_2 = -1/16$ cm^{-2}. Figure 5.4 corresponds to this case. Let the transducer be oriented along the Oz axis, and the coordinates of its ends are: points $S' - y = 0$ cm, $z = -1.00$ cm, points $S' - y = 0$ cm,

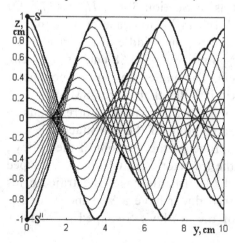

Fig. 5.4. Transformation of a wide SMSW wave beam of 3000 MHz frequency in a 'shaft'-type field. Trajectories are shown coming from different parts of the transducer.

z = 1.00 cm. The angle between the conductor of the transducer and the Oz axis is 10°, that is, for SMSW = 0°. The MSW frequency is 3000 MHz.

From Figure 4.5 it can be seen that the trajectories emanating from all sections of the transducer have a pseudo-sinusoidal character, and their amplitude is equal to the distance of the starting point from the axis. As it propagates along the Oy axis, the maximum width of the wave beam remains equal to the initial one (2 cm), but the beam undergoes several narrowings (pseudofocusings), the depth of which, due to the dephasing of individual trajectories, decreases as it increases. So, for the initial (at $y = 0$ cm) beam width equal to 2 cm, near $y = 1.5$ cm, the beam width is 0.30 cm, near $y = 4.5$ cm it is 0.5 cm, and near $y = 8.0$ cm it is 1.1 cm.

Thus, when propagating along the plane of the film, magnetized by a field of the 'shaft' type, the wave beam is preserved the better, the greater the distance travelled by it.

5.2. Method for analysis of amplitude-frequency and phase-frequency characteristics of transmission lines of SMSW

In the previous section (5.1), it was shown that the wave beam undergoes a spatial transformation during propagation in an inhomogeneously magnetized ferrite film. Let us now consider how this transformation affects the characteristics of the transmission line containing the emitting and receiving SMSW transducers superimposed on the film surface. The following describes a method for calculating the amplitude-frequency and phase-frequency characteristics (AFC and PFC) of a transmission line of the most general type, based on the concepts of geometric optics of inhomogeneous media and consisting in the study of partial wave beams of the SMSW. The method is applicable for any shape and relative position of the transducers, for cases of a uniformly and inhomogeneously magnetized film of any shape and with any inhomogeneities (thickness, magnetization, elastic stresses). The method can be generalized to the calculation of devices with layered structures based on ferrite films.

5.2.1. General scheme of the method for calculating the frequency phase responses

The amplitude-frequency and phase-frequency characteristics of the transmission line to the SMSW are determined by the combined effect of the following factors:

1) the efficiency of excitation of the SMSW by the emitting converter;

2) damping of the SMSW during propagation over the film, caused by the processes of relaxation of the precession of magnetization;

3) absorption of SMSW by obstacles on the path of proliferation;

4) the limited size of the receiving transducer;

5) the efficiency of receiving the SMSW by the receiving transducer;

6) the time phase of the arrival of the SMSW to the receiving transducer;

7) the mutual orientation of the conductor of the receiving transducer and the phase front of the SMSW arriving at this transducer.

The task of calculating the frequency response and phase response, taking into account the listed factors, is simplified by the fact that most of the factors act independently of each other and their contribution is calculated separately. The frequency dependence of the conversion efficiency (factors 1 and 5) and the dependence of the SMSW attenuation on the frequency (factor 2) are known [136, 139, 294, 415–417, 441]. These factors are not discussed further. Their accounting can be carried out by multiplying the results obtained below by the corresponding coefficients from the cited works. So, let's consider factors 3, 4, 6 and 7. In this case, we will consider factors 6 and 7 in aggregate, since the resulting phase of the total signal at the receiving converter is the sum of the SMSW phase delay on the path between the converters and the delay caused by the non-parallelism of the phase front of the wave and the conductor receiving transducer.

The successive steps of the proposed method for calculating the frequency response and phase response are as follows.

1. The emitting transducer is divided along its length into end sections, each of which is considered as an independent elementary emitter emitting a partial wave beam.

2. A frequency is selected in the frequency range of the existence of the SMSW.

3. The trajectories and temporal phases of all partial wave beams at the selected frequency are calculated.

4. Beams that leave the geometrical dimensions of the receiving transducer and hit obstacles are excluded from consideration.

5. Summation of the amplitudes is performed taking into account the phase shift of the beams that have reached the receiving transducer, which gives the amplitude–phase characteristic of the transmission at a given frequency.

6. The frequency changes in the specified range with the selected step, as a result of which the full frequency response and phase response are calculated for the transmission line.

The calculation method is illustrated in Figs. 5.5–5.7. Figure 5.5 shows a generalized diagram of a transmission line, where the plane of the drawing corresponds to the plane of the ferrite film, the letters S' and S'' denote the emitting transducer, and the letters R' and R'' denote the receiving one. In the general case, the field along the plane of the film is nonuniform, and the shape of the conductors of the transducers is arbitrary. The shaded areas show obstacles that the MSW cannot pass through. The emitting converter $S'S''$ is divided into sections of the same length (in this case $n = 15$), from the ends of which the trajectories of the MSW wave beams are constructed $n+1$ – in total (here 16 trajectories).

From Fig. 5.5 it can be seen that of the 16 trajectories leaving the emitting transducer, only four (2, 3, 6, 7) reach the receiving transducer, the rest go to the sides or are delayed by obstacles. Thus, the output signal at the receiving transducer is created only by wave beams corresponding to these four trajectories. In this case, if the

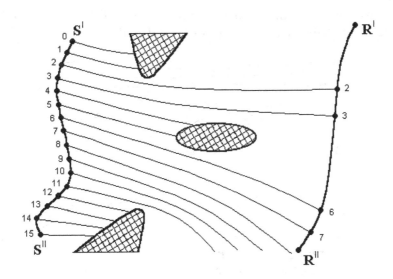

Figure 5.5. Generalized diagram of the transmission line to the PMSV

phases of the wave beams reaching the receiving transducer are the same, then the amplitude of the output signal will be 0.25 of the amplitude of the input signal, that is, the transmission coefficient, in this case equal to the ratio of the number of trajectories that have reached the receiving transducer to the initial their number is equal to 0.25. If the phases of the wave beams that have reached the receiving transducer are different, then the output signal must be calculated taking into account this phase difference and the transmission coefficient will be slightly less than in the first case.

The distribution of trajectories shown in Fig. 5.5 corresponds to one specific MSW frequency. At a different frequency, the distribution of trajectories will be different and the transmission coefficient will change. By setting the frequency values with a certain step and finding the transmission coefficients for them taking into account the phase, it is possible to construct the complete amplitude-frequency and phase-frequency characteristics (AFC and PFC) of any device under study. In the general case of curvilinear transducers and an inhomogeneous field, the partition grid of the emitting transducer should be so fine that the elementary sections of both transducers can be considered rectilinear, and the MSW phase front within the width of one beam between adjacent trajectories also remains rectilinear.

5.2.2. *Frequency response diagram*

The construction of the frequency response using the example of a device of two converters in a uniform field is illustrated in Fig. 5.6.

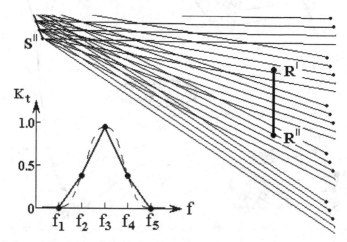

Fig. 5.6. Scheme of constructing the frequency response using the example of a device of two converters in a uniform field. 1–5 – wave beams of different frequencies.

Here, the emitting transformer $S'S''$ is divided into four equal segments, from the ends of which five trajectories, which are straight lines, proceed. The construction was carried out for five different frequencies f_1–f_5 such that $f_1 < f_2 < f_3 < f_4 < f_5$. The frequency f_1 corresponds to the group of trajectories 1, frequency f_2 to 2, frequency f_3 to 3, frequency f_4 to 4, frequency f_5 to 5. It can be seen from the figure that only two frequency trajectories, five frequency trajectories and two frequencies fall on the receiving $R'R''$ converter. The corresponding frequency dependence of the gain (excluding phase) is shown at the bottom left. In the same place, the dashed line shows the frequency response, built with a smaller frequency step. The observed frequency response has a well-pronounced bell-shaped character.

5.2.3. PFC construction scheme

Let us now consider the phase properties of the transmission line to the MSW. The total output signal at the receiving transducer is made up of signals from individual wave beams, one of which is schematically shown in Fig. 5.7. The ends of the elementary segment of the emitting transducer are designated by the letters B and C. The wave beam emanating from this segment is limited by the trajectories CN and BD, where points N and D correspond to the intersection of the trajectories originating from points C and B with the conductor of the receiving transducer. The line of equal phase of the beam passing through point N is denoted by NA, where point A belongs to the trajectory BD. By virtue of the above assumptions BC, ND, NA, AD are straight line segments. For the convenience of geometric constructions, the following coordinate systems have been introduced:

1) Oyz – the coordinate system associated with the film;
2) $O'y'z'$ – the coordinate system, the origin of which coincides with point N, the axis $O'z'$ is directed along the segment ND of the receiving transducer, and the $O'y'$ axis is perpendicular to it;
3) $O''y''z''$ – coordinate system, the origin of which coincides with point N, the $O''z''$ axis is directed along NA, and the $O''y''$ axis is perpendicular to NA;
4) $O'''y'''z'''$ is a coordinate system, the origin of which coincides with point N, and the axes are parallel to the axes of the Oyz system.

Using Fig.5.7, it can be shown that the signal at the receiving transducer, created by the wave beam emanating from the emitting transducer, is equal to:

$$S_i = \frac{W_s}{n} A_i \cos(\omega t - \phi_i),$$

(5.1)

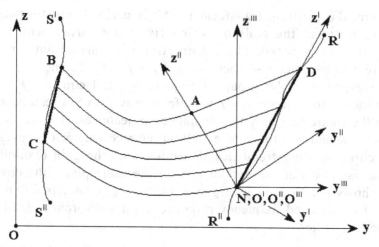

Fig. 5.7. Diagram explaining the calculation of the phase properties of the transmission line on the SMSW.

where: A_i is the signal amplitude in the ND section of the receiving transducer, taking into account the phase lag inside the triangle AND, φ_i is the phase incursion along the shortest trajectory of the i-th beam – CN, equal to:

...

$$\varphi_i = \int_{y_N}^{y_C} k(y,z)\sqrt{1+\left(\frac{dz}{dy}\right)^2}\,dy, \tag{5.2}$$

The expression for can be obtained from solving the AND triangle. It becomes especially simple when the $\angle NAD$ angle is close to 90°. In this case, A_i is reduced to an integral well known in geometric optics [481] and is equal to:

$$A_i = \frac{\sin\alpha_i}{\alpha_i}, \tag{5.3}$$

where $\alpha_i = \frac{1}{2}b_i k_i \sin\beta_i$, , $b_i = ND$, k_i is the wavenumber of the MSW inside the triangle ΔAND, β_1 is the angle $\angle AND$.

Let us now find the complete output signal at the receiving transducer. Let us denote by W_S the complete initial signal supplied to the emitting transducer. If this transducer is divided into sections of equal length, then the total signal at the receiving transducer is:

$$W_R = \sum_{i=1}^{n} S_i = \frac{W_S}{n} \sum_{i=1}^{n} A_i \cos\left(\omega t - \phi_i\right) \quad (5.4)$$

moreover, A_i and φ_i are determined by the formulas (5.2) and (5.3) for each bundle separately from the corresponding triangle ΔAND and trajectory CN.

The above value of the output signal (5.4) is obtained at only one predetermined frequency. To obtain full frequency response and phase response in the frequency range, one should set the frequency grid in this range and calculate the values on it. So, for example, if the original signal at the emitting transducer is described by the expression

$$W_S(\omega_j) = W_{S0}(\omega_j)\cos\omega_j t \quad (5.5)$$

and the output signal at the receiving transducer is

$$W_R(\omega_j) = W_{R0}(\omega_j)\cos\left[\omega_j t + \phi_R(\omega_j)\right], \quad (5.6)$$

then the AFC is found by calculating the transmission coefficients:

$$K_t(\omega_j) = W_{R0}(\omega_j)/W_{S0}(\omega_j) \quad (5.7)$$

The frequency dependence $K_t(\omega_j)$ is the desired AFC, and the PFC, determined by the phase difference between $W_R(\omega_j)$ and $W_S(\omega_j)$ is expressed by the relationship $\omega_R(\omega_j)$.

Comment. In order to simplify the problem and identify the main regularities of the considered phenomena, the amplitude–frequency characteristics presented below in the Sections 5.3–5.5 were obtained without taking into account the possible phase mismatch between individual partial beams of the SMSW. Generally speaking, taking into account such a mismatch can lead to some irregularity of the frequency response, so that the frequency response given below should be considered as 'envelopes' of the overall characteristics. The question of the formation of the phase–frequency properties of transmission lines is considered in more detail in the next chapter.

5.3. Amplitude-frequency characteristics of transmision lines on ferrite films magnetized by fields of different configurations

In the previous section (5.2), a general method is described that makes it possible to calculate the frequency response and phase response of the transmission of a transmission line on an SMSW propagating in an arbitrarily magnetized ferrite film. Let us now consider specific examples of calculating the frequency response, first for a homogeneous field, and then for the main cases of a longitudinally inhomogeneous field – linear, of the 'valley' and 'shaft' type.

5.3.1. Homogeneous field

Let's calculate the frequency response of a band-pass filter based on SMSWs propagating in a ferrite film, uniformly magnetized. The device diagram is shown in Fig. 5.8.

A ferrite film, the plane of which coincides with the plane of the pattern, is magnetized by a uniform field directed along the Oz axis. The saturation magnetization of the film is 1750 Oe, its thickness is 15 μm, and the field is 437.5 Oe. The emitting $S'S''$ and the receiving $R'R''$ converters are made in the form of straight

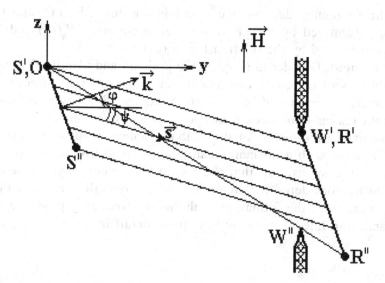

Fig. 5.8. Scheme of a band-pass filter based on an SMSW in a uniform field.

conductors 10 µm thick and 5 mm long. This length is sufficient to eliminate dispersive splitting of the wave beam (Section 1.3) practically in the entire range of existence of the SMSW. The angle between the conductor of the $S'S''$ transducer and the Oz axis is 30°. A diaphragm with a 5 mm long slot $W'W''$ is placed in front of the receiving transducer, the task of which is to ensure the constancy of the receiving transducer aperture at different angles of incidence of the SMSW on it. The diaphragm, together with the receiving transducer, can move along the film plane, and the orientation of the conductor of the $R'R''$ transducer is selected parallel to the phase front of the received wave beam, which ensures optimal reception conditions. In this case, the field is uniform, therefore the SMSW trajectories are straight lines, and the direction of the SMSW phase front does not change and remains parallel to the conductor of the emitting $S'S''$ converter, therefore the conductor of the $R'R''$ receiving converter is also parallel to $S'S''$, and makes an angle of 30° with the Oz axis. With the accepted parameters of the problem, the beginning of the SMSW spectrum corresponds to 2739 MHz, and the end to 3675 MHz. In the calculations, the emitting transducer was divided into 100 sections, and the frequency step was 10 MHz.

Figure 5.9 shows the change in the frequency response when the receiving transducer moves along the wave beam corresponding to a frequency of 3000 MHz with a step along the Oy axis equal to 1 cm. With a small distance between the transducers (curve 1), wave beams of almost all frequencies fall on the receiving transducer, as a result of which the AFC width (at half maximum) is close to the full width of the SMSW spectrum (at $\varphi = 30° - \sim 500$ MHz) and is 450 MHz. As the distance increases (curves 2–4), the beams of the first higher and then lower frequencies cease to fall on the receiving transducer, which leads to a narrowing of the frequency response first from the high side (curves 2, 3), and then low (curves 3, 4) frequencies. At a distance of more than 4–5 cm (curves 5, 6), the narrowing of the frequency response somewhat slows down and its width decreases to ~100 MHz.

Let us now consider the change in the frequency response when moving the receiving transducer along the coordinate axes. Figure 5.10 shows the frequency response corresponding to the movement of the transducer along the Oz axis with a step of 0.5 cm at a constant distance between the point R' and the Oz axis of 2.0 cm.

Figure 5.10 shows that most of the frequency response (1–3) have a bell-shaped, more or less symmetric form, the asymmetry of curve

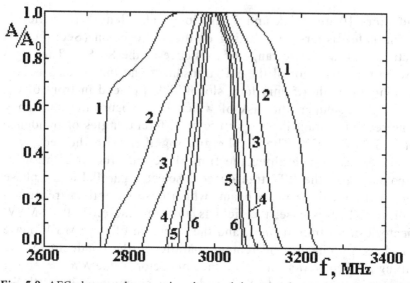

Fig. 5.9. AFC change when moving the receiving transducer along the 3000 MHz wave beam. The curve numbers correspond to the change in the y-coordinate of the upper end of the transducer from 1 to 6 cm in 1 cm steps. The curves correspond to the following coordinates of the upper end of the receiving transducer: 1 – = 1 cm, = –0.68 cm; 2 – = 2 cm, = –1.38 cm; 3 – = 3 cm, = –2.07 cm; 4 – = 4 cm, = –2.78 cm; 5 – = 5 cm, = –3.47 cm; 6 – = 6 cm, z_R = –4.17 cm.

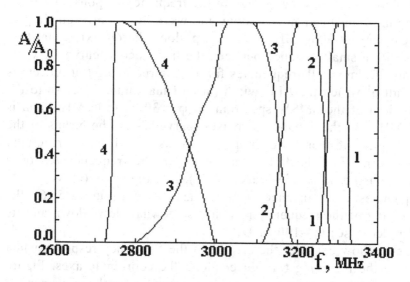

Fig. 5.10. Changing the frequency response when moving the receiving transducer along the Oz axis while maintaining the distance between the upper end of the transducer and the Oz axis equal to 2.0 cm. The numbers of the curves correspond to the y-coordinate of the end of the transducer: 1 – z'_R = –2.5 cm; 2 – z'_R = –2.0 cm; 3 – z'_R = –1.5 cm; 4 – z'_R = –1.0 cm.

4 is due to the limitation of the SMSW spectrum from below with a frequency of 2739 MHz. With the removal of the receiving transducer from the axis, the centre frequency of the frequency response is shifted towards higher frequencies, and the width of the frequency response decreases. The reason for this phenomenon is that with a small distance between the transducers (curve 4), only low frequency wave beams (2740–2950 MHz), which are least deviated from the Oy axis, fall on the receiving transducer. With an increase in this distance (curves 3, 2), wave beams of higher frequencies (2900-3200 MHz), more deflected from the Oy axis, fall on the receiving transducer, while low-frequency beams (2740–2850 MHz) pass by. At a large distance (curve 1), only the highest frequencies (3250–3350 MHz) wave beams hit the receiving transducer, which deviate most strongly from the Oy axis. As a result, the maximum AFC shifts from 2750 MHz (curve 4) to 3270 MHz (curve 1). The narrowing of the frequency response that accompanies such a displacement is due to an increase in the distance between the transducers, similar to the curves shown in Fig. 5.9. In this case, the frequency response narrows from ~300 MHz (curves 4, 3) to ~60 MHz (curve 1). An increase in the distance between the R' point and the Oz axis leads to a further narrowing of the frequency response. Thus, at $y_R = 4.0$ cm and $z_R = -2.5$ cm, the maximum AFC falls on 2920 MHz, and the width is 180 MHz; at $z_R = -5.0$ cm, the same values are equal to 3300 MHz and 30 MHz. The reason for this narrowing of the frequency response is the same as for the curves shown in Fig.5.9.

Figure 5.11 shows the frequency response corresponding to the movement of the transducer along the Oy axis with a step of 0.5 cm at a constant distance between the point R' and the Oy axis, equal to -1.0 cm.

It can be seen from the figure that most of the frequency response (1–3) have a bell-shaped, more or less symmetrical form, the asymmetry of curve 4 is due to the limitation of the SMSW spectrum from below by the frequency of 2739 MHz. With an increase in the y-coordinate of the end R' of te hreceiving transducer, the centre frequency of the frequency response shifts towards low frequencies, and the width of the frequency response increases (especially for curves 1, 2). The reason for this phenomenon is that with a small distance between the transducers (curve 1), only high-frequency wave beams (3100–3300 MHz), which are deviated from the Oy axis most strongly, fall on the receiving transducer. With an increase in this distance (curves 2, 3), the wave beams of

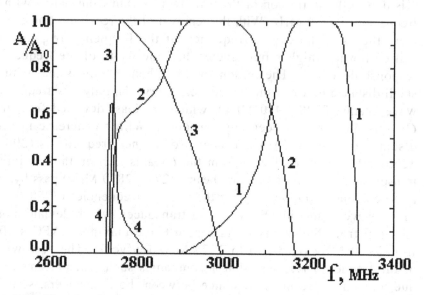

Fig. 5.11. Changing the frequency response when moving the receiving transducer along the Oy axis while maintaining the distance between the upper end of the transducer and the Oy axis equal to 1.0 cm. The curves correspond to the following values of the y-coordinate of the point R': $1 - y_R = 1.0$ cm; $2 - y_R = 1.5$ cm; $3 - y_R = 2.0$ cm; $4 - y_R = 2.5$ cm.

the lower frequencies (2750–3100 MHz), less deviated from the Oy axis, fall on the receiving transducer, while the high-frequency (3150–3300 MHz) beams pass by. At a large distance (curve 4), only the lowest frequencies (2740–2800 MHz) wave beams hit the receiving transducer, which deviate from the Oy axis least of all others. As a result, the maximum frequency response shifts from 3250 MHz (curve 1) to 2750 MHz (curve 4). At this offset, the frequency response width first increases from ~200 MHz (curve 1) to ~400 MHz, after which it decreases to 20 MHz (curve 4). The expansion of the frequency response is due to an increase in the frequency range of wave beams falling on the receiving transducer, and the subsequent narrowing is due to the limitation of the SMSW spectrum at a level of 2739 MHz. An increase in the distance between the R' point and the Oy axis leads to a further narrowing of the frequency response. Thus, at $z_R = -2.0$ cm and $y_R = 4.5$ cm, the maximum AFC falls on 3310 MHz, and the width is 60 MHz; at $y_R = 2.5$ cm, the same values are equal to 2830 MHz and 190 MHz. The reason for such a narrowing of the frequency response with increasing distance between the transducers is the same as in the previous case.

5.3.2. Linearly inhomogeneous field

Let us now onsider the frequency response of a band-pass filter based on SMSWs propagating in a ferrite film magnetized by a linearly inhomogeneous field. The scheme of the device and its main parameters are the same as before (Section 5.3.1), but here, due to the field inhomogeneity, the SMSW trajectories are V-shaped curves (Section 3.4). Suppose that the field is described by formula (3.2), where H_0 = 437.5 Oe, a_0 = 1, a_1 = 1/16 cm^{-1}, a_2 = 0.

Figure 5.12 shows the wave beams emanating from the $S'S''$ converter at six frequencies. The insets show the geometry of the emitting (left) and receiving (right) transducers. The $R'R''$ conductor of the receiving transducer is oriented parallel to the phase front of the wave at the receiving location. The type of trajectories completely coincides with that considered in Section 3.4 and is discussed there in sufficient detail. Here, the main focus will be on the resulting frequency response.

Figures 5.13 (a, b) and 5.14 (c, d) show the change in frequency response at some fixed values of the coordinate y: $a - y$ = 2 cm, $b - y$ = 4 cm, $c - y$ = 6 cm, $d - y$ = 8 cm, in the process of moving the receiving transducer along the Oz axis from -3.5 cm to 4.5 cm with a step of 0.5 cm. Note, first of all, that as the coordinate y increases, the interval z along, in which the AFC is nonzero, shifts upward along the coordinate z, which is associated with the general deviation

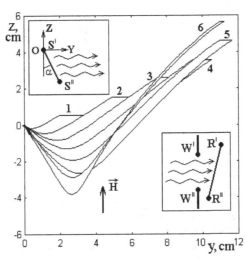

Fig. 5.12. Scheme of a band-pass filter based on an SMSW in a linearly inhomogeneous field. Beam 1 corresponds to a frequency of 2800 MHz, 2 to 2900 MHz, 3 to 3000 MHz, 4 to 3100 MHz, 5 to 3200 MHz, 6 to 3300 MHz. Insets – converter circuits.

increases with increasing from 3250 MHz (curve 4) to 3350 MHz (curve 8), the width is 20–50 MHz, and the maximum value reaches 0.8 (curve 5). As can be seen from Fig. 5.12, this peak is due to beam 6 (as well as others, not shown in the figure, whose frequencies are close to 3300 MHz). Due to its 'overlap' through the beams 5, 4 and partly 3, the maximum AFC corresponding to this beam is isolated and manifests itself at larger values of the coordinate than the maxima caused by beams of lower frequencies (2–5 in Fig.5.12). For the AFC corresponding to $y = 6$ cm (Fig. 5.14c), these tendencies are even more pronounced: the main maximum is now accompanied by an additional one for the greater part of the coordinate change range (curves 9–12), and at $y = 2.5$ cm it remains the only one. At $y = 8$ cm (Fig. 5.14 d), the additional maximum already exceeds the main one in amplitude (curves 13, 14) and remains the only one at $y > 3.5$ cm. These features are associated with an increase in the mentioned 'overlap' of the trajectories of high-frequency beams as the coordinate increases (Fig. 5.12).

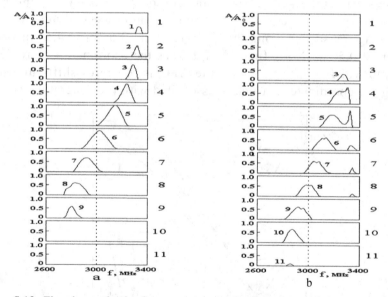

Fig. 5.13. The change in the frequency response when the receiving transducer is moved along the Oz axis at the y-coordinate of its upper end: $a – 2$ cm, $b – 4$ cm. The numbers of the curves correspond to a change in y from –3.5 to 1.5 cm with a step of 0.5 cm. Figures 1–17 correspond to the following values of the z-coordinates of the upper edge W' of the aperture $W'W''$: 1 – –3.5 cm; 2 – –3.0 cm; 3 – –2.5 cm; 4 – –2.0 cm; 5 – –1.5 cm; 6 – –1.0 cm; 7 – –0.5 cm; 8 – 0.0 cm; 9 – 0.5 cm; 10 – 1.0 cm; 11 – 1.5 cm; 12 – 2.0 cm; 13 – 2.5 cm; 14 – 3.0 cm; 15 – 3.5 cm; 16 – 4.0 cm; 17 – 4.5 cm..

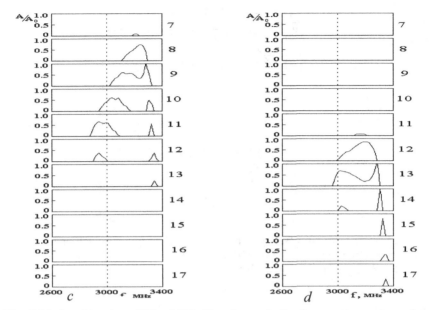

Fig. 5.14. (continuation of Fig. 5.13). The change in the frequency response when the receiving transducer moves along the Oz axis at the y-coordinate of its upper end: c – 6 cm, d – 8 cm. The numbers of the curves are the same as in Fig. 5.13

5.3.3. Valley-type field

Let us now consider the frequency response of a band-pass filter based on SMSWs propagating in a ferrite film magnetized by an inhomogeneous field of the 'valley' type. Let the field be described by formula (3.2), where $H_0 = 437.5$ Oe, $a_0 = 1$, $a_1 = 0$ cm^{-1}, $a_2 = 1/32$ cm^{-2}. In such a field, the SMSW trajectories, depending on the frequency, have a V-, L- or S-shaped character (Section 3.4). The diagram of the device and its main parameters are the same as before (Fig. 5.8). The coordinates of the ends of the emitting transducer: point S' $y = 0$ cm, $z = 2$ cm, point S'' $y = 0.25$ cm, $z = 1.567$ cm. The $R'R''$ conductor of the receiving transducer is parallel to the phase front of the wave at the receiving point.

Figure 5.15 shows the wave beams emanating from the S'S" converter at six frequencies. The shape of the beams is similar to that considered in Section 5.1.

Figure 5.15 shows that frequency beams below ~3150 MHz (1 and 2) are V-shaped, frequency beams above ~3350 MHz (5 and 6) are S-shaped, and beams with frequencies between 3150 and 3350 MHz (3 and 4) are split into two, one of which is V-shaped and the other is S-shaped. The splitting starts from ~ 4 cm, therefore, for

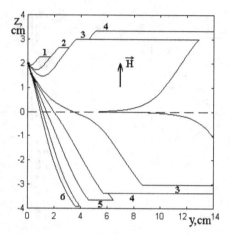

Fig. 5.15. Wave beams in a 'valley'-type field emanating from the emitting transducer at different frequencies: 1 – 3000 MHz, 2 – 3100 MHz, 3 – 3200 MHz, 4 – 3300 MHz, 5 – 3400 MHz, 6 – 3500 MHz.

the analysis of the frequency response, two values of this coordinate were chosen, lying on both sides of the indicated value, namely: 2 cm and 6 cm. The obtained frequency response is shown in Fig. 5.16 ($y = 2$ cm) and 5.17 ($y = 6$ cm). Each series of frequency response corresponds to the movement of the receiving transducer along the axis with a step of 0.5 cm. In this case, the coordinate of the upper end of the window of the receiving transducer W' varies from –3.75 cm to 3.25 cm.

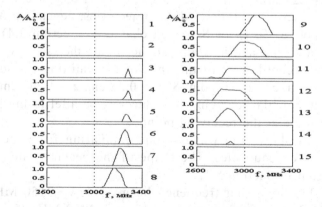

Fig. 5.16. AFC change when moving the receiving transducer along the axis at the -coordinate of its upper end equal to 2 cm. The numbers of the curves correspond to the following values of this coordinate: 1 – –3.75 cm; 2 – –3.25 cm; 3 – –2.75 cm; 4 – –2.25 cm; 5 – –1.75 cm; 6 – –1.25 cm; 7 – –0.75 cm; 8 – –0.25 cm; 9 – 0.25 cm; 10 – 0.75 cm; 11 – 1.25 cm; 12 – 1.75 cm; 13 – 2.25 cm; 14 – 2.75 cm; 15 – 3.25 cm.

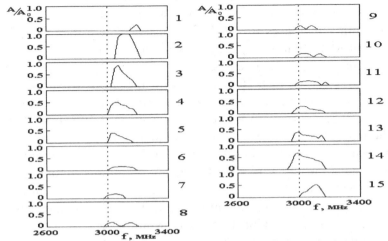

Fig. 5.17. The change in the frequency response when the receiving transducer moves along the Oz axis at the y-coordinate of its upper end equal to 6 cm. The numbers of the curves are the same as in Fig.5.16.

From Fig. 5.16 ($y = 2$ cm), it can be seen that as the window of the receiving transducer moves up, the maximum AFC shifts from a frequency of 3450 MHz (curve 3) to a frequency of 3100 MHz (curve 14), the width of the AFC first increases from 50 MHz (curve 3) to 300 MHz (curves 9–11), and then drops again to 50 MHz (curve 14). The maximum transmittance first increases from 0.45 (curve 3) to 1.0 (curves 8, 9), and then decreases to 0.15 (curve 14). Thus, the frequency of the maximum AFC falls all the time, and its maximum width and maximum transmission coefficient are achieved in the middle of the range of movement of the receiving transducer window, near the axis.

From Fig.5.17, corresponding to $y = 6$ cm, it can be seen that in this case, when the window of the receiving transducer is moved up, the maximum frequency response first moves slightly downward from the frequency of 3380 MHz (curve 1) to the frequency of 3250 MHz (curve 7), after which the frequency response is bifurcated to 3220 MHz and 3350 MHz (curve 8) and in this form almost does not change for some time (curves 9–11), after which the frequency response becomes uniform again and its maximum slowly shifts from 3210 MHz (curve 12) to 3180 MHz (curve 14), after which it tends to 3300 MHz. The frequency response width first rapidly increases from 50 MHz (curve 1) to 120 MHz (curve 2), after which it slowly decreases to 70 MHz (curve 7), when bifurcated, the width of each part remains within 50–70 MHz (curves 8–11), and after merging, it first smoothly increases to 220 MHz (curve 14) and then sharply drops to 100 MHz (curve 15). The maximum transmittance at first rapidly increases from 0.25 (curve 1) to 1.0 (curve 2), after which it gradually decreases to 0.20 (curve 7), in the bifurcation area it remains within 0.15–0.20 (curves 8–11), after which it gradually increases to 0.65 (curve 14) and drops sharply to 0.40 (curve 15). Thus, the main difference between the frequency response from the previous case ($y = 2$ cm) is its splitting when the receiving transducer window is near the Oy axis. The frequency of the maximum AFC decreases all the time, but much less than at $y = 2$ cm, the width of the AFC and its maximum value are large at the edges of the range of movement of the receiving transducer window, that is, far from the axis, and in the splitting region (near) are small.

The described behaviour of the frequency response is clearly explained using Fig. 5.15, from which it can be seen that at $y = 2$ cm the wave beams of different frequencies have not yet had time to noticeably disperse, therefore the behaviour of the frequency response is close to the case of a linear field with a small distance between the transducers (Fig. 5.13 a). At $y = 6$ cm, beams of different frequencies diverge along the sides of the 'valley' very strongly, therefore the behaviour of the frequency response in the case when the receiving transducer is below the 'bottom of the valley' resembles the case of a uniform field (Fig. 5.10), and in the case when it is above 'valley bottom' – the case of a linear field with a large distance between the transducers (Fig. 5.14 c, d, except for the phenomenon of 'overlap'). A decrease in the maximum and width of the frequency response, as well as its splitting when the receiving transducer is near the axis, are due to the rarefaction of the trajectories near the bottom of the 'valley', which coincides with the Oy axis. This rarefaction is clearly illustrated in Fig. 5.3, where the number of trajectories of beams 1 and 3 increases with distance from the Oy axis, while the number of trajectories is extremely small near this axis.

5.3.4. Shaft-type field

Let us now consider the frequency response of a band-pass filter based on SMSWs propagating in a ferrite film magnetized by an inhomogeneous field of the 'shaft' type. Let the field be described by formula (3.2), where $H_0 = 437.5$ Oe, $a_0 = 1$, $a_1 = 0$ cm^{-1}, $a_2 =$

$-1/8$ cm^{-2}. For the SMSW, such a field forms a channel (Section 3.4) parallel to the axis.

The boundaries of the SMSW spectrum corresponding to the field of the top of the 'shaft' (437.5 Oe at $z = 0$) are: 2739 and 3675 MHz. With distance from the top of the 'shaft', the frequency spectrum decreases. The coordinate z values at which the upper limit of the spectrum drops to 2739 MHz are 2.473 cm (the edge of the 'shaft'). At the edges of the 'shaft', the lower frequency of the SMSW spectrum is 1224 MHz.

Let the emitting transducer $S'S''$ be oriented along the Oz axis, and the z-coordinate of the S' point is $+2.5$ cm, and the S'' -2.5 cm, that is, it covers the entire channel. Sections of the emitting transducer located near the top of the 'shaft' ($z = 0$) excite the SMSW in the range of 2739–3675 MHz, and similar sections at the edge of the 'shaft' ($z = \pm 2.473$ cm) – in the range of 1224–2739 MHz. The sections of the transducer, located between the edge and the top of the 'shaft', give the SMSW spectrum in the intermediate range. Let the receiving transducer $W'W''$ also be oriented along the axis Oz, have a length of 0.5 cm, and move along the axis from one edge of the channel to the other. Obviously, due to the symmetry of the problem, it is sufficient to consider the movement of the receiving transducer from one of the channel edges to its middle, since then the whole picture is repeated symmetrically.

Let us first consider the trajectories of the wave beams emanating from the emitting transducer at different frequencies. The corresponding picture is illustrated in Fig. 5.18, which shows the trajectories of different frequencies emanating from the points with coordinates $y = 0$ cm, $z = \pm 2$ cm. Similar trajectories were discussed earlier in Section 3.4.

Figure 5.18 shows two typical types of trajetories. The first (curves 1-4) rush to the top of the 'shaft' the stronger, the higher their frequency, but do not reach the top of the 'shaft', breaking off at a point, the closer to the top of the 'shaft', the higher the frequency of the trajectory. The second ones (curves 5–10) also rush to the top of the 'shaft' the stronger the higher their frequency, but unlike the first ones they reach it, pass right through and then perform periodic oscillating (pseudosinusoidal (section 3.4)) oscillations. The frequency boundary between the trajectories of both types corresponds to the lower boundary of the SMSW spectrum for the top of the 'shaft' – 2739 MHz. Pseudo-sinusoidal trajectories near $y = 2$ cm experience 'pseudofocusing' (Section 3.4). Near $y = 6$ cm

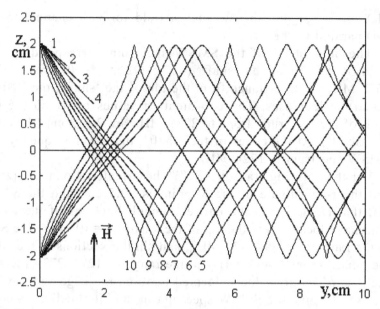

Fig. 5.18. Trajectories in a 'shaft'-type field originating from points with coordinates $y = 0$ cm, $z = \pm 2$ cm for frequencies: 1 – 2000 MHz, 2 – 2200 MHz, 3 – 2400 MHz, 4 – 2600 MHz, 5 – 2800 MHz, 6 – 2850 MHz, 7 – 2900 MHz, 8 – 2950 MHz, 9 – 3000 MHz, 10 – 3050 MHz.

for trajectories 5–9, a second region of 'pseudofocusing' is observed, which is much weaker. Near $y = 4.0$ cm, the trajectories move as far as possible from the top of the 'shaft', that is, their 'defocusing' occurs. At $y > 6$ cm, the trajectories are distributed more or less evenly over the cross section of the 'shaft', which corresponds to the steady state of the wave channel. To identify the strongest differences in frequency response, consider the movement of the receiving transducer along the axis in the immediate vicinity of the emitting transducer, as well as in the regions of pseudofocusing and defocusing.

Figure 5.19 shows the frequency response for the first case, when the distance between the transducers is 0.5 cm ($y = 0.5$ cm), when the -coordinate of the upper end of the transducer is changed from –2.00 to 0.25 cm with a step of 0.25 cm. Curve 1 corresponds to the location of the receiving transducer near the edges of the 'shaft', curve 10 – near its top. The figure shows that when the receiving transducer moves from the edge to the middle of the channel, the frequency response is shifted towards higher frequencies. So, when the receiving transducer is located near the edge of the channel, the

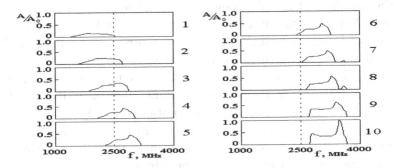

Fig. 5.19. Change in frequency response when moving the receiving transducer along the axis at = 0.5 cm. The numbers correspond to the -coordinate of the upper end of the receiving transducer W' equal to: 1 – –2.00 cm; 2 – –1.75 cm; 3 – –1.50 cm; 4 – –1.25 cm; 5 – –1.00 cm; 6 – –0.75 cm; 7 – –0.50 cm; 8 – –0.25 cm; 9 – 0.00 cm; 10 – 0.25 cm.

frequency response is in the range from 1200 to 2500 MHz, when it is located near the middle of the channel, these boundaries are approximately 2700–3700 MHz. In this case, the average value of the frequency response first increases by 2–3 times (curves 1–3), and then (excluding the high-frequency peak) remains more or less constant (curves 4–10). In the high-frequency region of the frequency response, a rise (one and a half to two times) is observed, the more pronounced the closer the receiving transducer is to the middle of the channel.

The observed features of the frequency response are explained as follows. As mentioned above, the emitting transducer emits wave beams in the range from 1224 to 3675 MHz, and the specific frequency range emitted by each element of the transducer is determined by the field at the location of this element: the further the element is from the middle of the channel, the lower its frequency interval. With a small distance between the transducers, only those wave beams fall onto the receiving transducer, the trajectories of which originate from the sections of the emitting transducer located opposite the receiving transducer with the capture of a certain 'periphery', the width of which is determined by the distance between the transducers and the smaller, the closer to each other the transducers are. Therefore, since the field at the ends of the emitting transducer is less than at its middle, the interval of frequencies passing through the entire system as a whole is higher, the closer the receiving transducer is to the middle of the channel, which explains the first of the above-mentioned features of the frequency response. Further, the tendency of the trajectories to the middle of the channel is manifested the stronger, the further from the middle of the channel is that element of the emitting transducer from which they emanate. Therefore, near the edges of the channel, fewer trajectories fall on the receiving transducer than near its middle, and this circumstance manifests itself noticeably only at the edges of the channel. This explains the increase in the absolute value of the frequency response with its subsequent stabilization when the receiving transducer is moved from the edges of the channel to the middle. High-frequency trajectories tend to the middle of the channel more strongly than low-frequency ones (Fig. 5.18), therefore the receiving transducer 'collects' the trajectories from the longer section of the emitting transducer, the higher the frequency, which explains the peak observed on the frequency response in the high-frequency region (curves 4–10).

Let us now consider the case when the receiving transducer in its movement across the channel passes through the first pseudofocusing region. This situation corresponds to the distance between the emitting and receiving transducers equal to 1.5 cm. The obtained frequency response is shown in Fig. 5.20.

From a comparison of this figure with Fig. 5.19, it can be seen that all AFCs are shifted to the frequency range above 2400 MHz, and when the -coordinate of the W' point changes from –2.50 cm to –1.50 cm (curves 1–3), there is no signal transmission. With further movement of the receiving transducer, the frequency response consists of two regions: low-frequency – from 2450 to 2900 MHz and

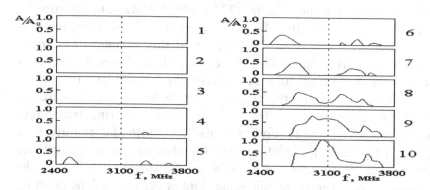

Figure 5.20. AFC change when the receiving transducer moves along the axis at = 1.5 cm. The numbers of the curves correspond to the same positions of the point W / as for Fig.5.19.

high-frequency – from 3100 to 3600 MHz (curves 4–7), which move towards each other and gradually merge, form one wide bandwidth from 2700 to 3900 MHz (curves 8-10), the maximum of which falls on approximately 2900–3100 MHz. There is also a small maximum at the high-frequency end of the frequency response around 3600-3700 MHz.

The observed features of the frequency response are also determined by the trajectories of the SMSW wave beams shown in Fig. 5.18. Indeed, wave beams of frequencies below 2739 MHz, which cannot propagate in the channel, break off before reaching the pseudofocusing region (curves 1–4 in Fig. 5.18), therefore, when the receiving transducer is located outside the pseudofocusing region, the corresponding AFCs are zero (curves 1–3 in Fig. 5.20). A small maximum in the frequency range of 2450–2739 MHz on curves 5–7 is caused by those SMSW trajectories that cannot propagate in the channel, but before the break they come closest to the top of the 'shaft'. When the receiving transducer enters the pseudofocusing region, it is first hit by the wave beams of the lowest and highest frequencies that can propagate in the channel (trajectories 10 and symmetric to 5, corresponding to 3050 MHz and 2800 MHz), which gives two maxima on the frequency response in the region of these frequencies (curves 7–8 in Fig. 5.20). The appearance of several maxima in the frequency range 3200–3600 MHz (curves 4–7) is due to the trajectories emanating from the central sections of the emitting transducer. When approaching the top of the 'shaft', the frequencies of the wave beams entering the receiving transducer approach the middle of the spectrum of frequencies propagating in the channel (trajectories 9 and symmetric to 6, and then 8 and symmetric to 7 in Fig. 5.18), as a result of which the maxima on the frequency response gradually approach, expand, and then merge into a single frequency response (curves 8–10 in Fig. 5.20). The maximum at this frequency response in the region of 2900–3100 MHz (curve 10) is due to the optimal conditions for the hitting of the receiving transducer located in the middle of the channel with those wave beams whose frequencies lie near the middle of the channel transmission spectrum (trajectories 6–9 and symmetric to them in Fig. 5.18). The small maximum at the high-frequency end of the frequency response (3600–3700 MHz) is due to the same reasons as the analogous maximum frequency response in Fig. 5.19. When the receiving transducer is moved along the Oz axis in the second pseudofocusing region ($y = 6.0$ cm), the AFCs change in a similar

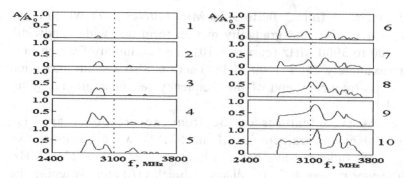

Fig. 5.21. AFC change when the receiving transducer moves along the Oz axis at $y = 4.0$ cm. The numbers of the curves are similar to those shown in Fig.5.19.

way, however, they are much more indented, which is due to the more sparse arrangement of the wave beam trajectories in the second pseudofocusing region compared to the first.

Let us now consider the movement of the receiving transducer across the channel within the same limits, but in the defocusing region, that is, at $y = 4.0$ cm. The obtained AFCs are shown in Fig.5.21.

From Fig.5.21 it can be seen that when the receiving transducer moves from the channel edge to its middle, transmission first appears near the middle of the spectrum of frequencies propagating in the channel (2900–3000 MHz, curves 1-4), and then appears also at higher frequencies (3200-3600 MHz – curves 2-5), after which both transmission regions merge, forming a single wide frequency response, covering all frequencies propagating in the channel (2739-3675 MHz).

The observed form of the frequency response is also explained using the trajectories of the SMSW wave beams shown in Fig.5.18. Indeed, when the receiving transducer is located near the channel edge, the trajectories of only those beams whose frequencies are close to the middle of the full spectrum of the SMSW fall on it (curves 7 and 8 in Fig. 5.18, having frequencies of 2900 and 2950 MHz, respectively). This case corresponds to the frequency response 1 and 2 in Fig.5.21. With further movement of the receiving transducer to the middle of the channel, trajectories 6 and 9, corresponding to frequencies of 2850 and 3000 MHz, fall on it, which causes the appearance of two maxima on the frequency response (curves 3–6 in Fig.5.21). The frequency gap between these maxima is gradually filled by trajectories emanating from the elementary sections of the emitting transducer located closer to the middle of the channel than those shown in Fig. 5.18, which ultimately leads to an expansion of the frequency response over the entire frequency range propagating in the channel (curves 9, 10 in Fig. 5.21). The presence of several not sharply expressed maxima is due to the uneven distribution of trajectories over the channel width.

The AFCs shown in Figs. 5.19–5.21 correspond to the length of the receiving transducer equal to 0.5 cm. With an increase in its length by two or three times, the general character of the AFC changes little, and the irregularity decreases noticeably, which is associated with the weakening of the influence of the local distribution of individual beams.

Thus, as a result of the performed studies, it has been shown that the frequency response of the waveguide channel formed by magnetizing the ferrite film by a 'shaft'-type field depends significantly on the location of the receiving transducer over the channel cross-section, and can vary by more than an order of magnitude (from 100 MHz for curve 1 to (see Fig. 5.21, up to 900 MHz for curves 10 in Figs. 5.19–5.21). In this case, the nature of the frequency response can be either continuous (curves 1–10 in Fig.5.19, 8–10 in Fig.5.20 and others), and discrete (curves 5–7 in Fig.5

5.4. Ampliture–frequency characteristics of a waveguard channel for SMSW formed by the inhomogeneous 'shaft'-type field

In the previous section (5.3.4), the frequency response of the filter based on a film magnetized by a field of the 'shaft' type was considered. At the same time, as shown in the sections 3.4 and 3.8, such a field creates a waveguide channel for MSWs with very low propagation losses. Let us consider here the frequency properties of such a channel. The geometry of the problem is similar to that considered in Section 5.3.4. The trajectories of the SMSW wave beams are shown in Fig. 5.18. Since further we will be interested in the channelizing properties of a field of the 'shaft' type, that is, it is supposed to answer the question of which SMSW frequencies propagate inside the channel and which ones do not, let us assume that both the emitting and receiving transducers are made in the form of straight conductors parallel to the Oz axis, moreover, the receiving transducer covers the entire channel as a whole, and the length of the emitting transducer will vary. In this case, we assume that the emitting transducer is relatively stationary and is always at $y = 0$ cm, and the receiving transducer can be moved along the Oy axis.

5.4.1. Changing the length of the channel

Let us first assume that both transducers completely cover the entire channel, that is, they have a length along the Oy axis from -2.5 cm to $+2.5$ cm, the emitting transducer is at the origin, and the receiving transducer moves along the Oy axis. The resulting frequency response for this case is shown in Fig. 5.22. The curves 1–5 correspond to the displacement of the receiving transducer from $y = 0.5$ cm to $y = 2.5$ cm with a step of 0.5 cm. With an increase of more than 2.5 cm, the frequency response does not change and coincides with curve 5.

Figure 5.22 shows that at any distance between the transducers, the frequency response has a maximum near the frequency of 2739 MHz, on both sides of which there are smooth drops. At distances between transducers less than 2.5 cm, the low-frequency roll-off is smooth, at large distances it is sharp. The shorter the distance, the lower the low frequency roll-off frequency is. The form of the high-frequency falloff is practically independent of the distance between the transducers.

The observed features of the frequency response are explained as follows. As mentioned above (section 5.3.4), the emitting transducer emits wave beams in the range from 1224 to 3675 MHz, and the

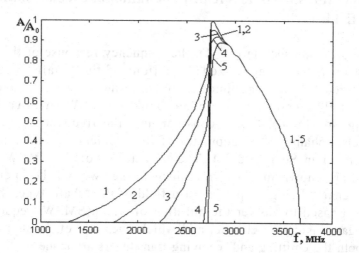

Fig. 5.22. AFC change when moving the receiving transducer along the Oy axis. The curve numbers correspond to the following values of the y-coordinate of the upper end of the transducer: 1 – 0.5 cm, 2 – 1.0 cm, 3 – 1.5 cm, 4 – 2.0 cm, 5 –] 2.5 cm.

frequency of the beam emitted by each individual element of the transducer is determined by the field at the location of this element: the farther it is from the axis of the 'shaft', the lower the frequency. With a small distance between the transducers (0.5 cm), almost all beams leaving the emitting transducer fall on the receiving one. In this case, the frequency response extends from 1224 to 3675 MHz (curve 1). When the receiving transducer is removed from the emitting one (0.5 cm $< y <$ 2.5 cm), the wave beams of lower frequencies, starting near the ends of the emitting transducer, are cut off (see Fig. 5.18), that is, they do not fall on the receiving transducer. In this case, the frequency response from the side of low frequencies narrows (curves 1–4). With an even greater distance from the receiving transducer, all the beams that do not pass through the top of the 'shaft' are cut off, and those passing further propagate in the channel, that is, they always fall on the receiving transducer. This situation corresponds to the frequency range from 2739 to 3675 MHz. In this case, the type of frequency response (curve 5) does not change. The critical distance, starting from which the type of frequency response remains unchanged, is determined by the end point of the 2739 MHz frequency trajectory outgoing from the end of the emitting transducer. This point lies on the axis of the 'shaft' and corresponds to a distance of about 2.5 cm. The maximum AFC near the frequency of 2739 MHz is due to the fact that the beams of this frequency are emitted along the entire length of the converter, the beams of lower frequencies are emitted by the regions near the ends of the the middle. Therefore, near the frequency of 2739 MHz, the largest number of SMSW trajectories emanates from the converter, which gives a maximum at the frequency response.

5.4.2. Changing the channel excitation conditions

Let us now consider the situation when the distance between the transducers is greater than the critical one, that is, only those wave beams that propagate inside the channel reach the receiving transducer. Let the receiving transducer cover the entire channel width and be located relative to the emitting transducer at a y-coordinate equal to 6 cm. Let us consider various variations of the excitation conditions, determined by the length and position of the emitting transducer.

5.4.2.1. Symmetrical arousal

We will first change the length of the emitting transducer symmetrically on both sides of the Oy axis. This case corresponds to Fig.5.23, where the y-coordinate of the receiving transducer is 6 cm, and the z-coordinates of the ends of the emitting transducer vary from 0.25 cm to 2.50 cm with a step of 0.25 cm. The minimum length of the emitting transducer is 0.5 cm (curve 1), the maximum is 5.0 cm (curve 10).

From Fig.5.23 it can be seen that all AFCs lie in the frequency range from 2739 to 3675 MHz, which corresponds to wave beams passing through the top of the 'shaft'. From the side of low frequencies, all the frequency response breaks off abruptly, and from the side of high frequencies they fall off smoothly, and with their drops, all curves (1–9) fit into the frequency response corresponding to the maximum length of the emitting transducer (curve 10), and between these drops they have a flat horizontal section, located the higher, the longer the length of the emitting transducer. This type of frequency response is explained by the fact that wave beams emanating from the entire length of the converter participate in the formation of the frequency response corresponding to the maximum

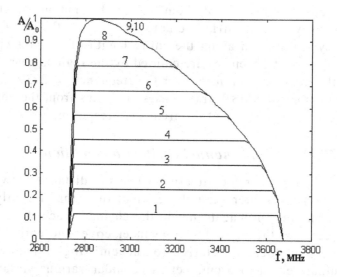

Fig, 5.23. The change in frequency response when changing the length of the emitting transducer is symmetric about the Oy axis. The numbers of the curves correspond to the following length: 1 – 0.25 cm, 2 – 0.50 cm, 3 – 0.75 cm, 4 – 1.00 cm, 5 – 1.25 cm, 6 – 1.50 cm, 7 – 1.75 cm, 8 – 2.00 cm, 9 – 2.25 cm, 10 – 2.50 cm.

length of the emitting transducer (curve 10), and with a decrease in the length of the emitting transducer, the sections located near its ends are gradually eliminated, which leads to a decrease in frequency response. The horizontalness of the central part of the frequency response reflects the fact that the amplitude of the wave beam emanating from each elementary section of the emitting transducer does not depend on the frequency within the limits determined by the field at the location of this element. The general smooth falloff of the frequency response from the high-frequency side, corresponding to the maximum length of the emitting transducer (curve 10), is due to the same reasons as the high-frequency falloff of the frequency response in Fig.5.18.

5.4.2.2. Asymmetrical excitement

Let us now consider the case when the length of the emitting transducer changes asymmetrically relative to the axis. This situation corresponds to Fig.5.24, where the coordinate of the lower end of

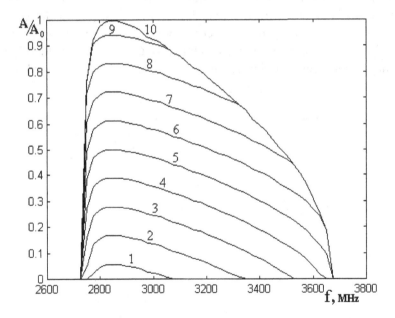

Fig. 5.24. The change in the frequency response at an asymmetric change in the length of the emitting transducer relative to the Oy axis. The numbers of the curves correspond to the following values of the z-coordinates of the upper end of the transducer: 1 – –2.0 cm, 2 – –1.5 cm, 3 – –1.0 cm, 4 – –0.5 cm, 5 – 0.0 cm, 6 – 0.5 cm, 7 – 1.0 cm, 8 – 1.5 cm, 9 – 2.0 cm, 10 – 2.5 cm.

the emitting transducer is fixed and equal to −2.5 cm, and the upper end changes from −2.0 cm to +2.5 cm with a step of 0.5 cm.

The obtained frequency response in many respects resembles the previous case, but now they all have a maximum near 2820 MHz, after which they smoothly fall towards high frequencies. The general appearance of the curves is explained by the action of the same factors as for Fig. 5.23. Here, however, the lower end of the emitting transducer (at z = −2.5 cm) is rigidly fixed, that is, the wave beams emanating from it, corresponding to the smaller field and the lowest frequencies propagating in the channel, participate in the formation of the frequency response at any length of the transducer. For this reason, all AFCs in the low-frequency region (near 2820 MHz) have a maximum.

5.4.2.3. Transverse shift of the emitting transducer

Let us now consider such a case when the length of the emitting transducer does not change, but the whole of it, as a whole, moves along the Oz axis. Due to the symmetry of the problem, it is enough to restrict the movement of the transducer only from one of the edges of the 'shaft' to its top. This case corresponds to Fig. 5.25, where the y-coordinate of the receiving transducer is 6 cm, the length of the emitting transducer remains fixed equal to 0.5 cm, the z-coordinate

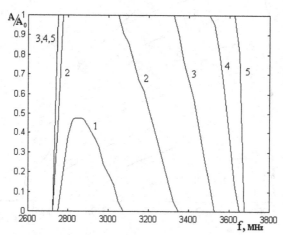

Fig. 5.25. The change in the frequency response when moving the emitting transducer with a length of 0.5 cm along the Oz axis at y = 6 cm. The numbers of the curves correspond to the following values of the coordinates of the endstransducer: 1 − z_1 = −2.5 cm and z_2 = −2.0 cm; 2 − −2.0 cm and −1.5 cm; 3 − −1.5 cm and −1.0 cm; 4 − −1.5 cm and −0.5 cm; 5 − −0.5 cm and 0.0 cm.

of the lower end varies from −2.5 cm to −2.0 cm, and the upper end - from −2.0 cm to 0.0 cm with a step of 0.5 cm.

From Fig.5.25 it can be seen that when the emitting transducer is located near the edge of the 'shaft', the frequency response has a maximum near the frequency 2820 MHz, and smooth drops on both sides of this frequency (curve 1). When the transducer is moved from the edge of the 'shaft' to its top, the frequency response expands, and the maximum value first increases, and then, starting from curve 2, is limited to a constant level, as a result of which a flat section appears at the top of the frequency response (curves 2–5), which As the transducer moves to the top of the 'shaft', it gradually expands, reaching the full width of the spectrum of frequencies that can propagate in the channel (curve 5), which is from 2739 to 3675 MHz. This type of observed AFC is explained by the fact that only those wave beams are involved in the formation of the AFC corresponding to the location of the emitting transducer near the edge of the 'shaft', the frequencies of which correspond to the field near this edge, that is, they are rather low. Curve 1 corresponds to this case. When the emitting transducer is moved to the top of the 'shaft', the field at its location increases, as a result of which beams of higher frequencies come into play, which leads to expansion of the frequency response (curves 2–5). Due to the fixed length of the emitting transducer, the number of wave beams emanating from its elementary sections and able to propagate in the channel does not change when the frequency changes starting from curve 2, as a result of which a flat section is observed on the frequency response shown by curves 2–5. The absolute value of the frequency response within this section is determined, as in the previous cases, only by the length of the emitting transducer.

Thus, as a result of the performed studies, it has been shown that the frequency response of the waveguide channel formed by magnetizing the ferrite film by a 'shaft'-type field significantly depends on the length and location of the emitting transducer over the channel cross section, and the total frequency response varies by more than an order of magnitude (from 200 MHz for curve 1 in Fig. 5.25 to 2450 MHz for curve 1 in Fig. 5.22).

5.5. Amplitude–frequency characteristics of the transmission line to the SMSW at an arbitrary orientation of the magnetizing field

In the previous sections (5.3 and 5.4), the frequency response was considered for the case of a constant field direction in the film plane. Let us now consider what happens to the frequency response when the direction of the field changes. As an example, we take a band-pass filter consisting of two rectilinear SMSW transducers superimposed on a film magnetized in the plane by a uniform field of an arbitrary direction.

5.5.1. The general geometry of two variants of the location of the transducers: mutually opposite and mutually shifted

The geometry of the problem is shown in Fig. 5.26. The plane of the drawing coincides with the plane of the film and the plane of the coordinate system. The emitting $S'S''$ and the receiving $W'W''$ converters are made in the form of segments of straight conductors of the same length, and these conductors are always parallel to each other and to the Oz axis, and the distance between them along the Oy axis can vary. The magnetizing field is oriented in the plane of the film, the angle between the direction of the field and the Oy axis is indicated by α.

In this geometry, the $S'S''$ emitting converter emits a SMSW beam, the direction of which is determined by the field orientation and the SMSW dispersion law, and the total width is determined by the length of the emitting converter. Due to the homogeneity of the field, the trajectories of individual wave beams emerging from the emitting transducer are straight lines parallel to each other, schematically shown in the figure by a dotted line (dashed lines). The angle between the direction of propagation of these wave beams and the Oy axis is indicated by ψ. In the same figure, dotted and dashed-dotted lines show the trajectories of wave beams at other angles α or frequencies.

Let us consider filters of two types – the first, in which the receiving transducer is located exactly opposite the emitting one (Fig. 5.26 a) and the second, in which the receiving transducer is shifted along the axis relative to the emitting one by a distance equal to the length of the emitting transducer (Fig. 5.26 b). For each type of filter, we set two different distances between the transducers:

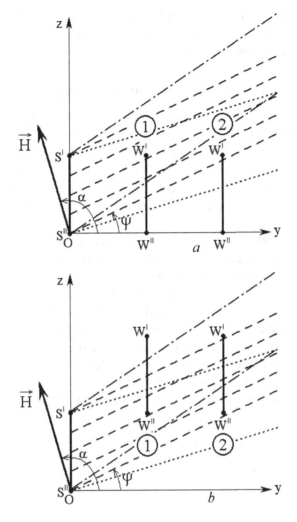

Fig. 5.26. General geometry of the problem. a – filter of the first type, mutually opposite geometry; b – filter of the second type, mutually shifted geometry. Dotted, dashed and dashed-dotted lines are the positions of the wave beam at different angles α.

equal to the length of the emitting (or receiving) transducer (1) and twice as long (2). For brevity, we will call the distance between the transducers equal to their single length – 'small', and the doubled length – 'large'. Note that the geometry of the transducers presented here was selected only for convenience and maximum clarity, since, obviously, the used method of wave beams in this regard has no limitations. As before, we take the film magnetization equal to 1750 G, and its thickness 15 μm. Field $H = 437.5$ Oe. In this case, the SMSW spectrum lies within 2739–3675 MHz. Let the length of the transducers be 0.5 cm and the distance between them is 0.5 cm (small) or 1.0 cm (large).

5.5.2. Filtration of the first type, mutually opposite geometry

Let us first consider the frequency response for a filter of the first type with a small distance between the transducers (Fig. 5.26 a – 1). Due to the symmetry of the problem with respect to the Oy axis, it is sufficient to consider only $\alpha > 90°$, since at $\alpha < 90°$ the whole picture is symmetrically repeated. The resulting frequency response is shown in Fig. 5.27, where the angle α changes from 90° to 110° in 2° steps.

Figure 5.27 shows that with an increase, the low-frequency boundary of the frequency response remains in place, and the high-frequency boundary decreases, which leads to a narrowing of the frequency response down to zero. In this case, the top of the frequency response, being at first almost horizontal, gradually turns into an oblique, leading to a change in the shape of the frequency response from rectangular to triangular, and the maximum value of the frequency response decreases from one to zero. So, at $\alpha = 90°$ (curve 1) the lower boundary of the frequency response is 2739 MHz, and the upper one is 3675 MHz, that is, the frequency response

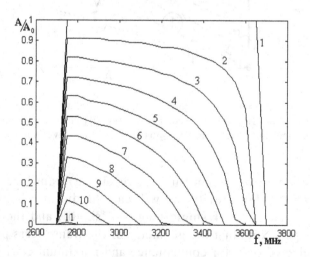

Fig. 5.27. AFC for a filter of the first type with a small distance between the transducers for different field orientations. 1 – 90°; 2 – 92°; 3 – 94°; 4 – 96°; 5 – 98°; 6 – 100°; 7 – 102°; 8 – 104°; 9 – 106°; 10 – 108°; 11 – 110°.

covers the entire spectrum of the SMSW and is almost rectangular. Its FWHM is 930 MHz. With an increase of 2° (curve 2), the low-frequency limit of the frequency response is still 2739 MHz, and the upper one decreases to 3650 MHz. At the same time, the AFC, although it decreases somewhat towards high frequencies, remains close to rectangular, and its maximum value decreases to 0.8–0.9 from the same value at $\alpha = 90°$. The frequency response width is reduced to 900 MHz. A further increase leads to a sharper narrowing of the frequency response and the transition from rectangular to triangular, and the maximum frequency response always remains near the lower boundary of the SMSW spectrum. At $\alpha = 106°$ (curve 9), the frequency response becomes almost triangular, its lower boundary remains close to 2739 MHz, and the upper one is 3095 MHz. In this case, the maximum frequency response falls at 2780 MHz, the height of the frequency response at the maximum is 0.22 of the maximum at $\alpha = 90°$, and the frequency response width approaches 160 MHz. At $\alpha = 110°$ (curve 11), the frequency response narrows to 40–50 MHz, and its value becomes less than 0.01 of the maximum at $\alpha = 90°$. With a further increase in the angle, there is no signal transmission from the emitting transducer to the receiving one.

The observed features of the frequency response are explained as follows. The emitting converter emits wave beams in the range of 2739–3675 MHz, and due to the homogeneity of the field, the trajectories of the wave beams of all frequencies are rectilinear. The emitting transducer is straight, therefore, the individual beams at each given frequency are parallel to each other, and the direction of their propagation (angle ψ in Fig. 5.26) is determined by the specific value of the angle α and the wave frequency. The larger the angle ψ, the larger the higher, the higher the SMSW frequency, the higher the angle being fixed. At $\alpha = 90°$, wave beams emanating from the entire length of the emitting transducer at all frequencies propagate parallel to the axis ($\psi = 0°$), as a result of which all without exception fall on the receiving transducer, which gives a rectangular frequency response of the maximum value (curve 1). At $\alpha > 90°$ wave beams of the same frequency, while remaining parallel to each other, do not propagate exactly along the Oy axis, but at an angle ψ to it, which is the greater, the higher the frequency. As a result, the frequency response is formed primarily due to beams of lower frequencies, which results in its triangular character. On the other hand, with an increase of α, the angle ψ also always increases, therefore, part of the beams of even the lowest frequencies

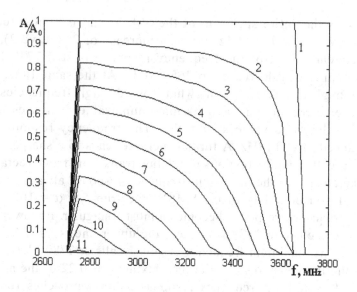

Fig. 5.28. Frequency response for a filter of the first type with a large distance between transducers for different field orientations. 1 – 90°; 2 – 92°; 3 – 94°; 4 – 96°; 5 – 98°; 6 – 100°.

pass by the receiving transducer, leading to a general decrease in the maximum magnitude of the frequency response. Beginning with a certain value α (somewhat larger than 110°), beams of any frequencies cease to reach the receiving transducer, which leads to a drop in the frequency response to zero.

Similar frequency response for a filter of the first type with a large distance between the transducers (Fig.5.26 a – 2) are shown in Fig.5.28. The angle step and curve numbers are the same as before.

Figure 5.28 shows that the general view of the frequency response for this case is similar to the previous one, however, the range of angles at which the filter passes the signal is approximately halved (from 90°–110° to 90°–100°). In this case, the frequency response, plotted with the same step as in Fig. 5.27 (2°), are now located much less frequently, that is, the criticality of the frequency response to the field orientation increases significantly. The observed behavior of the frequency response in comparison with the previous case is quite obviously explained by the course of the wave beams shown in Fig. 5.26 a, taking into account the increase in the distance between the transducers.

5.5.3. Filtering of the second type, mutually shifted geometry

Let us now consider the frequency response for a filter of the second type with a small distance between the transducers (Fig. 5.26 b – 1). Since in this case the receiving transducer is shifted along the Oz axis relative to the emitting one, there is no symmetry of the frequency response in the angle α and it is necessary to consider the change in this angle relative to the Oy axis in both directions. The resulting frequency response is shown in Fig. 5.29, where the angle α varies from 92° to 122° in 2° steps.

Figure 5.29 shows that, as in the previous cases, with an increase, the low-frequency boundary of the frequency response remains in place, and the high-frequency boundary decreases, which leads to a narrowing of the frequency response down to zero. In this case, the shape of the frequency response changes significantly. There is a critical value of the angle $\alpha = 110°$, below which the frequency response from the low-frequency side increases first stepwise, and then gently up to a maximum, after which it drops to zero. With an increase in the range from 92° to 110°, the low-frequency jump grows from 0.1 at $\alpha = 92°$ (curve 1) to unity at $\alpha = 110°$ (curve 10), the length of the flat section decreases from ~800 MHz (curve 1) to zero (curve 10), the length of the high-frequency falloff increases

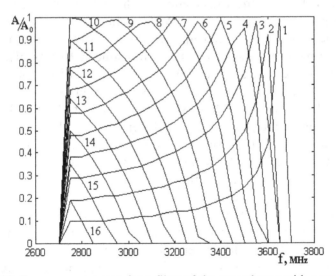

Fig. 5.29. Frequency response for a filter of the second type with a small distance between transducers for different field orientations. 1 – 92°; 2 – 94°; 3 – 96°; 4 – 98°; 5 – 100°; 6 – 102°; 7 – 104°; 8 – 106°; 9 – 108°; 10 – 110°; 11 – 112°; 12 – 114°; 13 – 116°; 14 – 118°; 15 – 120°; 16 – 122°.

from 50 MHz (curve 1) to 600 MHz (curve 10). The frequency response width at half height increases from 75 MHz (curve 1) to 500 MHz (curve 10). In this case, the frequency of the maximum decreases from 3650 MHz (curve 1) to 2750 MHz (curve 10), and the maximum magnitude of the frequency response in the entire interval $92° < \alpha < 110°$ remains close to unity. With an increase above the critical value of $110°$, the frequency response from the side of low frequencies at first still increases in a jump, reaches a maximum, after which it gradually decreases to zero. With an increase in the range from $110°$ to $122°$, the value of the low-frequency jump falls from unity (curve 10) to 0.1 (curve 16), while the maximum frequency response, always remaining near the beginning of the SMSW spectrum (2740–2750 MHz), decreases from unity (curve 10) to 0.1 (curve 16). The frequency of the frequency response to zero decreases from 3300 MHz (curve 10) to 2850 MHz (curve 16). In this case, the width of the frequency response decreases from 500 MHz (curve 10) to 90 MHz (curve 16).

The observed features of the frequency response are explained similarly to the case of a filter of the first type, however, here, when the field is oriented exactly along the axis, not a single wave beam emerging from the emitting transducer falls on the receiving one, since the upper (according to Fig. 5.26 b) edge of the emitting transducer S' falls to the lower edge of the receiving W'', and the beams go parallel to the Oy axis. With an increase in the angle α, all wave beams are deflected upward, as a result of which some of them begin to fall on the receiving transducer. Beams of higher frequencies are deflected more strongly, as a result of which a high-frequency maximum is observed on the AFC (curves 1–9). At the same time, the beams of the highest frequencies first of all go beyond the upper edge of the receiving transducer, as a result of which the upper frequency boundary of the frequency response in the entire range of increase gradually decreases (curves 2–16). The deviation of the beams from the horizontal axis occurs the stronger, the larger the angle. The critical value $\alpha = 110°$ corresponds to the situation when the wave beam of the lowest frequency of the SMSW spectrum (~2740 MHz) emerging from the lower edge of the emitting transducer (S') falls on the lower edge of the receiving transducer (W''). At $\alpha < 110°$, there is always such a frequency at which all the beams outgoing from the emitting transducer exactly fall on the receiving one, as a result of which the maximum AFC approaches

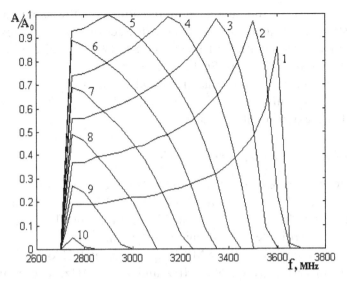

Fig. 5.30. Frequency response for a filter of the second type with a large distance between transducers for different field orientations. 1 – 92°; 2 – 94°; 3 – 96°; 4 – 98°; 5 – 100°; 6 – 102°; 7 – 104°; 8 – 106°; 9 – 108°; 10 – 110°.

unity (curves 1–10). At> 110o such a frequency is absent, therefore the maximum AFC decreases (curves 11–16).

Similar frequency response for a filter of the second type with a large distance between the transducers (Fig.5.26 a–2) are shown in Fig.5.30. The angle step and curve numbers are the same as before

The observed features in comparison with the previous case are similar to the features of the frequency response for a filter of the first type in the transition from a small distance between transducers to a large one. Here, there is also a narrowing of the permissible interval from 92°–122° to 92°–110°, and the frequency response, plotted with the same step as in Fig. 5.29 (2°), like Fig. 5.28, are located here much less frequently. All of the above features are also explained by the straightness of propagation of wave beams in combination with an increase in the distance between the transducers.

5.6. Experimental study of SMSW beams of finite width and amplitude–frequency characteristics

To test the theoretical concepts developed above, an experimental study of the propagation of wide wave beams of SMSWs, as well as the amplitude-frequency characteristics in nonuniformly magnetized ferrite films, was carried out. The experiments were carried out on the same setup, in the same geometry and with the same samples as the study of trajectories described in Section 3.8. As a result of studies of the propagation of beams of SMSWs of finite width, the main provisions outlined in Section 5.1 were confirmed. The study of the amplitude–frequency characteristics also confirmed the theoretical conclusions presented in Sections 5.3 and 5.4. Let us consider briefly some of the experimental results.

5.6.1. Linearly inhomogeneous field

The case of a linear field was investigated in the experimental situation described in Section 4.8 (Fig. 4.25). The length of the emitting transducer was 0.4 cm, its centre was located at the point with coordinates $y = 0$ cm, $z = 3.5$ cm, and the angle φ was 32°. The boundaries of the SMSW spectrum at the location of the emitting converter were 3693 MHz and 4479 MHz. The converter excited the SMSW in the 3700–4200 MHz frequency band (at −40 dB level from the maximum near 3750 MHz). The change in the frequency response was investigated when the receiving transducer moved along the axis at a fixed value of the coordinate. In the first position, $y = 0.5$ cm was set. In this case, for $z = 2.5$ cm, the frequency response boundaries at a level of −10 dB from the maximum were 4000–4100 MHz. At $z = 2.8$ cm, the same values were 3750-4150 MHz. At $z = 3.1$ cm, they were equal to 3700–4100 MHz. The second position corresponded to $z = 1.5$ cm. In this case, for $z = 2.3$ cm, the frequency response boundaries were 3960–4120 MHz. At $z = 2.7$ cm, the same values were 3860–3990 MHz. At $z = 3.2$ cm they were equal to 3760–3920 MHz. In the third position, $z = 2.5$ cm. In this case, for $z = 2.7$ cm, the frequency response boundaries were 3960–4120 MHz. At $z = 3.3$ cm, the same values were 3860–3990 MHz. At $z = 3.7$ cm, they were equal to 3800–3910 MHz.

Thus, when the receiving transducer was moved in the positive direction of the Oz axis, in all cases there was a shift in the frequency response towards low frequencies, the frequency response width was 100÷400 MHz, and its maximum was in the middle of the permissible range of variation of the z-coordinate of the receiving transducer. All these features fully correspond to the behaviour of the frequency response in a linear field, described in Section 5.3.2, Figs. 5.13 and 5.14. The detachment from the AFC of a separate peak in the high-frequency region, noted in Figs 5.13 b and 5.14 a, b, was also observed near 4200 MHz, and its amplitude was about −20 dB from the maximum, and its width was 50 MHz. Such a small amplitude was due to a drop in the efficiency of excitation of the SMSW by the emitting transducer with increasing frequency due to a decrease in the SMSW length.

5.6.2. Valley type field

The case of a 'valley'-type field was studied in an experimental situation similar to that described in Section 4.8, (Fig. 4.26), however, here, for a more visual identification of the splitting of the wave beam, the centre of the emitting transducer was located exactly at the bottom of the 'valley', and the origin of the coordinate system was aligned with the centre of the transducer. The angle was chosen equal to −5°. The boundaries of the SMSW spectrum at the location of the emitting converter were 3031 MHz and 3951 MHz. The converter excited the SMSW in the frequency band 3050–3840 MHz (at a level of −40 dB from the maximum near 3100 MHz). In the experiments, a bifurcation of the beam was observed, similar to that shown in Figs. 5.3 and 4.26, which is more pronounced, the lower the SMSW frequency. So, for a frequency of 3790 MHz, the beam boundaries at a level of −10 dB from the maximum at $y = 0.5$ cm were 0.1 cm and 0.7 cm, at $y = 1.0$ cm 0.7 cm and 1.3 cm, at $y = 1.5$ cm 1.1 cm and 1.9 cm, i.e. the beam width increased slightly, but there was no bifurcation. At the same time, for a frequency of 3200 MHz, the same values at $y = 0.5$ cm, when the bifurcation was still absent, were −0.4 cm and 0.5 cm, at $y = 1.0$ cm, when the bifurcation was just beginning to appear −0.5 cm and 0.9 cm, and at $y = 1.5$ cm, when the beam was already clearly bifurcated, the boundaries of individual beams were equal to −0.8 cm and −0.2 cm for the first, and 0.2 cm and 1.4 cm for the second. The study of the frequency response, carried out while moving the receiving transducer along the Oz axis, showed that, despite the splitting of the beam, the frequency response always remains the same and for a constant value $y = 0.5$ cm, at $z = -0.3$ cm, it is within 3100–3700 MHz, at $z = 0.0$ cm – within 3200–3750 MHz, and at $z = 0.5$ cm - within 3300–3800 MHz. At $y = 1.5$ cm, the frequency response at $z = -0.5$ cm lies within 3100–3200 MHz, at $z = 0.0$ cm, there is practically no signal transmission, at $z = 0.5$ cm, the signal reappears and the frequency response boundaries are 3100–3200 MHz, at $z = 1.0$ cm its boundaries are equal 3200 and 3600 MHz, and at $z = 1.5$ cm –

3400 and 3800 MHz. The observed shift of the frequency response towards high frequencies is due to the negative value of the initial angle. A drop in the frequency response to zero at $y = 1.5$ cm and $z = 0.02$ cm corresponds to the location of the receiving transducer in the gap between two separate wave beams into which the original beam was split (similar to curves 7–10 in Fig. 5.17). This behaviour of wave beams and frequency response in a 'valley'-type field corresponds to that described in Section 5.3.3.

5.6.3. Shaft-type field

The case of a 'shaft'-type field was investigated in the experimental situation described in Section 4.8 (Fig. 4.27). The boundaries of the SMSW spectrum at the top of the 'shaft' were equal to 2394 and 3486 MHz. Investigation of the signal distribution over the channel cross section showed that the profile of the full wave beam propagating in the channel, starting from the coordinate $y = 2$ cm, is not critical to the value of the initial angle within at least 20°, as well as to the value of the z-coordinate of the center of the emitting transducer within not less than 0.8 cm (Fig. 3.27). In all these cases, the -10 dB beam width was 0.6–0.8 cm. At $y < 2$ cm, an oscillation of the beam width was observed, similar to that shown in Fig. 5.4. The maximum oscillation amplitude of the SMSW beam width relative to the stationary value (at $y > 2$ cm) at a frequency of 2570 MHz was ~10%, and at a frequency of 3150 MHz ~50%.

The most important property of a 'shaft' type field is the suppression of any beam divergence, as a result of which the wave loss during propagation decreases significantly (Section 3.8). So, in a uniform field with an intensity equal to the field of the top of the 'shaft', at $\varphi = 0°$, the width of the SMSW beam increases as it propagates along the Oy axis, the faster the lower the MSSW frequency. At $y = 3$ cm at a frequency of 3150 MHz, it increases 1.4 times compared to the initial one, and the wave attenuation exceeds 27 dB, while in a 'shaft'-type field, the beam width remains 0.6 cm, and the attenuation is less than 19 dB. At the same value at a frequency of 2570 MHz, the beam width in a uniform field increases by a factor of 2.5, and the attenuation exceeds 17 dB, while in a 'shaft'-type field, the beam width remains 0.6 cm and the attenuation is less than 3 dB.

Thus, the experimentally observed behaviour of wide wave beams and the AFC in a 'shaft'-type field also corresponds to the theoretical principles developed in Sections 5.3 and 5.4.

Conclusions for chapter 5

The main issues discussed in this chapter are as follows.

1. The features of the propagation of wide wave beams of surface magnetostatic waves (SMSW) in nonuniformly magnetized ferrite films are considered. It is shown that under these conditions a wide wave beam is spatially transformed: its width and transverse profile change, and splitting can occur. When propagating over the plane of the film, magnetized by a linear field, the wave beam is preserved the better, the higher its frequency and the smaller the initial angle between the wave vector and the normal to the direction of the field. In the case of a 'valley'-type field, the wave beam is split into two clearly defined parts corresponding to different sections of the transducer, and the part corresponding to the section of the transducer closest to the bottom of the 'valley' is better preserved. In a 'shaft'-type field, the wave beam at the beginning of the path undergoes several constrictions (pseudofocusings), the depth of which decreases as the beam propagates and its width approaches the initial one, that is, in general, the wave beam is preserved the better, the greater the distance traveled by it.

2. The formation of the amplitude–frequency and phase–frequency characteristics (AFC and PFC) of transmission lines on magnetostatic waves is considered. To calculate the frequency response and phase response of such systems, the method of partial wave beams of MSWs, propagating in a nonuniformly magnetized ferrite film in accordance with the laws of geometric optics, is proposed. The method includes the analysis of the trajectories of individual wave beams emanating from the discrete sections of the emitting transducer, followed by summation of the resulting output signal at the receiving transducer. The method is applicable for any shape and mutual arrangement of transducers, for cases of a uniformly and inhomogeneously magnetized film of any shape and with any inhomogeneity of parameters, and also allows the consideration of transmission lines based on multilayer ferrite structures. A computer-based calculation algorithm has been developed, suitable for analyzing the frequency response and phase response of transmission

lines on ferrite films magnetized by a uniform and non-uniform field, linear, of the 'valley' and 'shaft' type.

3. The frequency response of transmission lines on ferrite films magnetized by fields of various configurations is considered. For the case of a uniform field, it is shown that all AFCs have a bell-shaped form, and with the distance of the receiving transducer from the emitting one, the AFC width decreases. With the removal of the receiving transducer from the axis of the transmission line, the center frequency of the frequency response shifts towards high frequencies, and the width of the frequency response decreases. With the removal of the receiving transducer from the emitting one along the axis of the transmission line, the center frequency of the AFC shifts towards low frequencies, and the width of the AFC first increases and then decreases. The maximum width of the frequency response corresponds to the full width of the SMSW spectrum, the minimum – 2–5% of the maximum.

4. For the case of a linearly inhomogeneous field, it is shown that when the receiving transducer is moved perpendicular to the axis of the transmission line, there are two characteristic options for changing the frequency response: the first corresponds to a small distance between the transducers, until the wave beams are turned towards a larger field; the second corresponds to a large distance between the transducers, after the said rotation. In the first case, as the receiving transducer approaches the axis of the transmission line, the frequency response, while maintaining a bell-shaped appearance, shifts towards low frequencies, gradually expanding. In the second case, with the same movement of the receiving transducer, the frequency response is split into two parts, one of which, low-frequency, behaves similarly to the first case, and the other, high-frequency, shifts towards high frequencies, gradually narrowing. The maximum width of the low-frequency part of the AFC approaches the full width of the SMSW spectrum, the maximum width of the high-frequency part is 5–10 times less.

5. In the case of a 'valley'-type field, with a similar movement of the receiving transducer, there are also two characteristic variants of changing the frequency response: the first corresponds to a small distance between the transducers, before the splitting of the wave beam into two, the second – to a large distance after splitting. In the first case, as the receiving transducer approaches the axis of the transmission line, the frequency of the maximum AFC decreases all the time, and its maximum width and maximum transmission

coefficient are achieved when the receiving transducer is located near the axis of the transmission line. In the second case, with the same movement of the receiving transducer, the frequency of the maximum AFC also decreases with time, but more slowly than in the first case, and when the receiving transducer is located near the axis of the transmission line, the frequency response width and transmission coefficient are minimal.

6. In the case of a field of the 'shaft' type, with a similar movement of the receiving transducer, there are also two characteristic options for changing the frequency response: the first corresponds to a small distance between the transducers, when the system has trajectories that do not reach the top of the 'shaft', the second – to a large one, at which only periodic trajectories passing through the top of the 'shaft'. In the first case, as the receiving transducer approaches the axis of the transmission line, the frequency of the maximum AFC increases, and its width approaches the full width of the SMSW spectrum. The lowest frequency boundary of the AFC is determined by the beginning of the SMSW spectrum in the field corresponding to the edge of the 'shaft', the upper – by the end of the SMSW spectrum in the field corresponding to the top of the 'shaft'. In the second case, with the same movement of the receiving transducer, the AFC width remains close to the full spectrum width all the time, the AFC is noticeably indented, and its frequency boundaries are determined by the boundaries of the SMSW spectrum in the field corresponding to the top of the 'shaft'.

7. The frequency properties of the waveguide channel for the SMSW formed by a field of the 'shaft' type are considered. It is shown that in the case when both transducers completely cover the entire channel, at any distances between them, the frequency of the maximum AFC is close to the lower frequency boundary of the SMSW spectrum corresponding to the field of the top of the 'shaft'. On both sides of the maximum, smooth drops are observed, and at small distances between the converters, the low-frequency drop is smooth, at large distances it is sharp, and the form of the high-frequency drop does not depend on the distance between the converters. The full bandwidth at large distances between the transducers approaches the width of the SMSW spectrum in the field of the 'shaft' top. A change in the length of the emitting or receiving transducers, as well as their relative position leads to a change in the frequency response, mainly from the high-frequency side, the position of the maximum, as a rule, remains near the lower

frequency boundary of the SMSW spectrum corresponding to the field of the top of the 'shaft', and the total the frequency response width changes by more than an order of magnitude.

8. The frequency response of the transmission line to the SMSW is considered for an arbitrary orientation of the magnetizing field. For the case of two identical rectilinear parallel-oriented transducers, it is shown that the form of the frequency response strongly depends on the orientation of the field direction relative to the transducers and can vary from rectangular, covering almost the entire spectrum of SMSW, to triangular with a width of an order of magnitude smaller. In this case, the frequency of the maximum AFC can be located anywhere in the SMSW spectrum, and the transmittance in the region of the maximum changes by more than an order of magnitude.

9. An experimental study of the propagation of wide beams of SMSWs, as well as the frequency response of transmission lines on inhomogeneously magnetized ferrite films, has been carried out. The main configurations of the field are considered: linear, of the 'valley' and 'shaft' type. A good agreement of the experimental results with the conclusions of the theory was revealed. So, in a linear field, when the receiving transducer moved perpendicular to the axis of the transmission line, both characteristic variants of the frequency response change were observed: corresponding to a small and large distance between the transducers. In both cases, a shift in the frequency response toward low frequencies was observed, and at a large distance, a detachment from the frequency response of a separate peak in the high-frequency region was observed. In a field of the 'valley' type, a splitting of the wave beam into two, two frequency regions of the location of the frequency response corresponding to these beams and a sharp drop in the transmittance when the receiving transducer was located near the bottom of the valley were observed. In a field of the 'shaft' type, clearly expressed waveguiding properties were observed, manifested in the formation of a channel for the SMSW and independence of the frequency response from the location of the receiving transducer along the length of this channel, complete suppression of the dispersive divergence of the wave beam and record low losses of SMSW during propagation in the channel, constituting dB/cm.

6

Amplitude–frequency properties of transmission lines on magnetostatic waves taking into account the phase run

This chapter presents the results of a study of the amplitude-frequency characteristics of transmission lines on the SMSW, formed by wave beams of finite width. On the basis of taking into account the phase incursion of individual components of the full wave beam, a detailed account of the phase relations providing signal interference at the receiving transducer is made. The deformation of the plane wavefront of surface magnetostatic waves propagating in ferrite films magnetized by a linearly inhomogeneous field is investigated. The experimentally observed features of the frequency response and phase response of real transmission lines are noted, and recommendations for their optimization are given.

In this chapter, we used the results of works [118, 274, 275, 282, 397, 582, 585, 615, 616, 617, 627–633]. The rest of the links are given in the text.

6.1. General characteristics of typical transmission lines to SMSW

Let us first consider the main general characteristics of typical transmission lines on SMSW. The traditional transmission line to the SMSW contains a YIG film, on the surface of which there are emitting and receiving transducers, made in the form of thin wire antennas. The film is magnetized in the plane by a constant field generated by an external magnet. Signal processing is carried out due to a general or local change in the magnitude, direction or time dependence of the bias field, as well as by proper choice of the shape or relative position of the transducers. For a transmission line to SMSW, the typical frequency range is 1–20 GHz, the range of magnetizing fields is 100–10000 Oe, the length of the conductors of the transducers is 2–20 mm, the distance between them is of the same order. A typical medium for the propagation of SMSW is YIG films with a thickness of 5–20 μm, having a magnetization of 1750 G.

Under these conditions, the length of the SMSW is from 50 μm to 5 mm, that is, as a rule, it is much less than the distance between the transducers; therefore, the phase incursion of the wave during its propagation from the emitting transducer to the receiving one is quite large and can be tens and hundreds of radians. On the other hand, the length of the SMSW, as a rule, is much less than the length of the transducers; therefore, the waves excited by different parts of the emitting transducer arrive at the receiving transducer in different phases and interfere with each other, leading to distortion of the output signal. Thus, the phase incursion of the wave along the path of its propagation can radically change the transmission characteristics of the transmission line.

In the previous chapter, the frequency response of transmission lines was obtained by the method of partial wave beams under various conditions. The analysis carried out there presupposes the complete inphase of the signals induced by the SMSW arriving at individual sections of the receiving transducer. Let us now consider a more general case when such an inphase is absent due to the difference in the phase incursion for individual partial wave beams.

6.2. General case of waves in a magnetic medium

Let us first consider the general case of a homogeneous magnetic medium without damping with a dispersion law of the form and assume that the geometric dimensions of the transducers significantly (not less than an order of magnitude) exceed the wavelength in the medium. This assumption makes it possible to consider the problem in the approximation of geometrical optics, that is, further any possible diffraction effects, as well as the dispersive splitting of wave beams, will not be taken into account. Let us also assume that the attenuation of the wave during propagation in the medium between the transducers is sufficiently small, that is, it can be neglected.

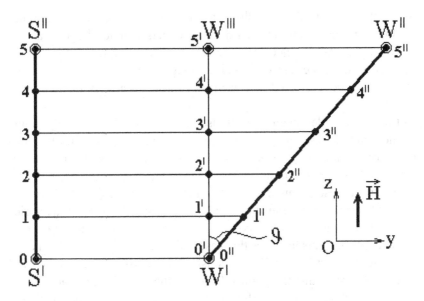

Fig. 6.1. General geometry of the total phase incursion problem.

To solve such problems, in the previous chapter, the method of partial wave beams was proposed and successfully tested, which consists in analyzing the trajectories of individual wave beams emanating from discrete sections of the emitting transducer, followed by summing the resulting output signal at the receiving transducer [615, 616].

Let's use this method here to solve the problem, the general geometry of which is shown in Fig. 6.1. Let the plane of the ferrite film coincide with the plane of the drawing. We will use a Cartesian coordinate system $Oxyz$ the plane of which coincides with the plane of the film, and the axis is perpendicular to it. The film is magnetized in the plane by a uniform field applied along the Oz axis. The conductor of the emitting converter $S'S''$ is oriented parallel to the Oz axis, and the conductor of the receiving $W'W''$ makes an angle ϑ with this axis.

In such a geometry, the waves excited by the emitting transducer propagate rectilinearly along the axis in the form of a parallel beam, the width of which is equal to the length of the transducer. Let us further assume that the receiving transducer is located exactly opposite the emitting one, and its length is such that the coordinates of the points S' and W', as well as the points S'' and W'' coincide in pairs. Under this condition, the receiving transducer intercepts the entire width of the wave beam excited by the emitting transducer.

We will assume further that the phase incursion of the wave in the medium during propagation between the transducers is much larger than the phase incursion of the electromagnetic signal along the length of each transducer, that is, the phase incursion along the length of an individual transducer can be neglected and each transducer separately operates in a single phase.

Let us now divide the emitting transducer into equal sections by points 0–5. The number of sections equal to five is chosen here only for the convenience of the drawing, in real calculation the more points, the more accurate the result obtained. Let us construct the trajectories of the wave beams outgoing from these points. Due to the uniformity of the field and the orientation of the emitting transducer parallel to the field, these trajectories are straight lines parallel to the axis. The trajectory outgoing from the point S' of the emitting transducer, in this case, falls into the point W' of the receiving one. From the point W'' we construct an auxiliary segment W"W'" parallel to $S'S''$ such that the point W''' lies on the line $S'W'$. Obviously, S'" $SW'''W'$ is a rectangle, that is, all trajectories come into the section $W''W'''$ in the same phase. Therefore, to calculate the resulting signal on the receiving transducer, it is sufficient to take into account the phase incursion only on the segments of the trajectories enclosed within the triangle $W''W'''W'$. From the point W'' we construct an auxiliary segment $W''W'''$ parallel to $S'S'''$ such that the point W'''' lies on the line $S''W''$. Obviously, $S''S'W'W''''$ is a rectangle, that is, all trajectories enter the section $W'''W'$ in the same phase. Therefore, to calculate the resulting signal at the receiving transducer, it is sufficient to take into account the phase incursion only on the segments of the trajectories enclosed inside the triangle $W'W''''W'$.

Suppose that the signal in the section $W''W'$ is

$$U_s = A_0 \cos \omega t. \tag{6.1}$$

where A_0 is the constant initial amplitude, ω is frequency, t time. In this case, within the triangle $W''W'''W'$, a wave of the following form propagates:

$$U_w = A_0 \cos(\omega t - ky), \tag{6.2}$$

where the coordinate y is measured from the section $W'''W'$.

Let us first consider the phase incursion of the wave beam. outgoing from point 1. The length of the section of the trajectory of this beam inside the triangle $W''W'''W'$ is equal to:

$$\Delta y = \Delta z \, tg \, \vartheta \qquad (6.3)$$

where Δz is the length of the section of the emitting transducer between points 0 and 1. The phase incursion of the wave on the section of the trajectory with the length is equal to Δy:

$$\Delta\varphi_1 = -k\Delta y = -k \, \Delta z \, tg \, \vartheta, \qquad (6.4)$$

where k is the wave number determined by the field, frequency and dispersion law for waves in the medium. Assuming:

$$\Delta z = L/N, \qquad (6.5)$$

where L is the length of the emitting converter $S'S''$, N is the number of sections into which the emitting converter is divided, we obtain the phase incursion from the first section of the converter in the form:

$$\Delta\varphi_1 = -\frac{kL tg \vartheta}{N}. \qquad (6.6)$$

Similarly, the phase incursion of the beam emerging from the point numbered is:

$$\Delta\varphi_n = -\frac{kL tg \vartheta}{N}.n \cdots \qquad (6.7)$$

The total signal at the receiving transducer is the sum of signals of the form:

$$U_n = \frac{A_0}{N+1}.\cos(\omega t + \Delta\varphi_n). \qquad (6.8)$$

Converting this expression, we get:

$$U_n = \frac{A_0}{N+1}\left[\cos\omega t \cdot \cos(\Delta\varphi_n) - \sin\omega t \cdot \sin(\Delta\varphi_n)\right]. \qquad (6.9)$$

Summing over all the segments into which the emitting transducer is divided, we obtain the complete signal at the receiving transducer in the following form:

...

$$U = \frac{A_0}{N+1} \cdot \left[\cos \omega t \sum_{n=0}^{N} \cos\left(\frac{kL \, tg \, \vartheta}{N} n\right) + \sin \omega t \sum_{n=0}^{N} \sin\left(\frac{kL \, tg \, \vartheta}{N} n\right) \right]$$
(6.10)

Let's introduce the notation:

$$B = \frac{A_0}{N+1} \sum_{n=0}^{N} \cos\left(\frac{kL \, tg \, \vartheta}{N} n\right) \quad (6.11)$$

$$D = \frac{A_0}{N+1} \sum_{n=0}^{N} \sin\left(\frac{kL \, tg \, \vartheta}{N} n\right) \quad (6.12)$$

The sums included in the expressions for B and D can be transformed using the well-known formulas [630]:

$$\cos \alpha + \cos 2\alpha + \cos 3\alpha + \ldots + \cos n\alpha = \frac{\cos\left(\frac{n+1}{2}\alpha\right) \cdot \sin\left(\frac{n\alpha}{2}\right)}{\sin\left(\frac{\alpha}{2}\right)}$$
(6.13)

$$\sin \alpha + \sin 2\alpha + \sin 3\alpha + \ldots + \sin n\alpha = \frac{\sin\left(\frac{n+1}{2}\alpha\right) \cdot \sin\left(\frac{n\alpha}{2}\right)}{\sin\left(\frac{\alpha}{2}\right)}$$
(6.14)

Thus, we get:

$$B = \frac{A_0}{N+1} \cdot \left\{ 1 + \frac{\cos\left[\frac{(N+1) \, kL \, tg \, \vartheta}{2N}\right] \cdot \sin\frac{kL \, tg \, \vartheta}{2}}{\sin\frac{kL \, tg \, \vartheta}{2N}} \right\} \quad (6.15)$$

$$D = \frac{A_0}{N+1} \cdot \frac{\sin\left[\frac{(N+1) \, kL \, tg \, \vartheta}{2N}\right] \cdot \sin\frac{kL \, tg \, \vartheta}{2}}{\sin\frac{kL \, tg \, \vartheta}{2N}} \quad (6.16)$$

Now the complete signal at the receiving transducer can be written as:

$$U = B \cdot \cos \omega t + D \cdot \sin \omega t = R \cdot \cos(\omega t + \theta), \quad (6.17)$$

where:

$$R = \sqrt{D^2 + B^2}, \quad (6.18)$$

$$\theta = arctg\frac{D}{B} \quad (6.19)$$

The obtained expressions (6.15) and (6.16) include the number of partition points, while it is obvious that the final expressions (6.18) and (6.19) should not depend on this number. Indeed, the calculation should be the more accurate, the more, that is, and should tend to a certain limit. Using the well-known rules for finding the limits [631], at we obtain:

$$B = A_0 \frac{\sin(kL\, tg\vartheta)}{kL\, tg\vartheta}; \quad (6.20)$$

$$D = A_0 \frac{1 - \cos(kL\, tg\vartheta)}{kL\, tg\vartheta} \quad \ldots \quad (6.21)$$

Substituting (6.20) and (6.21) into (6.18) and (6.19), we obtain:

$$R = A_0 \frac{\sin\left(\dfrac{kL\, tg\vartheta}{2}\right)}{\dfrac{kL\, tg\vartheta}{2}}; \quad (6.22)$$

$$\theta = \frac{kL\, tg\vartheta}{2} \quad \ldots \quad (6.23)$$

As a result, from (6.17), (6.22) and (6.23) we obtain the complete signal at the receiving transducer in the form:

$$U = A_0 \frac{\sin\left(\dfrac{kL\, tg\vartheta}{2}\right)}{\dfrac{kL\, tg\vartheta}{2}} \cdot \cos\left(\omega t + \frac{kL\, tg\vartheta}{2}\right) \quad (6.24)$$

The relative amplitude of this expression, depending on the wave number, behaves like a function and tends to unity at, then, as it increases, passes through a series of zeros and maxima and tends to zero at $k \to \infty$. The wave number values corresponding to zeros (6.24) are determined by the expression:

$$k_m = \frac{2\pi m}{L\, tg\beta}, \quad (6.25)$$

where $m = 1, 2, ...$ is an integer.

The phase incursion of the signal at the receiving transducer increases linearly with increasing wavenumber k, the greater the angle between the transducers ϑ and the width of the wave L beam specified by the length of the emitting transducer.

The substitution of the obtained expressions for B and D into the formula U for performed here makes it possible to replace the cumbersome summation with a relatively simple calculation of trigonometric functions. Such a replacement, along with clarity, gives great convenience in the construction of machine algorithms for calculating the characteristics of transmission lines on the SMSW.

6.3. The case of surface magnetostatic waves (SMSW)

The obtained expressions include the signal frequency and the parameters of the magnetic medium through the wave number k determined by the law of wave dispersion in the medium. For the case of a ferrite film with a free surface, along which SMSWs propagate, the dispersion relation has the form [118]:

$$\beta - 1 - 2\mu\alpha \cdot cth(kd\alpha) = 0, \qquad (6.26)$$

or, allowed relative to the wavenumber:

$$k = \frac{1}{2d\alpha} \cdot \ln\frac{\beta - 1 + 2\mu\alpha}{\beta - 1 - 2\mu\alpha}, \qquad (6.27)$$

where:

$$\alpha = \sqrt{\frac{\sin^2\phi}{\mu} + \cos^2\phi} \; ; \qquad (6.28)$$

$$\beta = (v^2 - \mu^2 + \mu)\cos^2\phi - \mu \; ; \qquad (6.29)$$

$$\mu = 1 + \eta; \qquad (6.30)$$

$$\eta = \frac{\Omega_H}{\Omega_H^2 - \Omega^2} \; ; \qquad (6.31)$$

$$v = \frac{\Omega}{\Omega_H^2 - \Omega^2} \; ; \qquad (6.32)$$

$$\Omega_H = \frac{H_0}{4\pi M_0}; \quad (6.33)$$

$$\Omega = \frac{\omega}{4\pi\gamma M_0}; \quad (6.34)$$

φ is the angle between the direction of the wave vector $\vec{k}$ and the Oy axis, measured from the axis towards the Oz axis (in our geometry $\varphi = 0$), d – the thickness of the ferrite film, γ – the gyromagnetic constant. For given magnetic parameters, the frequency boundaries of the SMSW spectrum are determined by the formulas [118]:

– bottom:
$$\Omega_b = \sqrt{\Omega_H(\Omega_H + 1)}, \quad (6.35)$$

– upper:
$$\Omega_t = \Omega_H + 1/2 \quad (6.36)$$

Now, to obtain a complete signal at the receiving transducer, it is sufficient to substitute (6.27) in view of (6.280–(6.34) into (6.24). Thus, we have all the formulas necessary to calculate the amplitude-frequency characteristics of the transmission line on the SMSW, taking into account the signal phase, which we will proceed to in the next section.

6.4. Amplitude transmission line characteristics and its different geometric parameters

Let us first consider how the amplitude transmission characteristics of the transmission line depend for different geometric parameters of the problem, namely, when the mutual orientation of the transducers and the total width of the SMSW beam are changed. First, let's trace the change in the amplitude of the transmitted signal, and then its phase. The resulting dependences are illustrated in Fig. 6.2. The magnetic parameters of the problem are as follows: the thickness of the magnetic film is 15 μm, the magnetization is $4\pi M_0 = 1750$ G, the field in the plane of the film is 437.5 Oe.

6.4.1. Dependence of the amplitude of the transmitted signal on frequency when changing the relative orientation of the transducers

Figure 6.2 shows the dependences of the relative amplitude of the signal passing through the transmission line A/A_0 on its frequency ω for different values of the angle ϑ between the transducers. The length of the emitting transducer is 0.5 cm. Accordingly, the width of the SMSW beam is also equal to 0.5 cm and is unchanged, and

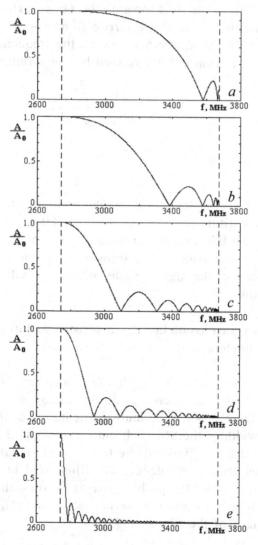

Fig. 6.2. Dependence of the amplitude of the output signal on the frequency at different values of tϑhe angle between the transducers: $a - \vartheta = 1°$, $b - \vartheta = 2°$, $c - \vartheta = 5°$, $d - \vartheta = 10°$, $e - \vartheta = 40°$.

the length of the receiving transducer is chosen so that it completely overlaps the entire SMSW beam. The vertical dashed lines in the figure show the boundaries of the SMSW spectrum determined by the formulas (6.35) and (6.36), for the given magnetic parameters of the problem equal to 2740 MHz and 3675 MHz.

It can be seen from the figure that all dependences have a highly indented multi-lobed character, and the width of the petals decreases with increasing frequency, and the greater the angle, the greater the number of well-resolved petals. Obviously, the multi-lobe nature of the curves is due to the interference of signals arriving at the receiving transducer from different parts of the emitting one, and the increase in the number of allowed lobes and their narrowing with increasing angle is due to the fact that with an increase in the indicated angle, the path difference between areas of the emitting transducer increases. On the whole, the dependence A/A_0 on frequency reflects the course of the function sin x/x (formula (6.22)), and near the lower boundary of the SMSW spectrum is equal to unity, and near the upper boundary it tends to zero. The first lobe, adjacent to the lower boundary of the SMSW spectrum, stands out in terms of magnitude: its maximum is equal to unity, while the next lobe barely exceeds 0.2, and the magnitude of all subsequent ones is even less. At each fixed value of the angle, the frequency width of the intervals between adjacent zeros or maxima decreases with increasing frequency. This behaviour is due to the fact that the values of the wave number corresponding to the zeros of the function are multiples of a series of consecutive natural numbers (formula (6.25)), while the dependence of the wave number on frequency for the SMSW is nonlinear [118] and near the upper boundary of the SMSW spectrum the wave number tends to infinity, which leads to the concentration of zeros and maxima of the function as the frequency approaches this boundary. Since the width of the main lobe of the function decreases with increasing angle, it is interesting to trace the quantitative course of this dependence. The answer to this question is given in Fig. 6.3, which shows the dependences of the frequency of the decay of the main lobe on the angle at different levels from the maximum. The relative position of the curves reflects the gradual decay of the main lobe as the frequency increases.

It can be seen from the figure that all curves fall off as the angle ϑ increases, and at small values ϑ (less than 10°) the decline is quite sharp, and as the angle ϑ increases, the decline slows down. Such their behaviour becomes understandable if we take into account that

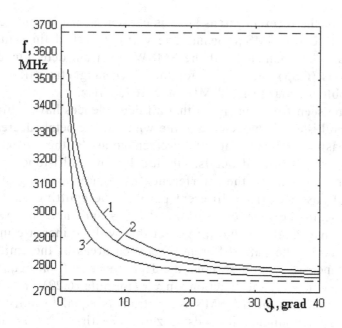

Fig. 6.3. Dependence of the frequency width of the main lobe of the AFC on the angle between the transducers at different levels from the maximum: 1 – 0.2; 2 – 0.5; 3 – 0.8.

the phase incursion during wave propagation between the transducers in accordance with formula (6.23) is proportional, that is, at small values it is small, but sharply increases with increasing. Therefore, when small, the path difference between the extreme rays of the total wave beam is small and the main lobe remains wide, and as the path increases, the path difference increases the more it approaches 90°, as a result of which the lobe sharply narrows and, in the limit, its width tends to zero.

6.4.2. Dependence of the amplitude of the transmitted signal on the frequency with a change in the width of the wave beam

Let us further consider how the amplitude transmission characteristics of the transmission line depend upon a change in the total width of the SMSW beam. Let us assume that the length of the emitting transducer is specified, and the length of the receiving transducer is chosen in such a way that it completely covers the entire width of the SMSW beam. In the chosen geometry (Fig. 6.1), the total beam width of the SMSW is equal to the length of the emitting converter. We take the magnetic parameters of the problem the same as in the previous case. The resulting dependences are illustrated in Figs 6.4 and 6.5.

Figure 6.4 shows the dependences of the relative amplitude of the signal passing through the transmission A/A_0 line on the signal frequency for different widths L of the SMSW beam. The value of the angle ϑ between the transducers is constant and equal to 3°. The vertical dashed lines in the figure show the boundaries of the SMSW spectrum determined by the formulas (6.35) and (6.36), for the given magnetic parameters of the problem equal to 2740 MHz and 3675 MHz.

It can be seen from the figure that, like the previous case, all dependences have a highly indented multi-lobed character, and the width of the lobes decreases with increasing frequency, and the total number of lobes is the greater, the larger the beam width. The multi-lobe nature of the curves observed here is also due to the interference of signals arriving at the receiving transducer from different parts of the emitting converters. As before, the dependence on frequency reflects the course of the function (formula (6.22)), and near the lower boundary of the SMSW spectrum is equal to unity, and near the upper boundary it tends to zero. The remaining features of the curves are similar to the previous case, which becomes clear if we take into account that the SMSW beam width enters into formula (6.22) in the same way as and, that is, both systems of curves are similar up to the tangent function, which grows faster than the values of its argument, therefore, the 'compression' of the curves towards low frequencies in Fig. 6.4 occurs less sharply than in Fig. 6.2.

Figure 6.5 shows the dependences of the main lobe fall off of the frequency on the SMSW beam width L at various levels on the maximum. The relative position of the curves reflects the gradual decay of the main lobe as the frequency increases.

It can be seen from the figure that all curves fall off as the beam width increases; moreover, at small values (less than 0.5 cm), the falloff is rather sharp, and as the beam increases, the falloff slows down. This behaviour, up to the tangent function, is completely analogous to the course of the curves shown in Fig.6.3.

6.4.3. Dependence of the amplitude of the transmitted signal on the relative orientation of the transducers at a fixed signal frequency

Let us now consider what happens to the received signal of a fixed frequency when the relative orientation of the transducers changes.

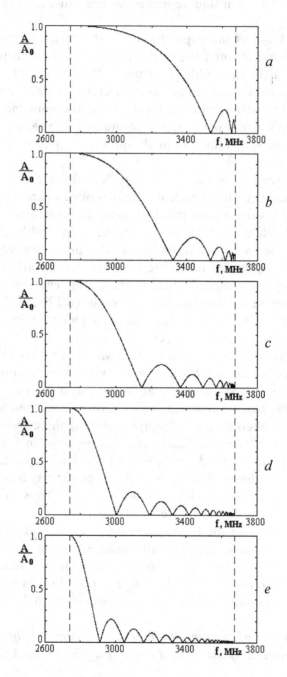

Fig. 6.4. Frequency dependence of the output signal amplitude for different widths of the SMSW beam: (*a*) 0.2 cm, (*b*) 0.4 cm, (*c*) 0.7 cm, (*d*) 1.2 cm, (*e*) 2.0 cm.

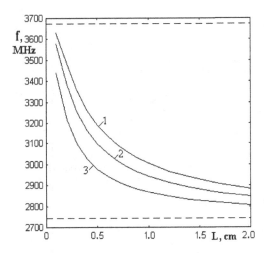

Fig. 6.5. Dependence of the frequency width of the main lobe of AFC on the width of the SMSW beam on different levels from maximum: 1 – 0.2; 2 – 0.5; 3 – 0.8.

Let the length of the emitting transducer be given, and the length of the receiving one such that it covers the entire width of the SMSW beam. The magnetic parameters of the problem remain the same as before. The resulting dependences are illustrated in Figs 6.6 and 6.7. Figure 6.6 shows the relationship between the relative amplitude of the signal passing through the transmission line A/A_0 and the angle ϑ between the transducers at different frequencies. The SMSW beam width is constant and equal to 0.5 cm.

It can be seen from the figure that, like the previous cases, all dependences have a highly indented multi-lobed character, and as the angle ϑ increases, the width of the petals decreases, and the total number of petals increases. The observed features of the curves are caused by the same reasons as in the previous cases, that is, the dependence A/A_0 on the angle ϑ reflects the course of the function $\sin x/x$ (formula (6.22)). Some difference is due to the fact that Figs. 6.2 and 6.4 show the dependences of the transmitted signal on the frequency included in formula (6.22) implicitly through the wave number k and dispersion law of the SMSW (6.26)–(6.27), and here (Fig. 6.6) the angle ϑ enters into formula (6.22) in explicit form under the tangent sign.

It is interesting to note that the decrease in the main lobe of the curves occurs the sharper, the higher the SMSW frequency. This feature is illustrated in more detail in Fig. 6.7, which shows the dependences of half (at $\vartheta > 0$) of the angular width of the main lobe on frequency at various levels on the maximum. The relative

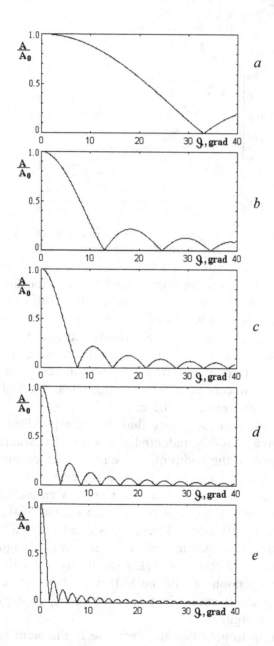

Fig. 6.6. Dependence of the output signal amplitude on the angle between the transducers at different frequencies: a – 2800 MHz, b – 2900 MHz, c – 3000 MHz, d – 3150 MHz, e – 3400 MHz.

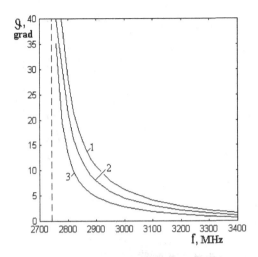

Fig. 6.7. Dependence of half (at $\vartheta > 0$) the angular width of the main lobe of the AFC on frequency at different levels from the maximum: 1 – 0.2; 2 – 0.5; 3 – 0.8.

position of the curves reflects the narrowing of the main lobe with increasing frequency.

It can be seen from the figure that all curves fall off with increasing frequency, and at frequencies below 2900 MHz, the decline is quite sharp, and then it slows down and the width of the lobe tends to zero. This behaviour of the curves is due to the dispersion law of the SMSW, due to which, when the frequency approaches the upper boundary of the SMSW spectrum (3675 MHz), the wavenumber k, and with it the wave phase incursion, increase indefinitely and the coefficient at tg ϑ in formula (6.22) also tends to infinity.

6.4.4. Dependence of the phase of the transmitted signal on frequency when changing the relative orientation of the transducers

Let us now consider how the phase of the signal passing through the transmission line to the SMSW changes when its geometric parameters change. From the formula (6.23) it is easy to see that the phase incursion linearly depends on the width of the SMSW beam, as well as on the tangent of the angle ϑ between the transducers. The dependence of the phase of the transmitted signal on its frequency is much more complicated, since it is determined by the dispersion relation (6.26)–(6.27), which gives the frequency dependence of the wave number k. This dependence can be obtained in explicit form, for which it is sufficient to substitute (6.28) and (6.29) in (6.27),

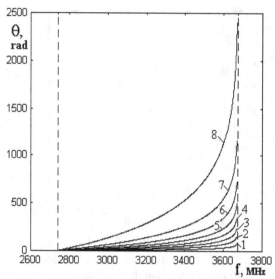

Fig. 6.8. Dependence of the phase of the output signal on the frequency at different values of the angle between the converters: 1 –10°, 2 – 20°, 3 – 30°, 4 – 40°, 5 – 50°, 6 – 60°, 7 – 70°, 8 – 80°.

and then (6.27) in (6.23). The results of this procedure are illustrated in Figs. 6.8 and 6.9.

Figure 6.8 shows the dependence of the output signal phase on the frequency for different values of the angle between the converters. The curves are plotted under the assumption that the SMSW beam width is constant and equal to 0.5 cm, and the receiving transducer covers the entire wave beam. Vertical dashed lines show the boundaries of the SMSW spectrum, determined by the formulas (6.35) and (6.36), equal to 2740 MHz and 3675 MHz.

It can be seen from the figure that all the curves have an upward character, the steeper the larger the angle ϑ, which reflects the proportionality of the value θ to the tangent ϑ (formula (6.23)). The nonlinear growth of the phase incursion versus frequency is due to the nonlinearity of the dispersion law of the SMSW, due to which, when the frequency approaches the upper boundary of the spectrum, the wavenumber tends to infinity, providing the same tendency to infinity for the phase incursion.

Figure 6.9 shows the dependences of the phase incursion of the output signal on the angle ϑ between the transducers at different frequencies. It can be seen from the figure that as the angle ϑ increases, all curves smoothly increase in accordance with formula

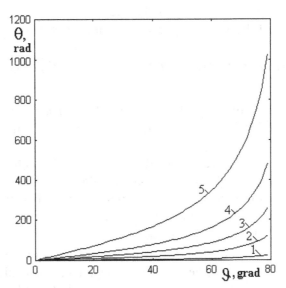

Fig. 6.9. Dependence of the phase of the output signal on the angle between the transducers at various frequencies: 1 – 2800 MHz, 2 – 3000 MHz, 3 – 3200 MHz, 4 – 3400 MHz, 5 – 3600 MHz.

(6.23), where the phase incursion is proportional to tg ϑ. It is also seen that the higher the frequency, the higher the corresponding curve is. Their arrangement corresponds to the SMSW dispersion law, according to which the wave number, and with it the phase incursion, increases with increasing frequency.

6.5. Effect of the phase run on AFC

In the previous sections, examples of the simplest frequency response of transmission lines on the MSSW are given, provided that the receiving transducer is located exactly opposite the emitting one, and its length is such that the z-coordinates of the end points of the emitting transducer coincide in pairs with the z-coordinates of the end points of the receiving transducer. That is, the receiving transducer intercepts the entire width of the SMSW beam excited by the emitting transducer.

Let us now consider a more realistic but also more complicated case – the location of the receiving transducer with an offset along the Oz axis relative to the emitting transducer. In this case, the length of the receiving transducer may differ from the length of the emitting transducer. Thus, the receiving transducer intercepts only that part of the full SMSW beam emanating from the emitting transducer, which falls on its length between the start and end points.

6.5.1. Geometry of the problem with relative mutual displacement of transducers

The general geometry of the problem for an arbitrary mutual displacement and relative length of the transducers is illustrated in Fig. 6.10.

The figure shows the plane of the ferrite film. The Cartesian coordinate system is chosen so that the plane coincides with the plane. film, and the Ox axis is perpendicular to it (directed towards the reader). The Oz axis is oriented along the direction of the constant field $\vec{H}$ lying in the plane of the film.

Emitting and receiving transducers are straight sections of conductors. The ends of the emitting transducer are designated by the letters A and B, the receiving ones by the letters C and D. The origin of the Cartesian coordinate system coincides with the upper end of the emitting transducer, that is, with point B. The emitting transducer AB is deflected from the field direction, that is, the Oz axis by an angle φ. Receiving transducer CD is parallel to the Oz axis, its y-coordinate is indicated by y_S.

In this geometry, the wave vector of the emitted wave, perpendicular to the length of the emitting transducer AB, makes exactly the same angle φ as the emitting transducer with the Oz axis. The group velocity vector $\vec{s}$ makes an angle ψ with the Oy

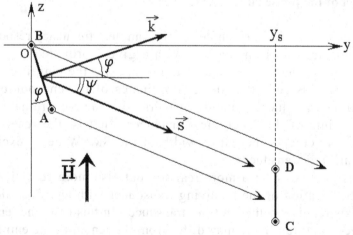

Fig. 6.10. The relative position of the emitting (AB) and receiving (CD) transducers at their mutual displacement.

axis determined by the dispersion law. The wave beam created by the emitting transducer propagates along the direction of the vector, and in the situation shown in the figure, the lower edge of this beam falls on the inner part of the receiving transducer (above point C, but below point D), and the upper edge passes by it (above point D). When the frequency changes, the direction of propagation of the wave beam, determined by the angle ψ, changes, so that the region of the receiving transducer covered by the wave beam also changes, which leads to the formation of a frequency-dependent frequency response.

6.5.2. Formation of the amplitude–frequency characteristic

Sections 5.3–5.5 show various options for the formation of frequency response in non-uniform fields – linear, 'valley' and 'shaft' type. The AFCs given there assume that the partial wave beams arriving at the receiving transducer are in phase, that is, if the wavefront is nonlinear, the shape of the receiving transducer must repeat the configuration of such a front. In the case of a mismatch in the configurations on the receiving transducer, the interference of the wave beams takes place, leading to the transformation of the initially specified configuration of the frequency response.

Let us consider such a transformation using the example of a transmission line based on a free ferrite film with a magnetization of 1750 G and a thickness of 15 μm, magnetized by a uniform field with an intensity of 437.5 Oe.

Let the emitting transducer have a length of 0.50 cm and is deflected from the Oz axis at an angle of 30°. The receiving transducer is 0.45 cm long and is parallel to the Oz axis. The distance between the transducers along the Oy axis is 1.00 cm. The coordinates of the ends of the emitting transducer: $y_A = 0.29$ cm, $z_A = -0.50$ cm; $y_B = 0.00$ cm, $z_B = 0.00$ cm. The coordinates of the ends of the receiving transducer: $y_C = 1.00$ cm, $z_C = -1.13$ cm; $y_D = 1.00$ cm, $z_D = -0.68$ cm.

To form partial beams, the emitting transducer was divided into 100 segments, respectively 0.005 cm long. The calculation of the amplitude and phase incursion of the signal on the receiving transducer was carried out according to the method described in Section 6.2 using the formulas (6.18) and (6.19). Due to the incomplete overlapping of the length of the receiving transducer by the outgoing beam, the use of the shortened formulas (6.15) and (6.16) was unacceptable, therefore the calculation was carried out by direct summation according to the original formulas (6.11) and (6.12), in which only those partial beams that fell on receiving transducer, that is, passed between points C and D in Fig. 6.10.

The results are illustrated in Fig. 6.11. In this figure, curve 1 shows the frequency response without taking into account the phase incursion, that is, similar to those obtained in sections 5.3-5.5, and curve 2 – the frequency response taking into account the phase incursion, obtained by the method of section 6.2. Figure 6.11 a shows both frequency responses on a linear scale. Figure 6.11 b shows the same curves on a logarithmic scale along the vertical axis, allowing one to see the details in more detail.

It can be seen from the figure that, firstly, the frequency response, taking into account the phase, has a strongly indented character,

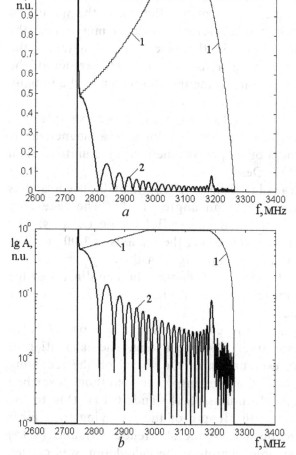

Fig. 6.11. Amplitude–frequency characteristics of a transmission line from two mutually displaced converters without taking into account (1) and taking into account (2) the phase incursion. a – linear scale, b – logarithmic scale.

and secondly, that its amplitude, except for a relatively small initial section, is significantly (an order of magnitude or more) less than the signal amplitude without taking into account the phase.

The reason for the irregularity of the frequency response is obviously the same as the irregularity of the frequency response discussed in Section 6.4, and in appearance reflects the same function sin x/x of the species. A small overshoot near 3200 MHz is random and does not repeat itself when the parameters change. The observed significant decrease in the amplitude of the frequency response is the result of the interference of a large number of partial beams arriving at the receiving transducer in different phases.

A sharp spike in amplitude at the initial frequency of the SMSW spectrum, that is, near the frequency of 2740 MHz, is quite characteristic. In this case, the normalized amplitude in both cases, both without taking into account the phase, and with it, tends very closely to unity.

Such an outlier is due to the fact that in this region the SMSW length is quite large and can be several centimeters, that is, it becomes comparable to or even greater than the distance between the transducers, so that in this case the phase incursions for different partial beams differ little from each other and there is no noticeable interference. going on.

6.5.3. Formation of the phase-frequency response

Let us now consider the phase incursion of the SMSW, corresponding to different frequencies within the total frequency response, for which we turn to Fig. 6.12, which shows the phase-frequency characteristic with the same problem parameters as the AFC in Fig. 6.10.

From Fig. 6.12 a, one can see that the phase response is a frequently-periodic sawtooth curve, the upper and lower ends of the teeth of which fall on (1.57), which corresponds to a continuous phase incursion, which consists of successively following segments along. As the frequency increases, the slope of the saw teeth increases, that is, the phase incursion becomes the greater, the higher the frequency. On a larger scale in frequency, the structure of the teeth is shown in Figs. 6.12 b and 6.12 c, corresponding to the low-frequency end of the phase response (Fig. 6.12 b) and its high-frequency end (Fig. 6.12 c). It can be seen that the period of the teeth decreases with increasing frequency, that is, the phase incursion also increases with increasing frequency. This behaviour of

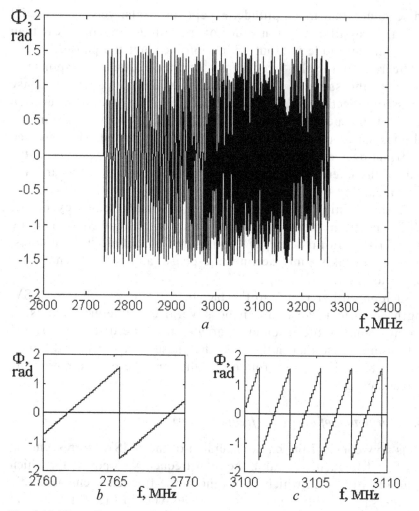

Fig. 6.12. Phase-frequency characteristics of a transmission line from two converters. All parameters are the same as in Fig. 6.11. *a* – phase response in the full range of filter transmission frequencies; *b, c* – PFC fragments in the low-frequency (*b*) and high-frequency (*d*) regions of the full transmission frequency range.

the phase incursion corresponds to the acceleration of the growth of the curves shown there as the frequency increases observed in Fig. 6.8 (Section 6.4.4) and is explained by the same reasons.

6.5.4. The influence of the length of the transducers on the structure of the frequency response

Let us now consider to what extent the change in the length of the transducers affects the formation of the frequency response. In the accepted geometry (Fig. 6.10), such a change in length leads to a change in the number and conditions of interference of partial beams falling on the receiving transducer, which affects the formation of the frequency response.

In further consideration, we restrict ourselves to the case of reducing the length of the transducers, since an increase in the limit should lead to a complete overlap by the receiving transducer of the wave beam emanating from the emitting transducer, which has already been discussed in Sections 6.2–6.4.

Let us refer to Fig. 6.13, which shows the frequency response without taking into account (curves 1) and taking into account (curves 2) the phase incursion with a decrease in the length of the emitting (*a, b*) and receiving (*c, d*) converters in linear (*a, c*) and logarithmic (*b, d*) scale. For convenience of comparison, the reduction in length in both cases is performed in a close ratio: forthe emitting transducer by 4 times, for the receiving one by a factor of 3.75.

From Fig. 6.13 it can be seen that in both cases, with a decrease in the length of both the emitting and receiving transducers, the frequency response width decreases. So, with the original length of the converters, the frequency response is located between 2740 and 3260 MHz, that is, its width at the minimum level is 520 MHz. With a small length of the emitting converter, the frequency response boundaries of the frequency response are 2895 MHz and 3260 MHz, that is, the width is 365 MHz. With a small length of the receiving transducer, the frequency response is between 2740 and 3105 MHz, that is, its width is the same 365 MHz. Thus, in both cases, the frequency response is narrowed by a factor of 1.42. With a decrease in the length of the emitting transducer, the frequency response is narrowed from the low-frequency end, with a decrease in the length of the receiving transducer, the frequency response is narrowed from the high-frequency end.In both cases, a sufficiently intense peak remains at the frequency of the beginning of the SMSW spectrum – 2740 MHz, similar to that in Fig. 6.11. It must be assumed that the nature of such a peak is the same as in Fig. 6.11 and is due to the significant length of the SMSW near this frequency.

With a small length of the emitting converter (Fig. 6.13) between the peak of the beginning of the SMSW spectrum at a frequency of 2740 MHz up to a frequency of 2895 MHz (that is, over 155 MHz), a wide dip is observed in the frequency response. Such a dip in the frequency response in this case is due to the fact that in this

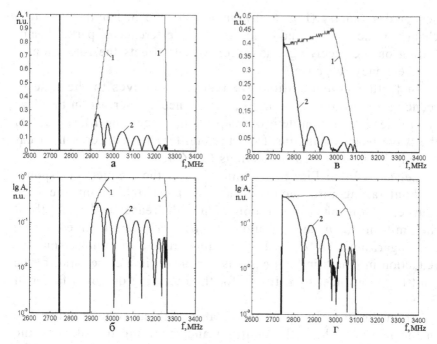

Fig. 6.13. Transformation of the frequency response when changing the length of the emitting (*a, b*) and receiving (*c, d*) converters. 1 – frequency response without taking into account the phase incursion, 2 – taking into account the phase incursion. *a, b* – decrease in the length of the emitting transducer, its length is 0.125 cm (initial 0.50 cm), the *z*-coordinate of the upper end is preserved (z_B = 0.00 cm); *c, d* – decrease in the length of the receiving transducer, its length is 0.12 cm (initial 0.45 cm), the coordinate of the upper end is preserved (z_B = –0.68 cm). All other parameters are the same as in Fig. 6.11. *a, c* – linear scale; *b, d* – logarithmic scale.

interval the angle ψ is so small that the wave beam emanating from the emitting transducer completely passes above the upper end of the receiving transducer (above point *D*).

A characteristic property of the frequency response taking into account the phase (curves 2) with a decrease in the length of one or another transducer is a decrease in the irregularity of the corresponding characteristic, so that the width of each peak between adjacent minima increases several (3–5) times. Such a broadening of peaks is generally due to the properties of a function of the type sin *x/x*, where, according to (6.22) (Section 6.2), the quantity $kL\mathrm{tg}\,\vartheta/2$, where *L* is the width of the wave. As the length of one or another transducer decreases, the width of the beam falling on the receiving transducer decreases, so that the interval between the first minima of the function increases. Thus, with a decrease in 4 times, the mentioned interval also increases approximately 4 times (the first minimums from 4.5 go to 17.5). Such an increase in the intervals between the minima occurs over the entire width of the frequency response, which is manifested as a decrease in its irregularity.

6.6. Deformation of the wave front of surface magnetostatic waves in ferrite films magnetized by linearly inhomogeneous field

In Sections 6.26.5 it was shown that the frequency response of transmission lines, depending on the frequency, wavenumber and width of the beam of magnetostatic waves, as well as the mutual orientation of the transducers, have a multi-lobe character due to the interference between SMSWs arriving at different parts of the receiving transducer. At the same time, for a number of applications it is desirable to have an AFC having only one lobe, that is, it is desirable to exclude interference phenomena. For a homogeneous magnetizing field, this problem is solved by using two rectilinear transducers parallel to each other. The situation becomes much more complicated in the case of an inhomogeneous field. In this case, obviously, to eliminate interference, the conductor of the receiving transducer should be located in those places where the SMSWs arrive with the same phase incursion, that is, along the lines of the constant phase of the wave. If in this case the emitting transducer is chosen to be straight, then the receiving transducer can be bent in a rather complicated manner, determined by the inhomogeneous configuration of the magnetizing field. To clarify the shape of the receiving transducer, it is necessary to investigate the structure of the constant phase lines of the SMSW emitted by the emitting transducer. Further sections are devoted to the consideration of this issue for the case of a linearly inhomogeneous field of constant direction.

In this consideration, we used the materials of works [282, 397, 582, 585, 615-617, 627, 632, 633].

6.6.1. General geometry of the problem

Let the general geometry of the problem have the form shown in Fig. 6.14. A rectangular frame with edge hatching shows the plane of the ferrite film on which the transducers are located: AB – emitting, CD – receiving. In the selected coordinate system Oyz, the plane coincides with the plane of the film, its Ox axis is perpendicular to it and is directed towards the reader in the drawing. Let the

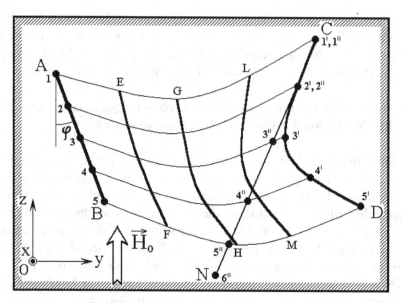

Fig. 6.14. General geometry of the problem.

magnetizing field be directed along the Oz axis.

Let us further consider a 'longitudinally inhomogeneous' field, the direction of which does not change along the axis, and the intensity depends linearly on the coordinate:

$$\ldots \; H_z(z) = 4\pi M_0 \left(a_0 + a_1 z \right) \qquad (6.37)$$

Setting a fixed frequency value determines the range of fields in which SMSWs can exist. In an inhomogeneous field, this interval gives the spatial configuration of the regions in which the SMSW propagation is possible. For a ferrite film with a free surface, the field boundaries of the existence of SMSWs have the form:

...

$$H_1 = \frac{\omega}{\gamma} - 2\pi M_0. \qquad (6.38)$$

...

$$H_2 = \sqrt{\left(\frac{\omega}{\gamma}\right)^2 + \left(2\pi M_0\right)^2} - 2\pi M_0 \qquad (6.39)$$

For a longitudinally inhomogeneous linear field with $a_1 > 0$, we find that SMSWs can propagate within a strip, the boundaries of which are parallel to the axis and intersect the axis at the points:

$$z_1 = \frac{\frac{\omega}{\gamma} - (2\pi M_0 + 4\pi M_0 \cdot a_0)}{4\pi M_0 \cdot a_1} \qquad (6.40)$$

$$z_2 = \frac{\sqrt{\left(\frac{\omega}{\gamma}\right)^2 + (2\pi M_0)^2} - (2\pi M_0 + 4\pi M_0 \cdot a_0)}{4\pi M_0 \cdot a_1} \qquad (6.41)$$

From (6.38)–(6.41) it is seen that $H_1 < H_2$ and $z_1 < z_2$. The above expressions do not impose any restrictions on the value H_0; however, it is obvious that the saturation field of the film must exceed H_S. Suppose that the entire portion of the film plane, bounded by the frame in Fig. 6.14, lies between z_1 and z_2, and consider the operation of the transmission line shown. According to the method of partial wave beams [615, 616], we divide the emitting transducer AB into segments by points 1–5 and construct the SMSW trajectories outgoing from these points, shown by thin solid lines 1–1'–5–5', where the points 1–5 correspond to the ends of the trajectories arriving at the receiving transducer. In the figure, the points of the emitting and receiving transducers lying on the same trajectory of the SMSW are shown by large round dots. The thickened solid lines EF, GH, and LM, crossing the SMSW trajectories 1–1'–5–5', show the lines on which the SMSW propagating along any of the trajectories have the same phase incursion. We will further call such lines 'constant phase lines' of the wave.

It can be seen from the figure that the SMSWs emanating from different points of the rectilinear emitting transducer, due to the inhomogeneity of the magnetizing field, are curvilinear and the lines of the constant phase are also curvilinear. We can say that the line of constant phase represents the wave front at a fixed moment in time. An emitting transducer, being rectilinear, creates a flat wavefront. When the SMSW propagates in an inhomogeneous field, this front is curved. This means that if the receiving transducer, like the emitting one, is also straightforward, then the SMSWs emanating from different parts of the emitting transducer arrive at the receiving one in a different phase.

Indeed, let the conductor of the receiving transducer be located on a straight line CN passing through the points 1″ and 2″ coinciding with points 1′ and 2′. In this case, all SMSWs emerging from section 1–2 of the emitting transducer fall on section 1″–2″ of the receiving transducer in the same phase, and SMSWs emerging from sections

2–3, 3–4 and 4–5, falling into sections 2″–3″, 3″–4″ and 4″–5″, the higher the section number, the more late in phase. They do not come to section 5″–6″ of the receiving converter at all. Due to the difference in the phases of the signals in different parts of the receiving transducer, the total signal taken from the receiving transducer is generally distorted. For undistorted reception, all waves emerging from any parts of the emitting transducer must arrive at the receiving transducer in the same phase, for which the receiving transducer must be bent in the form of a line of constant phase of the wave, as is done for the transducer denoted by the letters CD.

6.6.2. Various cases of orientation of the emitting transducer

Now that the general provisions of the problem have been qualitatively clarified, we will consider quantitative solutions for a number of special cases, namely, we will show what happens to the lines of constant phase when the orientation of the emitting transducer, the signal frequency, and the gradient of the magnetizing field change.

6.6.2.1. Orientation corresponding to $\varphi = 30°$

The general character of the lines of constant phase at different orientations of the emitting transducer is illustrated in Figs. 6.15 and 6.17. Here and everywhere below, the magnetization of the ferrite film $4\pi M_0 = 1750$ G, its thickness $d = 15$ µm.

Figure 6.15 a shows the SMSW trajectories (small dashed line) and lines of the constant phase of the wave (solid thickened) at $\varphi = 30°$, field parameters $a_0 = 1/4$; $a_1 = 1/8$ cm^{-1} ($H_0 = 4\pi M_0 (1/4+z/8)$), for frequency $f = 3000$ MHz. With the indicated parameters, the SMSW fields can propagate in a band from $z_1 = -1.1020$ cm to $z_2 = 0.3238$ cm. The initial point of transducer A is located near the upper edge of this band and corresponds to $y = 0.00$ cm, $z = 0.29$ cm. The transducer length is chosen equal to 1 cm. the end point of the transducer B corresponds to $= 0.5$ cm, $= -0.71$ cm. When calculating, the individual trajectories were constructed from the points of the emitting transducer spaced apart from each other with a coordinate step of 0.01 cm, along a coordinate $z - 0.017321$ cm, with a step along the length the transducer was 0.02 cm. The length of the emitting transducer (1 cm) was 50 steps. The figure shows every fifth of the calculated trajectories. In this case, only those parts

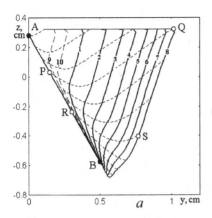 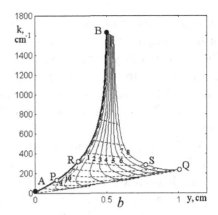

Fig. 6.15. SMSW trajectories and the dependence of the wavenumber on the coordinate, as well as the lines of constant phase at = 30°. *a* – trajectories (small dotted line), lines of constant phase (solid lines); *b* – wavenumber (small dotted line), constant phase lines (solid lines).

of the trajectories are presented that correspond to a phase incursion of no more than 160 rad, the remaining parts of the trajectories are not shown. The lines of the constant phase of the wave are plotted in the interval from 20 to 160 rad with a step of 20 rad. The numbers on the curves correspond to the phase incursion. To thicken the grid, the dotted line (with elongated strokes) in the same figure shows the constant phase lines for the incursion of 5 and 10 rad – curves 9 and 10, respectively.

Figure 6.15 a shows that the region in which the SMSWs propagate with a phase incursion not exceeding a certain specified value has the shape of a curvilinear triangle with its apex downward. In this case, one of the sides of this triangle is rectilinear and coincides with the upper boundary of the region of existence of the SMSW, the other is also rectilinear and coincides with the conductor line of the emitting transducer, and the third is in the upper part, which is up to 80–90% of the total length, almost (accurate to units of percent) is rectilinear, and at the bottom (no more than 20–40% of the length) deviates to the side with a phase incursion of less than 60–100 rad - from the emitting transducer, and with a phase incursion of more than 100 rad – to it.

In [397, 582, 585], it was shown that in the described geometry all trajectories should have a V-shaped character and end at the upper boundary of the region of existence of the SMSW (6.40) at $z_2 = 0.3238$ cm. From Fig. 6.15 a it can be seen that this is typical of the first four trajectories, starting at the upper section of the emitting transducer, denoted by the letters AP, have a character, but on these trajectories the phase incursion is always less than 160 rad. The PQ trajectory is the first on which a phase incursion of 160 rad is reached at the upper boundary of the region of existence of the SMSW. All trajectories starting on the emitting transducer below the point P gain phase equal to 160 rad, not reaching the indicated upper limit – on the curve indicated by the number 8.

All lines of constant phase in the upper part have an almost rectilinear section (for line 8 – from point Q to point S), the length of which increases as the phase incursion increases. The slope of this section is positive, the angle between the lines of constant phase and the Oz axis is slightly less than the angle φ of the emitting transducer, is about 25° and practically does not depend on the value of the phase incursion. Below the rectilinear section, the constant phase lines are bent first towards the emitting transducer, and then away from it, as a result of which their slope at first, being positive, slightly increases (up to 30–40°), and then decreases, for lines with a phase incursion less than 100 rad, changing negative.

It can be seen from Fig. 6.15 a that as the phase incursion increases, the lines of constant phase, built with the same phase step, gradually thicken: so the distance between the straight sections of the lines 1 and 2 is about 0.15 cm, and between the lines 7 and 8 less than 0.05 cm.

The above features of the lines of constant phase can be explained by considering the change in the wave number during the propagation of the SMSW in an inhomogeneous field, illustrated in Fig. 6.15 b. In Fig. 6.15 b, the solid lines correspond to the lines of the constant phase of the wave on the plane, the small dotted line – the dependences of the wave number on the coordinate on the SMSW trajectories. All curves are plotted with the same problem parameters that were used to plot the curves in Fig. 6.15 a. The points A, B, P, Q, R, S correspond to the points marked with the same letters in Fig. 6.15 a. The numbers on the curves also correspond to those shown in Fig. 6.15 a.

Figure 6.15b shows that transducer AB, being straight in Fig. 6.15 a, is transformed into a curved line AB in Fig. 6.15 b. Such a transformation is easy to understand if we take into account the dispersion law of the SMSW, according to which, at a constant frequency, larger fields correspond to smaller wave numbers and vice versa, and the closer the field is to the lower field boundary of the region of existence of the SMSW, the more the wave number changes. In the case under consideration, the boundaries of the region of existence of SMSWs in the field are equal to H_1 = 196.4286 Oe, = 508.3236 Oe, which corresponds to the boundaries along the coordinate z_1 = −1.1020 cm, z_2 = 0.3238 cm. In this case, point A, for which z = 0.29 cm is located near the upper boundary along the coordinate and along the field, and point B, for which z = −0.71 cm, approaches the lower boundary. Therefore, the wave number k at point A is close to zero, and at point B is more than 1600 cm^{-1}. The nonlinear growth of the wavenumber k with a linear change along the AB curve in Fig. 6.15 b reflects the nonlinear character of the SMSW dispersion.

Further, the sequential arrangement of trajectories from top to bottom in Fig. 6.15 a corresponds to the sequential arrangement of the curves (dashed line) in Fig. 6.15 b from bottom to top. Such a reverse motion is due to a decrease in the field in Fig. 6.15 a from top to bottom and the associated increase in the wavenumber k in accordance with the dispersion law of the SMSW, which gives rise to the curves in Fig. 6.15 b. The upward narrowing of the region enclosed between the AB curve and the line of constant phase 8 in Fig. 6.15 b reflects the downward narrowing of a similar region in Fig. 6.15 a. Note that this narrowing reflects the requirement to preserve the constancy of the integral phase incursion, determined through the product of the wavenumber and the coordinate along the SMSW trajectory, that is, an increase in the wavenumber due to a decrease in the field leads to a shortening of the trajectory in this field.

The thickening of the lines of constant phase as the coordinate increases in both Figs. 2 and 3 is due to the fact that the formation of these lines with a large phase incursion (with an increase in the number of curves) involves trajectories located in the region of smaller fields (below in Fig. 6.15 a), that is, where the wavenumbers are larger and the specified phase incursion is reached sooner.

Thus, the dependences for the wavenumber shown in Fig. 6.15 b, taking into account the dispersion relation for SMSW, make it possible to explain all the features of the curves in Fig. 6.15 a.

6.6.2.2. Other orientations

Let us now consider other orientations of the straight-line emitting transducer. Let φ = 0°, that is, the conductor of the transducer is elongated along the Oz axis. In this case, all other parameters of the transmission line will remain the same. This case is illustrated in Fig. 6.16, which (*a*) shows the trajectories (dashed line) and lines of constant phase (solid), as well as (*b*) dependences of the wavenumber $\varphi = 0°$ on the coordinate *y* (dashed line), and lines of constant phase in coordinates. All letters and numbers on the curves are completely analogous to those shown in Fig. 6.15. The phase step for the curves 1–8 is the same 20 rad, that is, curve 8 corresponds to a phase incursion of 160 rad. Here, however, the length of the emitting converter *AB* is extended over the entire width of the region of existence of the SMSW, which could not be done for Fig. 6.15 due to the narrowing of the SMSW spectrum at an angle other than zero.

In general, Figs. 6.16 a and 6.16 b are in many ways similar to Figs. 6.15 a and 6.15 b, so we will not repeat ourselves and dwell only on noticeable differences.

So, in Fig. 6.16 a, the SMSW propagation region with a phase incursion not exceeding 160 rad is a curved triangle with a right angle at apex *A*, while in Fig. 6.15 a a similar angle is acute. This circumstance is caused only by the orientation of the emitting transducer aligned along the *Oy* axis. Further, all the trajectories here depart from the emitting transducer at a right angle parallel to the *Oy* axis, and then turn upward – in the direction of increasing the field along the *Oz* axis. Such their behaviour is also known and was noted earlier in works [397, 582, 585]. Similarly to Fig. 6.15 a, here there is a critical trajectory *PQ*, which has the same properties: on all trajectories above it, the phase incursion is less than 160 rad,

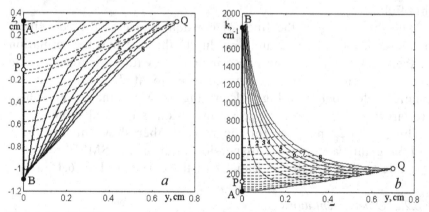

Fig. 6.16. SMSW trajectories and the dependence of the wavenumber on the coordinate *y*, as well as the line of constant phase at $\varphi = 0°$. *a* – trajectories (small dotted line), lines of constant phase (solid lines); *b* – wavenumber (small dotted line), constant phase lines (solid lines). The numbers on the curves correspond to the same phase incursion as in Fig. 6.15.

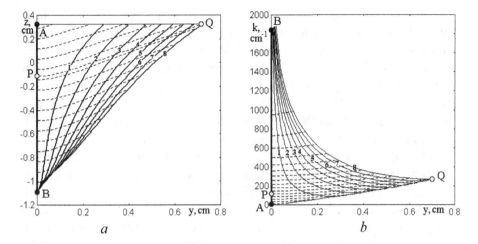

Fig. 6.17. SMSW trajectories and the dependence of the wavenumber on the coordinate, as well as the lines of constant phase at $\varphi = -30°$. *a* – trajectories (small dotted line), lines of constant phase (solid lines); *b* – wavenumber (small dotted line), constant phase lines (solid lines). The numbers on the curves correspond to the same phase incursion as in Figs. 6.15 and 6.16.

on all trajectories below it, more than 160 rad. Note that here, in contrast to Fig. 6.15 a, the lines of constant phase 1–8 are straighter than similar lines in Fig. 6.15 a, however, a clearly pronounced rectilinear section, similar to the segment QS in Fig. 6.15a, is absent and the entire line BQ is curved at nearly equally.

In Fig.6.16 b, the region between the emitting transducer AB and the line of constant phase 8 is limited to the left by a vertical straight line, which is due to the fact that here all the trajectories start on the axis. Otherwise, the behaviour of the curves in this figure is similar to the curves in Fig. 6.15 b.

Let us now consider the orientation of the straight-line emitting transducer, corresponding to $\varphi = -30°$, while keeping all the other parameters of the problem the same. This case is illustrated in Figure 6.17.

All designations are completely similar to those shown in the previous figures. Here, the length of the emitting converter AB, as in Fig. 6.15a, is limited by the limits of the SMSW spectrum, corresponding to $\varphi = -30°$.

Figure 6.17 a shows that the SMSW propagation region with a phase incursion not exceeding 160 rad is a curved triangle with an obtuse angle at the apex A, which corresponds to the corresponding orientation of the emitting transducer. Further, all the trajectories here depart from the emitting transducer at an acute angle upward and immediately tend to the upper boundary of the region of existence of the SMSW, on which the boundary of the region of propagation of the SMSW AQ lies. This behaviour of the trajectories was also noted in [397, 582, 585]. The consequence of such a rapid tendency of the trajectories to the AQ line is a sharp narrowing of the SMSW propagation region in comparison with similar regions in Figs. 6.15 a and 6.16 a. Similarly to Figs. 6.15 a and 6.16 a, there is a critical trajectory PQ, which has the same properties of dividing the trajectories into two groups by the magnitude of the phase incursion. As in Fig. 6.16 a, here the lines of constant phase 1–8 are straighter than similar lines in Fig. 6.15 a, however, a clearly pronounced rectilinear section, similar to the segment QS in Fig.6.15 a, is also absent.

In Fig. 6.17 b the region between the radiating transducer AB and the line of constant phase 8 is bounded on the left by a curved line bent in the same direction as the line of constant phase 1–8. Such a course of it is due to the location of point B of the emitting transducer, corresponding to a smaller field and a larger wavenumber, to the left along the coordinate than point A, corresponding to a larger field and a smaller wavenumber. Otherwise, the behavior of the curves in Fig. 6.17 a is similar to the analogous curves in Fig. 6.15 b and 6.16 b.

6.6. General character of transformation of the area of distribution of SMSW when various parameters of the structure change

Summarizing the above results, we can conclude that the general configuration of the SMSW propagation region with a phase incursion not exceeding a given value is a curvilinear triangle with a downward apex. Let us consider the transformation of the SMSW propagation regions when changing various parameters in more detail.

6.6.1. Changing the orientation of the emitting transducer

When the orientation of the emitting transducer is changed, the upper edge of the SMSW propagation region with a phase incursion not exceeding a given value remains in place and corresponds to the upper boundary of the SMSW existence region, its left edge follows the emitting transducer, and the right edge connects the lower end of the emitting transducer with the end point of the critical trajectory at the top border of the area.

The general configuration of the SMSW propagation region with a phase incursion not exceeding a given value on the plane y–k is a curved triangle with the apex facing up. When the orientation of the emitting transducer is changed, the left edge of the specified area is determined by the line corresponding to the emitting transducer and follows it, and the right edge is determined by the line corresponding to the given phase incursion, the lower edge is always located in the area of small values of the wavenumber (no more than 300 cm^{-1}) and, slightly increasing along the coordinate y, changes little on the whole.

6.6.2. Frequency change

Let us now consider what happens to the lines of constant phase when the frequency of the propagating SMSWs changes. This change is illustrated in Fig. 6.18 which shows the boundaries of the SMSW propagation region with a phase incursion of 160 rad for different frequencies. The emitting transducer is oriented along the axis Oz, that is, $\varphi = 0°$, the field parameters are as follows: $a_0 = 1/4$; $a_1 = 1/8$ cm^{-1} (that is $H_z = 4\pi M_0(1/4+z/8)$).

Figure 6.18a shows that all areas have the shape of a right-angled triangle facing one of the vertices downward, which is disassembled in detail when considering Fig. 6.16 a. As the frequency increases, the regions shift upward along the Oz axis almost unchanged, which reflects, in this frequency range, the nearly linear character of the dependence of the resonance field on frequency. Indeed, in

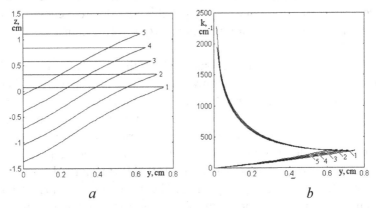

Fig. 6.18. Boundaries along the coordinate (*a*) and wave number (*b*) of the SMSW propagation regions with a phase incursion of 160 rad for different frequencies: 1 – 2800 MHz, 2 – 3000 MHz, 3 – 3200 MHz, 4 – 3400 MHz, 5 – 3600 MHz.

accordance with the linear law, the field increases linearly along the coordinate z, which leads to a similar increase in the resonance frequency and an increase in the region of propagation of the SMSW along the same coordinate. In the considered frequency range 2800-3600 MHz with magnetization $4\pi M_0 = 1750$ G, the dependence of the frequency on the field in accordance with the formulas (7) and (8), at $\omega/\gamma \gg H$ when approaches the linear form of the type $\omega \sim H$. Some shortening of the region along the Oy axis (from 0.75 cm at 2800 MHz to 0.60 cm at 3600 MHz) reflects a slight deviation from the linearity of the dependence of the SMSW frequency on the field, described by formula (8).

Figure 6.18 b shows the wavenumber boundaries for different frequencies under the same conditions. It can be seen from the figure that the boundaries of the region with respect to the wavenumber hardly change, which reflects the retention of the similarity of the dispersion curve of the SMSW described by formula (1) with a change in the frequency or field. Some shortening of this region along the coordinate with increasing frequency is due to the same reasons as a similar change in a similar region in Fig. 6.18 a.

6.6.3. Changing the gradient of the field

Let us now consider what happens to the lines of constant phase when the gradient of the magnetizing field changes. This change is illustrated in Fig. 6.19.

Figure 6.19 a shows the boundaries of the SMSW propagation region with a phase incursion of 160 rad for various field gradients. The emitting transducer is oriented along the Oz axis, that is, $\varphi = 0°$, frequency $f = 3000$ MHz. The field parameters are as follows: $a_0 = 1/4$; a_1 – changes (that is $H_z = 4\pi M_0(1/4 + a_1 \cdot z)$.

Figure 6.19a shows that all regions 3–5 have the shape of a right-angled triangle facing one of the vertices downward, which is disassembled in detail when considering Fig. 6.16a. An increase in these regions along the Oy axis with a decrease in the field gradient reflects the expansion of the region of existence of the SMSW, described by the formulas (6.40) and (6.41). Their elongation along the Oy axis under the same conditions reflects a slower set of the required phase incursion (160 rad) at lower field gradients, since with a decrease in the field gradient, the wave passes an increasingly significant part of the trajectory with a field close to the initial one and relatively slowly falls into the region of small fields where the phase incursion is greatest. The regions 1 and 2 degenerate into a straight line parallel to the

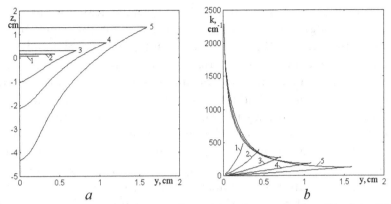

Fig. 6.19. Boundaries along the coordinate (*a*) and wave number (*b*) of the SMSW propagation regions with a phase incursion of 160 rad for different field gradients: 1 – 1/2 cm^{-1}; 2 – 1/4 cm^{-1}; 3 – 1/8 cm^{-1}; 4 – 1/16 cm^{-1}; 5 – 1/32 cm^{-1}.

Oy axis, since with such large gradients at a given frequency (3000 MHz), a phase incursion of 160 rad is not achieved (the wave tends so strongly towards a larger field that it does not have time to gain a sufficient phase).

Figure 6.19 b shows the wavenumber boundaries of the same SMSW propagation regions for different field gradients, the same as in Fig. 6.19 a. From this figure it can be seen that the course of the curves is basically the same as that discussed in detail in the discussion of Fig. 6.16 b. The increasing portions of the curves correspond to the horizontal boundaries of the regions in Fig. 6.19 a, where the specified phase incursion (160 rad) is not achieved. The descending portions of curves 3–5 are similar to those shown in Fig. 6.16 b and their nature is due to the same reasons. There are no such sections for curves 1 and 2, since the phase incursion on them is insufficient. The different lengths of the curves along the axis correspond to the same lengths along this axis of the SMSW propagation regions shown in Fig. 6.19 a.

6.7. Recommendations for optimizing the parameters of the transmission line of the SMSW

The results obtained make it possible to develop certain recommendations for optimizing the parameters of the transmission line on the SMSW with a given phase incursion operating under conditions of inhomogeneous magnetizing. Let us consider the possibilities of optimizing these parameters using the curves shown in Fig. 6.15 as an example.

Suppose that we want to have a receiving transformer in the form of a straight conductor with a phase incursion equal to 160 rad. In this case, on the line of constant phase corresponding to the indicated incursion (in the figure it is curve 8), it is necessary to select a straight section of the maximum length. Note that the requirement for the maximum length of the segment is due to the desirability of minimizing the effects of dispersive splitting of the wave beam [282, 314–318]. The segment of curve 8 that satisfies the specified requirements is the segment enclosed between points Q and S. It can be seen from the figure that only those trajectories that leave the emitting transducer in the segment between points P and R fall on it. It can be said that the trajectories PQ and RS are, as it were, critical: only those trajectories that are enclosed between these two fall from the emitting transducer to the receiving one with a given phase incursion. All trajectories above PQ go to the upper boundary of the region of existence of the SMSW, where they end. All paths below RS pass by the straight-line section of the receiving transducer and do not participate in the formation of the output signal of the transmission line. Thus, the sections AP and RB of the emitting transducer can be removed without affecting the parameters of the device.

If, by condition, the form of the receiving transducer can differ from the rectilinear one, then it is advisable to choose it along the line of constant phase. In this case, of the two critical trajectories, only one PQ remains, that is, the emitting transducer does not make sense to continue above the point P and its total length should be only the segment PB. The receiving transducer must extend from point Q to the lower apex of the curved triangle.

Similar recommendations can be given for all the other variants of the transmission line to the SMSW discussed above.

Conclusions for chapter 6

The main issues discussed in this chapter are as follows.
1. The role of the wave phase incursion in the formation of the output signal of the transmission line on SMSWs propagating in a homogeneous magnetic medium without attenuation under the condition of a small wavelength compared to the size of the transducers and the distance between them is considered. The method of partial wave beams is used to calculate the total signal at the receiving transducer for various geometric parameters of the

transmission line. It is shown that the amplitude characteristics of the signal transmission depending on the frequency, wavenumber, mutual orientation of the transducers and the width of the SMSW beam have a multi-lobe character, similar to a function, taking into account the dispersion law of the SMSW determined by the magnetic parameters of the medium. The phase characteristic of the signal transmission is proportional to the wave number, the width of the wave beam and the tangent of the angle between the transducers. The observed features of the transmission characteristics are explained on the basis of the geometry of the problem and the dispersion properties of the SMSW.

2. The role of the wave phase incursion in the formation of the output signal of the transmission line on the SMSW, containing two converters of arbitrary length, with their mutual displacement relative to each other in a direction parallel to the constant field, so that the receiving converter intercepts only part of the wave beam emanating from the emitting converter. It is shown that, in the general case of mutual orientation of the transducers, the frequency response, taking into account the phase, has a strongly indented character, and its amplitude, except for a relatively small initial section, is significantly (an order of magnitude or more) less than the signal amplitude without taking into account the phase. It is noted that the reason for the irregularity is the properties of the forming AFC function of the form, which, in addition to the main maximum, has many side maxima alternating with minima. The interference of a large number of partial beams arriving at the receiving transducer in different phases is proposed as a reason for the small amplitude. At the initial frequency of the SMSW spectrum, a sharp peak is noted on the AFC, for the interpretation of which a hypothesis has been proposed about the decisive role of an increase in the SMSW length at the beginning of the spectrum, which weakens the role of interference in the formation of the AFC.

3. The formation of the phase-frequency characteristic in the same transmission line is considered. It is shown that the phase response is a frequency-periodic sawtooth curve corresponding to a continuous phase incursion with increasing frequency. It is noted that the phase incursion within the SMSW spectrum is accelerated the more the higher the frequency becomes. ...

4. The transformation of the frequency response when changing the length of the emitting and receiving transducers is considered. It is shown that with a decrease in the length of the emitting transducer, the AFC narrows from the low-frequency end of the SMSW spectrum,

and in the case of a decrease in the length of the receiving transducer, the AFC narrows from the high-frequency end. In both cases, the narrowing is accompanied by a decrease in the frequency response irregularity due to the expansion of the maxima of the function of the form, and such an expansion occurs in proportion to the magnitude of the decrease in the length of the transducers.

5. The lines of constant phase of surface magnetostatic waves (SMSW) propagating in a ferrite film magnetized in the plane by a linearly inhomogeneous field of constant direction are calculated by the method of partial wave beams. It is shown that, under these conditions, the regions of SMSW propagation with a given phase incursion have the shape of a curvilinear triangle facing downward. Cases of changing the orientation of the emitting transducer, frequency and gradient of the magnetizing field are considered. The transformations of the SMSW propagation region with the indicated changes are explained by the nature of the behavior of the wave number in accordance with the dispersion law during the SMSW propagation in an inhomogeneous field.

6. On the basis of the results obtained, practical recommendations were developed for optimizing the parameters of the transmission line to the SMSW.

7

Use of magnetostatic waves in inhomogeneously magnetic ferrite films for information processing devices and other technical applications

The properties of magnetostatic waves in nonuniformly magnetized ferrite films described in the chapters 3–6 open up rich possibilities for various technical applications. Let us list briefly some of the most striking of them, which directly follow from the results obtained.

7.1. Brief overview of possible technical applications

The use of multilayer magnetic structures (Sections 2.5–2.6) makes it possible to form the dispersion law of MSWs (magnetostatic waves) with predetermined properties. The dispersion of the MSW can be controlled by changing the magnetization and thickness of the magnetic layers, as well as the thickness of the dielectric layer between the ferrite and the metal. Based on such structures, it is possible to create dispersionless delay lines, as well as with a linear or other desired dependence of the delay time on frequency. The use of sections with a horizontal tangent on the dispersion curve transitions from forward to backward waves (Section 3.1.1) makes it possible to dramatically increase the delay time for sufficiently long MSWs, the attenuation of which is relatively small, which gives hope to create delay lines with a long delay time (5–10 µs and more) and low attenuation (no more than 10–20 dB).

The anomalous laws of reflection and refraction of MSWs at the interface between two multilayer media with different parameters (Section 3.2) make it possible to perform spatial separation of MSW wave beams of different frequencies, which can become the basis for creating multi-channel microwave filters. The strong dependence of the angles of reflection and refraction on the field makes it possible to switch channels by changing the field.

Film magnetization by a linearly inhomogeneous field (sections 4.4–4.8), as a result of which the trajectories acquire a V-shaped character, provides a strong spatial separation of wave beams of different frequencies, which makes it possible to create a multichannel frequency-separating filter with up to 10–12 non-overlapping channels.

Even more possibilities in this regard are provided by biasing the film by a 'valley'-type field (Sections 4.4–4.8), which separates the frequency beams above and below the critical on opposite sides of the bottom of the 'valley'.

Film magnetization by a 'shaft'-type field (Sections 4.4–4.8) makes it possible to create a highly efficient waveguide channel for the MSW with very low losses (up to 1 dB/cm), which allows the MSW to be transmitted without significant losses of tens of centimeters. On the basis of such a channel, it is possible to create delay lines with a long delay time (tens of microseconds). A wide range of MSW speeds allows one to vary the length of the delay lines from a few millimeters to tens of centimeters, make taps and place additional signal processing devices along the length of the delay line.

An anomalously large phase shift in inhomogeneously magnetized structures of the ferrite–metal type (Section 4.5) makes it possible to create a phase shifter with a large phase rotation, controlled by a small field.

Using the strong dependence on the field of the shape of the trajectories of backward MSWs in inhomogeneously magnetized FDM structures (Section 4.6), it is possible to create devices that perform spatial switching of wave beams up to the reversal of the direction of wave propagation.

The method of partial wave beams (Section 5.2) makes it possible not only to calculate the form of the frequency response and phase response for a wide variety of transducer and magnetizing field configurations, but also to synthesize the shape of the transducers for a given field configuration. Complemented by methods for calculating magnetic systems, it will optimize the parameters of the magnetic system for the given characteristics of the device.

Along with information processing devices, the results obtained make it possible to develop fundamentally new methods for measuring the parameters of films with a complex character of anisotropy.

This list is far from complete. In this chapter, some specific examples of the implementation of the mentioned application possibilities will be given.

We will mainly use literature sources [634–654].

7.2. Wave guiding structures for SMSW on ferrite films magnetized by a shaft-type field

Most devices for analog information processing in the microwave range are based on waveguide structures, usually made in the form of strip lines. A common disadvantage of such structures is their rather large size due to the large length of the electromagnetic wave. Another disadvantage is the impossibility of changing the initial predetermined geometric configuration of the structure during operation, which complicates the implementation of various kinds of switching functions. The use of surface magnetostatic waves propagating in yttrium–iron garnet (YIG) films as an information carrier makes it possible, on the one hand, due to the small length of the SMSW, to sharply reduce the dimensions of the waveguide systems, and on the other hand, due to the strong dependence of the direction of propagation of the SMSW on the magnitude and direction of the field, facilitate the design of switching devices. At the same time, the use of a YIG film magnetized by a uniform field as a waveguide structure for the SMSW is ineffective because of the large divergence and splitting of the SMSW beam, which leads to unjustifiably high losses and distortion of the transmitted signal.

In the Sections 4.4–4.8, it was shown that a structure consisting of a YIG film magnetized by a 'shaft'-type field has effective waveguiding properties. In such a structure, there is no divergence of the SMSW beam, as a result of which the signal loss sharply decreases and no more than a few decibels per centimeter can be made.

In accordance with the foregoing, the object of investigation of this section was the waveguiding properties of YIG films magnetized by a field of the 'shaft' type. Such a field was created by a flat magnet made in the form of a thin strip, the plane of the wide face of which was parallel to the plane of the YIG film, and the magnetic poles were located on the narrow side faces of the strip. In this case, the field component in the plane of the YIG film was oriented almost perpendicular to the long strip size. In the experiments, such a magnet was made using the film technology. In the course of the work, a computer program was compiled, which makes it possible to calculate the structure of the fields of such magnets of complex shapes, including Y-shaped magnetic systems, which make it possible to create branched waveguiding structures. Comparison of the calculation data with the experimentally measured values of the field for film magnets made of $SmCo_5$ showed that the results agree within 10%.

The study of the field component in the plane of the YIG film, created by the magnet described above, showed that two extreme cases are possible: the distance from the plane of the strip magnet to the plane of the YIG film is much less than the width of the strip of the magnet; the same distance is much larger than the width of the magnet strip. In the first case, in the plane of the YIG film, two parallel magnetic 'shafts' with a 'dip' between them are created, corresponding to the location of the poles on the lateral edges of the strip magnet. In the second case, there is only one magnetic 'shaft', on either side of which the field decreases. This 'shaft' is elongated in the middle along the axis of symmetry of the magnet. When the distances between the planes of the magnet and the YIG film are comparable to the width of the magnet strip, a smooth transition is observed between these two cases.

The analysis of the SMSW trajectories, calculated by the method described in Section 4.3.2, showed that in both cases the SMSW propagate along trajectories close to sinusoids, oscillating along the axes of the corresponding magnetic 'shafts'. In this case, in the first case, there are two parallel waveguide channels corresponding to two magnetic 'rolls', and in the second case – one waveguide channel corresponding to a single magnetic 'shafts'.

Along with magnets in the form of straight strips, magnets made in the form of branched Y-shaped strips were also considered. For the first case (a small distance between the planes of the magnet and the YIG film), such a magnet forms two parallel magnetic 'shafts' up to the branching area, similar to a simple strip magnet. After the branching area, each arm of the magnet forms its own system of two parallel magnetic 'shafts', and the magnetic 'shafts' from the unbranched part smoothly pass into the magnetic 'shafts' of the branched part, corresponding to the outer edges of the branched arms of the magnet. The magnetic 'shafts' corresponding to the inner edges of the branched arms of the magnet merge together at the branch point and have no continuation in the unbranched part.

In the second case (a large distance between the planes of the magnet and the YIG film), a single magnetic 'roller' from the unbranched part of the magnet smoothly branches into two and goes into magnetic "rollers" corresponding to each arm of the branched part of the magnet. In both cases, some (up to 20%) decrease in the magnetic field strength at the tops of the magnetic 'shafts' is observed in the branching area, which is due to the inevitable expansion of the magnet strip in the branching area and the associated mutual distance between opposite magnet poles located in this area.

The analysis of the MSSV trajectories performed for a branched magnet showed that, depending on the distance between the planes of the magnet and the YIG film, it is possible to separate two initial waveguide channels of each into its own arm, or to branch one waveguide channel into two, which corresponds to the geometry of the magnetic 'shafts'

The calculation results described above were verified in experiments. In this case, the propagation of SMSW frequencies of 2–4 GHz in YIG films magnetized by film magnetic systems made of $SmCo_5$ was studied. Film magnets were strips 500-700 µm thick, 5–6 mm wide and 4–5 cm long, deposited on copper substrates. Branched Y-shaped magnets were made of the same strips, diverging at the branch at an angle of 50°–70°. All magnetic stripes were magnetized in a plane perpendicular to their maximum size. The distance between the plane of the magnet and the plane of the YIG film was varied within 1–5 mm. The fields created by magnets in the plane of the YIG film at the tops of the magnetic 'shafts' far from the branching area were 200–500 Oe, in the branching area the field decreased by 50–100 Oe.

Investigation of the SMSW propagation using mobile antennas showed good agreement of its results with the calculations. Both the propagation of the SMSW along one or two parallel channels on simple strip magnets and the branching of one beam of the SMSW into two or the merging of two beams into one on branched Y-shaped magnetic systems were observed. The attenuation of the MSSW in the waveguide channels was significantly less than in a uniformly magnetized film and was, depending on the frequency, no more than 1–6 dB/cm. In the branching area, the attenuation increased by 57 dB additionally.

The results obtained allow us to hope for the possibility of developing a new class of microwave devices using the waveguiding properties of the described unbranched and branched magnetic structures.

7.3. Optimization of the shape of SMSW converters for devices on inhomogeneous magnetized ferrite films

An important problem arising in the design of devices based on SMSW is miniaturization, which requires a reduction in the size of the magnetic system. This decrease leads to an inhomogeneity of the constant magnetic field in the area of the location of the rectilinear SMSW transducers. The efficiency of the transducers decreases, since their various sections, located in different magnetic fields, excite and receive SMSWs with different wavefront lengths and orientations. This leads to undesirable interference phenomena that significantly degrade the frequency response of the devices. This section describes an algorithm for calculating such a mutual arrangement and shape of transducers, in which there are no interference phenomena.

The physical principle of the calculation is based on the placement of the conductors of the emitting and receiving transducers on the lines of the same phase of the SMSW propagating in a non-uniformly magnetized ferrite film. In this case, all emitted SMSWs will arrive at the receiving transducer in the same phase and there will be no signal interference.

Let us consider the algorithm for calculating the transducers using the example of the relative position of the magnet and the YIG film, shown in Fig. 7.1. Here 1 is the plane of the ferrite film, 2 is the magnet, 3 and 4 are the emitting and receiving transducers of the SMSW. The origin of the $Oxyz$ coordinate system is on the plane of the film, the coordinate plane Oyz coincides with the plane of the film. The film is cut in the form of a rectangle with a size along the Oy axis 12 mm, along the Oz axis 8 mm. The magnet is made of $SmCo_5$ in the form of a rectangular parallelepiped with edges a, b and c (where $a \ll b < c$), parallel to the Ox, Oy, Oz axes, and, accordingly, with: $a = 1$ mm, $b = 6$ mm, $c = 10$ mm. The centre of the wide edge of the magnet is aligned with the Ox axis. The distance

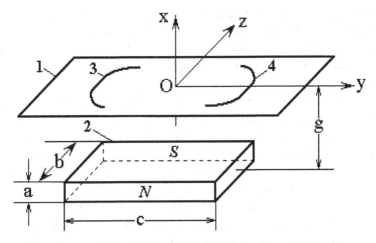

Fig. 7.1. General diagram of the device. 1 – YIG film, 2 – magnet, 3, 4 converters.

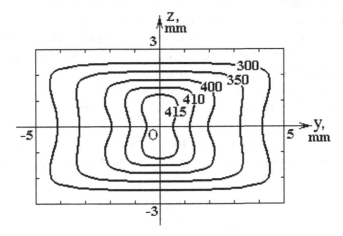

Fig. 7.2. Bias field configuration. The numbers on the curves are the values of the z-component of the field in oersteds.

between the centre of the magnet and the plane of the film is $g = 1.7$ mm. The magnet is magnetized parallel to the short edge b of the wide edge, that is, along the Oz axis. The poles of the magnet are on the side faces parallel to the plane.

Such a magnet creates a field in the plane of the film, illustrated in Fig. 7.2, which shows the lines of constant values of the field component in the plane. The numbers on the curves correspond to the field strength. It can be seen that the field is highly nonuniform and varies from the centre to the edges of the film by more than 25%; therefore, the expected shape of the transducers should be strongly curved.

The field is symmetric about the Oz axis; therefore, we assume that the wave beams coming out of different parts of the emitting transducer cross this axis at a right angle and their phase on this axis is the same. Further, the field is symmetric about the Oy axis, therefore the transducers must be symmetrical about this axis. Thus, as a reference line, it is sufficient to select an Oz axis segment from 0 to z_m, the value of which will determine the film section through which the SMSWs will pass on their way from the emitting transducer to the receiving one. Dividing this segment into equal sections, drawing from the splitting points the SMSW trajectories, intersecting the Oz axis at right angles, and laying a specified number of SMSW lengths along the length of these trajectories, we obtain a constant phase line along which the transducer conductor should be placed. Due to the symmetry of the field about the Oz axis, the pattern of trajectories, lines of constant phase and conductors of both transducers are also symmetrical about this axis.

The calculation results for a frequency of 2850 MHz are illustrated in Fig. 7.3, which shows the contours of the transducers (solid lines) and some SMSW trajectories (dashed line) plotted for z_m = 2 mm. The numbers on the curves correspond to the number of SMSW lengths that fit along the halves of the trajectories corresponding to $y > 0$.

The intersection of trajectories near y = 2 mm leads to the appearance of loops on the lines of constant phase, therefore, when

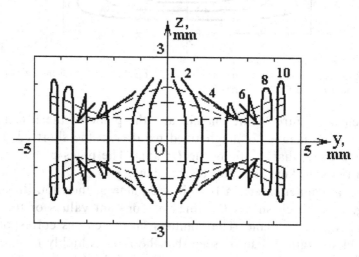

Fig. 7.3. Transducer contours (solid lines) and SMSW trajectories (dashed line). The numbers on the curves are the number of SMSW lengths at half the trajectory length.

calculating the transducer contour, the number should be set small, no more than 2–3, when loops are not yet formed. For other frequencies, the contours of the transducers are obtained differently. So for a frequency of 3150 MHz at $N = 2$, both the middle and the ends of the converters are shifted to the axis by one and a half to two times, and to varying degrees. The signals arriving at the receiving transducer from different parts of the radiating transducer are in phase only at a frequency of 2850 MHz, and when detuning from this frequency, skewing occurs, leading to a drop in the signal amplitude. Therefore, the frequency response of the transmission of the entire system as a whole has clearly pronounced frequency-selective properties, and its width decreases with increasing.

According to the calculation data at $N = 2$, a bandpass filter was manufactured for a frequency of 2850 MHz with a bandwidth of about 200 MHz. Compared to a similar design using straight-line converters, the passband loss has decreased by 2–3 dB, and outside this bandwidth it has increased by 12–15 dB. In a similar filter using focusing converters, taking into account the field inhomogeneity led to a narrowing of the passband from 50 to 20 MHz and a decrease in losses at the centre frequency by 10–12 dB.

7.4. Multi-channel filter on ferrite film magnetized by a valley-type field

One of the important tasks of information processing in the microwave range is frequency-selective channel separation. A convenient tool for solving this problem is the propagation of magnetostatic waves in a ferrite film magnetized by a 'valley'-type field, described in Sections 4.4–4.8. Such a field divides the wave beams of frequencies above and below the critical into two groups, propagating on different sides relative to the bottom of the 'valley', and in addition, within each group, spatially separates beams of different frequencies over considerable distances.

Thus, to construct a multi-channel frequency selective filter, it is sufficient to install a separate receiving transducer in the path of each of the beams of a given frequency. The length of the emitting transducer must be sufficient to eliminate the divergence phenomena, and the receiving transducers must be installed at a sufficient distance from each other to avoid overlapping beams. At the same time, with an increase in the length of the trajectories, the spatial separation of the channels improves, but the wave attenuation also increases, that is, the requirements for the maximum number of channels and the minimum attenuation are contradictory.

In the experiments it was found that in the frequency range 2–4 GHz, the length of the emitting converter should be at least 3–4 mm, and the maximum number of channels with an attenuation level of no more than 40–50 dB and mutual isolation between adjacent channels of at least 20 dB is 12–15. Four filters for 2, 4, 5 and 7 channels were made in the form of working models. As an example, consider the design of a five-channel filter. The mutual arrangement of the YIG film and the magnet coincided with that shown in Fig. 7.1. The 'valley'-type field configuration was created by making a magnet from three $SmCo_5$ strips parallel to the axis and having different thicknesses. In this case, thicker stripes were located at the edges, and thinner ones – in the middle. The length of the strips was 25 mm, the width of each strip was 6 mm, the thickness of the outer strips was 3 mm, and the average thickness was 2 mm. The strips folded together constituted a magnet with external dimensions of 3 18 25 mm. The distance between the centre of the magnet and the plane of the film was 2.5 mm. Such a magnet created a 'valley'-type field in the film plane with a bottom parallel to the Oy axis. The field on the Oy axis was 510 Oe, at a distance of 6 mm from this axis, 550 Oe.

The layout of the converters for a five-channel filter is shown in Fig. 7.4. Here 1 is the plane of the ferrite film, 2 is the emitting transducer, 3–7 are the receiving transducers. The horizontal dashed line shows the bottom of the 'valley'. Thin solid lines show trajectories for different frequencies.

The transducers were placed on a polycor plate with dimensions of 20×30×0.5 mm, metallized on one side, superimposed on a ferrite film. Emitting transducer 2 was made in the form of a straight conductor 3 mm long and 12 μm thick. The microwave signal was applied to one of its ends (the lower one in the figure) using a microstrip line (shown by a wide blackened strip). The other end of the transducer was grounded through a hole in the polycor to the metal layer on the other side of the plate (the ground is shown with a closed circle). The design of the receiving converters 3–7 was similar.

All frequency responses of individual channels are bell-shaped. The centre frequency of channel 1 (corresponding to converter 3) is 3280 MHz, channel 2 (converter 4) is 3340 MHz, channel 3 (converter 5) is 3400 MHz, channel 4 (converter 6) is 3460 MHz,

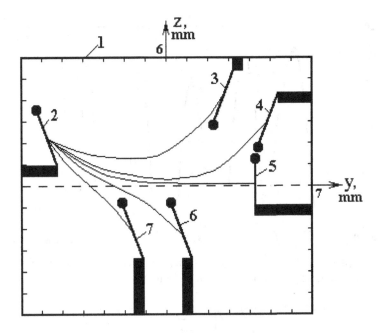

Fig. 7.4. The layout of the transducers of the multichannel filter. 1 – ferrite film, 2 – emitting converter, 3–7 – receiving converters.

channel 5 (converter 7) is 3520 MHz. The channel bandwidth varies from 40 MHz (channel 1) to 15 MHz (channel 5), the mutual isolation between adjacent channels is more than 20 dB, the attenuation is from 16 to 20 dB. The overall dimensions of the device together with the microwave connectors are 20 30 40 mm. The parameters of the other filters were similar.

7.5. Multichannel filter on packed ferrite structures

In the previous section 7.4, we considered the construction of a multichannel filter due to the separation of the trajectories of wave beams in the film plane. At the same time, the magnetic system creates a field not only in the plane of the film, but also on both sides of it. By placing additional films in the area of action of this field above and below the main one in the form of a plane-parallel package, it is possible to obtain a significant increase in the number of channels, as many times as the number of films will be installed. At the same time, the dimensions of the magnetic system, which

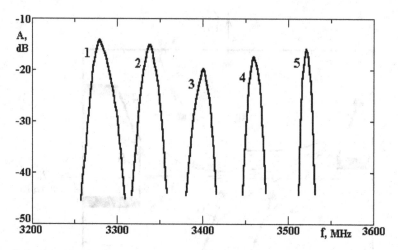

Fig. 7.5. Amplitude–frequency characteristics of a multichannel filter on a 'valley'-type field. 1–5 various frequency channels.

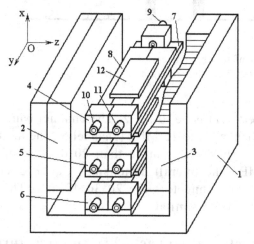

Fig. 7.6. Scheme of a multichannel filter on packet structures. 1 – yoke, 2, 3 – magnets, 4–6 – separate filters, 7 – brass plate, 8 – polycor plate, 9 – input connector, 10, 11 – output connectors, 12 – YIG film.

determine the overall dimensions of the device, remain practically the same.

In the experiments, such a filter was made on a package of three ferrite structures located one above the other. The device diagram is shown in Fig. 7.6. The constructive basis of the filter is a magnetic system consisting of a U-shaped yoke 1 made of soft iron, on the inner walls of which magnets 2 and 3 of $SmCo_5$ are fixed, made in the form of normally magnetized square plates with the dimensions of 30×30×8 mm. The gap between the wide edges of the magnets is 17 mm. In this gap, three two-channel filters 4, 5 and 6 are installed, each of which is based on a brass plate 7, on which a plate of polycor 8 is located, with one emitting and two receiving transducers made on it, connected respectively to one input 9 and two output 10 and 11 microwave connectors. On top of the transducers, a YIG 12 film is applied. The transducers are made in the form of wire antennas 3 mm long and 12 μm thick, the distance between the emitting and receiving transducers is 4–8 mm, all YIG films are identical, have a magnetization of 1750 G, the FMR line width is 0.6 Oe, dimensions 10×15 mm and thickness 10 μm.

The centre of the used coordinate system is in the middle of the distance between magnets 2 and 3 in the centre of their wide edges (in the figure, the coordinate system is shown separately in the inset for convenience of the drawing). The plane coincides with the planes of the magnets, the Oz axis is perpendicular to them, and the Ox and Oy axes and are parallel to the edges of the magnets. The planes of the ferrite films are parallel to the coordinate plane, and the direction of the field in the plane of the films is close to the axis.

In the described magnetic system, along the Oy axis at $y = 0$, the field z-component is maximum in the centre of the system and decreases towards the edges. So at $x = 0$ it is equal to 2200 Oe, at $x = \pm 10$ mm – 2000 Oe, at $x = \pm 20$ mm 0 – 1000 Oe. Along the direction of the Oz axis in the plane Oxz at $x = 0$ mm – the field z-component within the range 6 mm $< z <$ 6 mm is practically uniform and amounts to 2200 Oe, at $0 < x <$16 mm it has a 'valley' profile, at $x = 16$ mm it is also practically uniform and is equal to 1420 Oe, and at $x > 16$ mm it has a 'shaft' profile. The field of the 'valley' type, for example, at $x = 15$ mm at its 'bottom' is equal to 1620 Oe, and at $z = \pm 6$ mm it increases to 2100 Oe. The field of the 'shaft' type, for example, at $x = 20$ mm at its 'top' is equal to 860 Oe, and at $z = \pm 6$ mm it falls to 500 Oe. Thus, in the magnetic system along the coordinate there are three regions, the location in which the waveguide structure in the plane is favourable for the propagation of the SMSW: near its centre, as well as when moving away from the center in one direction or another by distance more than 16 mm. In these areas, the field is uniform or has a 'shaft' profile, therefore the attenuation of the SMSW during propagation is small, while between these regions, where the field has a 'valley' profile, the SMSW energy is scattered to the sides and the attenuation increases significantly. Thus, when the emitting and receiving transducers are oriented along the axis and the distance between them is 3 mm, the signal attenuation (at the frequency where it is minimal) at −5 mm < x <5 mm is equal to 15 dB, at −25 mm <<16 mm and 16 mm < x <25 mm − equal to 10–12 dB, and between these areas - 20-30 dB or more. Accordingly, one of the YIG films was located exactly in the centre of the magnetic system at $x = 0$, and the other two − at $x = -17$ mm and $x = 19$ mm. A two-channel filter was made on each film. In these filters, the channel separation was carried out due to the location of the emitting transducer at an angle of 20°–30° to the Oz axis, which ensured spatial separation of the SMSW beams of different frequencies. In the process of tuning, the locations of the two receiving transducers were pre-selected with the help of movable antennas so as to ensure optimal reception of each of the channels, after which the transducers were installed permanently

The obtained amplitude–frequency characteristics of the transmission of the entire filter as a whole are shown in Fig. 7.7. Here, AFCs 1 and 2 correspond to the YIG film set at $x = 19$ mm, 3 and 4 to the film set at $x = -17$ mm, and 5 and 6 to the film set at $x = 0$ mm. It can be seen that the frequencies of each pair of channels are lower the further from the centre of the system the corresponding YIG film is. The frequency distance between the lower and middle pairs is close to 1350 MHz, between the middle and higher ones - 2450 MHz, within each pair the channels are separated by ~100

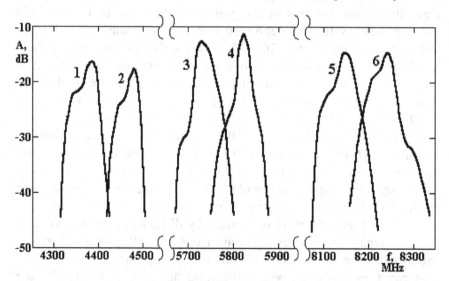

Fig. 7.7. Amplitude–frequency characteristics of the filter on packaged structures. 1–6 − different frequency channels.

MHz. The centre frequency of the lowest channel is 4380 MHz, the highest is 8245 MHz. The bandwidth of all channels is 25–70 MHz, the attenuation at the centre frequency is 12–17 dB. The isolation between channels at the frequency of their overlap inside each pair is 30–40 dB, and between the pairs it exceeds 70 dB.

Further improvement of the filter should go along the path of optimizing the size of the magnetic system in order to reduce the distances between individual ferrite films and increase their total number, as well as optimize the design of each single filter to increase the number of channels while maintaining good isolation between them. If these requirements are met, a noticeable increase in the number of channels is possible, so each single filter can be made four-channel with the total number of such filters at least five, that is, the total number of channels can reach 20 or more with the overall overall dimensions of the system not exceeding 40×40×40 mm.

7.6. Microwave signal delay line on a ferrite film magnetized by a shaft-type field

Along with filtering the most important analog signal processing, time delay. Magnetostatic improvement waves are a convenient means of solving this problem due to the low speed of electromagnetic waves. In most of the known designs of the delay lines at the SMSW. catastrophically increasing at long delay times (more than 100 ns). At the same time, as shown in the Sections 4.4–4.8, an inhomogeneous field of the 'shaft' type, due to the formation of a waveguide channel for the SMSW, completely suppresses any divergence and provides very small losses of wave energy during propagation.

This section uses such a field to plot a pulsed microwave signal. A YIG film with a magnetization of 1790 G and a thickness of 12 µm, cut in the form of a rectangle with dimensions of 15×68 mm, was used as a waveguide medium. The magnet was made in the form of a $SmCo_5$ rectangular parallelepiped with dimensions of 7×15×70 mm, magnetized along a size of 15 mm. The mutual arrangement of the layers and the magnet coincided with that shown in Fig.7.1. The distance between the centre of the magnet and the plane of the film was 9 mm. The magnet created a magnetic 'shaft' in the plane of the film, the field at the top of which was 750 Oe, and at a distance of 5 mm from the top, it dropped to 580 Oe. This field provided a waveguide channel about 6 mm wide. The emitting and receiving SMSW transducers were made in the form of straight antennas 4

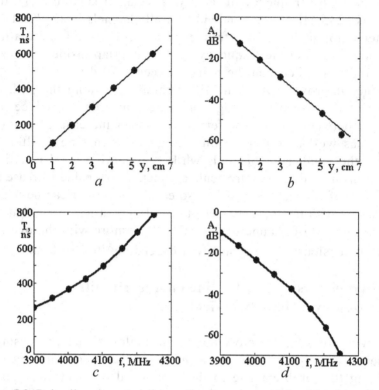

Fig. 7.8. Delay line characteristics. *a, b* – dependence of the delay (*a*) and signal attenuation (*b*) on the distance between the transducers; *c, d* – similar dependences on the signal frequency.

mm long and 12 μm in diameter, oriented along the axis (Fig. 7.1). The microwave frequency was 3800–4400 MHz, the pulse duration was 100–200 ns.

The resulting line characteristics are shown in Fig. 7.8.

Figure 7.8 shows the determination of the delay time of the 4100 MHz signal from the distance between the emitting and receiving transducers. Figure 7.8 b shows the dependence of the signal attenuation during propagation for the same conditions. It can be seen that with a change between converters, the time increases, and the signal attenuation increases. Both dependences are close to linear with slopes of 100 ns/cm and 8 dB/cm, respectively. Such a low attenuation makes it possible to remove the receiving transducer from the emitting one by more than 6 cm at an attenuation of less than 60 dB. In this case, the delay time reached 600 ns. In a uniform field, the attenuation at the same frequency is 17–25 dB/cm and does not allow the transducers to be moved apart by more than 2–3 cm, as well as to obtain a delay time of more than 150–200 ns. Figure 7.8 of the signal for establishing the delay time from the microwave frequency in the range 3900-4300 MHz with a distance between the transducers equal to 5 cm. Figure 7.8d shows the dependence of the signal attenuation during propagation for the same conditions. It can be seen that the time increases with frequency conversion. This circumstance is associated with a decrease in the group velocity of the SMSW with increasing frequency. At 3900 MHz, the slope of the delay growth is 0.7 ns/MHz, and at 4200 MHz – 2.7 ns/MHz. This nonlinear character of the given dependence reflects the nonlinear form of the dispersion law of the SMSW. In this case, the increase in attenuation is due to the shortening of the SMSW length, which also corresponds to the dispersion law.

The use of a 'shaft' type field allows obtaining a frequency of 4240 MHz with a signal attenuation of less than 70 dB for such a long time as 800 ns. In this case, the device between the transducers is 5–6 cm, which makes it possible to connect additional taps and signal processing along the wave path.

7.7. Measurements of parameters of yttrium iron garnet films with a complex anisotropy character

For the successful design of information processing devices on YIG films, a fuller knowledge of these films is possible. The methods used in this work for measuring the parameters of YIG films and the results obtained are already presented in Section 4.8, therefore, here, a comparative analysis of the efficiency of the microwave and magneto-optical method for measuring the parameters of YIG films with a complex character of anisotropy is presented. Below are the data that were directly measured in the experiments, as well as the parameters of the films measured from these measurements. All orientation dependences are removed when the film is rotated around the normal to its plane.

1) Orientational dependences of the initial frequency of the SMSW, the frequency of a uniform FMR. These dependences were used to determine the location of the projections of the cubic anisotropy axes onto the film plane, and also using the technique described in [651, 653], the value $4\pi M_0 - H_A$.

2) The frequency of the beginning of the SMSW spectrum, corresponding to the frequency of uniform FMR, when the film is

magnetized by a field perpendicular to its plane. This frequency was used to determine the value $4\pi M_0 - H_A$.

3) Orientation dependences of the dispersion curve for SMSW. These dependences were used to determine the location of the projections of the cubic anisotropy axes on the film plane, the orientation and the angle of deviation from the normal to the film plane of the uniaxial anisotropy axis, the cubic anisotropy field, and the film thickness using the technique described in [368].

4) Orientation dependences of the field of disappearance of the domain structure. From these dependences, using the technique described in [651, 653], the location of the projections of the axes of the cubic anisotropy on the film plane.

In addition, as control, for the studied films, there were data on the orientation of the axes of crystallographic anisotropy obtained from X-ray diffraction analysis, as well as the values of the film thickness obtained by the interference method.

Comparison of the above methods showed that the domain structure is very sensitive to the parameters of the film, as well as the FMR and dispersion of SMSWs in the presence of a domain structure. At the same time, the construction of a qualitative model of anisotropy (its type and orientation of the axes) is not difficult, however, due to the lack of an adequate model of the domain structure, it is difficult to obtain quantitative results. FMR and MSW dispersion in the saturation region are much less sensitive to small variations in film parameters. Based on these data, a qualitative model can be satisfactorily constructed only for cubic anisotropy, while the exact orientation of the uniaxial anisotropy axis is much more difficult to find. It is also impossible to separate uniaxial anisotropy from magnetization. In [368], it is argued that such a separation is feasible in the MSW dispersion; however, in the present work, the use of this method did not give good results. At the same time, if a qualitative model of anisotropy has already been built (for example, by studying the domain structure), then, due to the simplicity of the resonance model and the high accuracy of microwave measurements, good quantitative results can be obtained for the magnitude and cubic anisotropy.

7.8. Study of the spatial distribution of the magnetic field with the help of the sensor on the SMSW

For the design of microwave devices based on ferrites, the problem of local measurement of the magnetic field in small volumes is extremely important. The dimensions of traditional field sensors (for example, Hall) are quite large, so the measurement accuracy is low. In this work, we investigated the possibility of using beams of surface magnetostatic waves (SMSW) for this purpose, whose trajectories in an inhomogeneous field are strongly curved (Chapter 4). The field sensor contains a 15 µm thick yttrium iron garnet (YIG) film on a disk substrate 5 cm in diameter and 0.5 mm thick. On the surface of the film, there are SMSW transducers in the form of straight conductors 3 mm long and 12 µm thick, which can rotate and move over the entire area of the film. The thickness of the entire structure is 3 mm. The sensor is placed in the area of space where the field distribution is measured, after which the trajectory of the SMSW beam is recorded at a frequency of 3000 MHz. The process of measuring the trajectory parameters is similar to that described in Section 4.8. The field is determined by solving the problem inverse to the problem of integrating a system of three nonlinear first-order differential equations that determine the parameters of the SMSW wave beam trajectory (Section 4.3.2). The solution process consists in dividing the trajectory with a certain step into final sections and finding the field at the starting point of each section, assumed to be linear, and the field on it – homogeneous. A necessary condition for the unambiguousness of the solution is the setting of three parameters: the coordinates of the initial and final points of the investigated section and the direction of the phase velocity vector. The phase of the wave or the delay time of the pulse signal can also be taken as the third parameter.

The measurement results are illustrated in Fig. 7.9. Figure 7.9 a shows the SMSW trajectory (1) and the corresponding dependence of the angle between the phase velocity vector and the field direction (2). Figure 7.9 b shows the dependence of the field on coordinate (1) measured along the trajectory and the same dependence measured by the Hall sensor (2). The field values obtained with the SMSW sensor coincide with the data on the Hall effect with an accuracy of 2%. The spatial resolution at the 1% level is about 0.5 mm.

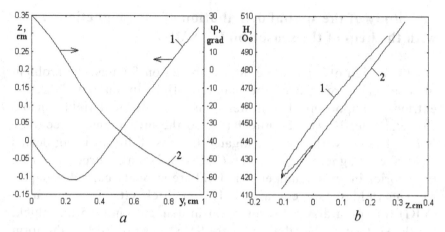

Fig. 7.9. Measurement of the field using a sensor on the SMSW, a – trajectory of the SMSW (1) and the coordinate dependence of the angle between the phase velocity vector and the direction of the field (2); b – dependence of the field on the coordinate, measured along the SMSW trajectory (1) and using a Hall sensor (2).

7.9. Use of the transmission line to SMSW to determine the orientation of the magnetic field

For the design of small-sized microwave devices on ferrite films, it is important to study the local orientation of the magnetic field. The dimensions of traditional field sensors (for example, Hall) are quite large, so the measurement accuracy is low. In this work, we investigated the possibility of using a surface magnetostatic wave (SMSW) transmission line for this purpose, which forms the basis of most ferrite film devices.

The transmission line diagram is shown in Fig. 7.10 a–c. The coordinate plane coincides with the plane of the ferrite film. SMSW transducers – emitting I'–I'', and receiving W'–W'', are oriented along β. Vector $\vec{H}$ is the magnetic field, vector $\vec{S}$ is the direction of propagation of the SMSW beam, α is the angle between the vector $\vec{H}$ and the axis, and β is the angle between the vector $\vec{S}$ and the Oy

In Fig. 7.10 a, the field is parallel to the conductors of the transducers. In hi case α = β = 0. In Fig. 7.10 b, the field is slightly deviated from the axis, in Fig. 7.10 c, the deviation is greater. In both cases β > α, and with increasing α the angle β increases faster than, which increases the sensitivity of measurements. When the field is parallel to the Oz axis, the entire wave beam created by the emitting transducer hits the receiving one (Fig. 7.10 a). In this case,

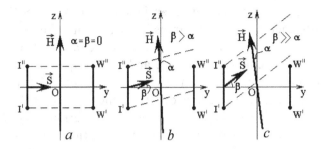

Fig. 7.10. Diagram of the path of a wave beam in a transmission line when the field orientation is changed. *a, b, c* – successive increase in the deviation of the field from the *Oz* axis.

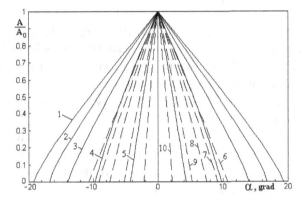

Fig. 7.11. Dependences of the amplitude of the output signal of various frequencies on the angle α at a field of 437.5 Oe and a distance between the transducers of 5 mm (solid lines) and 10 mm (dashed line). Curves 1 and 6 correspond to 2800 MHz, 2 and 7 to 3000 MHz, 3 and 8 to 3200 MHz, 4 and 9 to 3400 MHz, 5 and 10 to 3600 MHz.

the signal at the receiving transducer is maximum. When the field deviates from the *Oz* axis, part of the beam passes by the receiving transducer (Fig. 7.10 b) and the signal on it decreases. With an even greater deviation of the field, the entire beam goes to the side (Fig. 7.10 c) and the signal on the receiving transducer drops to zero.

The dependence of the relative amplitude of the output signal on the angle α is shown in Figure 7.11. The solid curves 1–5 correspond to the distance between the transducers equal to 5 mm, dashed curves 6-10–10 mm.

To measure the field orientation, it is necessary to measure the signal amplitude at the receiving transducer, and then find it using calibration curves similar to those shown in Fig. 7.11. The sensitivity of the field orientation measurement is determined by the slope of curves 1–10 and increases with an increase in the SMSW frequency and the distance between the transducers. So at a frequency of 2800 MHz and a distance of 5 mm, the signal changes from zero to a maximum when changing by 20°, and at a frequency of 3600 MHz and a distance of 10 mm, when changing by only 2°. In this case, if the measurement accuracy of the output signal is 1%, then the minimum recorded change in the field orientation is 0.02 degrees. Note that the Hall sensor under similar conditions allows registering an angle of at least 8o, which is 400 times worse than the given value.

Conclusions for chapter 7

The main issues discussed in this chapter are as follows.

1. The possibilities of using magnetostatic waves in nonuniformly magnetized ferrite films for information processing devices and other technical applications are considered. A number of promising directions have been identified, namely: the use of multilayer magnetic structures to form a dispersion law with predetermined properties, on the basis of which it is possible to create delay lines – dispersionless, with a linear or other desired dependence of the delay time on frequency, with a long delay time (5–10 μsec or more) and low attenuation (no more than 10–20 dB); the use of anomalous laws of reflection and refraction of MSWs at the interface between two multilayer media with different parameters to create multichannel filters and switches for microwave channels; the use of non-uniform fields – linear and of the 'valley' type to create multichannel frequency separation filters with the number of channels up to 10–12 and more; the use of a non-uniform field of the 'shaft' type to create a highly efficient waveguide channel with very low losses (up to 1 dB/cm), as well as delay lines with a long delay time (tens of microseconds) and a large number of taps (5–10 or more); the use of an anomalously large phase incursion in inhomogeneously magnetized structures of the ferrite–metal type to create a phase shifter with a large phase rotation (tens and hundreds of radians), controlled by a small field (units of oersted); the use of the strong dependence on the field of the shape of the trajectories of reverse MSWs in inhomogeneously magnetized FDM structures to create devices for spatial switching of wave beams; using the method of partial wave beams for calculating the frequency response and phase response of devices, synthesizing the shape of transducers

and optimizing the parameters of magnetic systems; use of the results obtained to measure the parameters of films with a complex character of anisotropy, as well as to determine the parameters of the magnetizing field.

2. The properties of waveguide structures for SMSWs on ferrite films magnetized by a 'shaft'-type field created by a flat magnet in the form of a thin strip, the plane of the wide face of which is parallel to the plane of the YIG film, and the poles are located on narrow lateral faces, have been experimentally investigated. It is shown that, depending on the distance between the planes of the magnet and the film, one or two parallel magnetic 'shafts' can be created in the film plane, forming, respectively, one or two parallel channels for the SMSW. Magnets made in the form of branched Y-shaped stripes are also considered. It is shown that with the help of such a system, it is possible to form either two parallel waveguide channels diverging after branching the magnet in different directions, or one channel branching into two channels corresponding to different arms of the magnet. At a microwave frequency of 2–4 GHz for $SmCo_5$ magnets, the attenuation of the SMSW in the waveguide channels was 1–6 dB/cm and in the branching region it additionally increased by 5–7 dB. The performed calculation of the fields of magnetic systems and the parameters of wave beams of SMSWs propagating in the channels showed agreement with experiment within 10%.

3. Based on the method of partial wave beams, an algorithm has been developed for calculating and optimizing the shape of SMSW transducers for devices using inhomogeneous magnetization of ferrite films. The shape of transducers for a magnet is calculated in the form of a rectangular plate, the wide face of which is parallel to the plane of the film, and the poles are located on opposite narrow faces. It is shown that as the distance between the transducers increases, their shape takes on the form of wide arcs facing outward bulges, after which loop-like sections appear at the ends of the arcs, where the conductors of the transducers experience self-intersection. Strong frequency-selective properties of such a system are revealed, which increase with the distance of the converters from each other. Based on the calculation data, mock-ups of band-pass filters for frequencies of 2–3 GHz with a bandwidth from 20 to 200 MHz were made. Comparison with filters based on linear converters revealed a decrease in losses in the passband by 2–3 dB and an increase in losses outside this band by 12–15 dB.

4. Implemented new principles for creating multichannel filters: on a non-uniform field of the 'valley' type and on packaged ferrite structures. It is shown that in a filter on a 'valley'-type in the frequency range 2–4 GHz, it is possible to place at least 12–15 channels with an attenuation level of no more than 40–50 dB and a mutual isolation between the channels of at least 20 dB. In a laboratory model of a filter for five channels, the bandwidth of each channel was 20 ± 5 MHz with a mutual isolation between the channels of more than 20 dB and an attenuation of 16 ± 4 dB. In the filter on packaged structures in the range of 2–10 GHz, it is possible to place 20 or more channels with an attenuation level of no more than 30-40 dB and a mutual isolation between channels of at least 20 dB. In a laboratory model of a filter for six channels, made on the basis of a package of three two-channel ferrite structures in the range 4.3–8.4 GHz, the bandwidth of each channel is 25–70 MHz with a mutual isolation between channels of more than 30–40 dB and an attenuation of 12–17 dB.

5. Implemented a signal delay line on a ferrite film magnetized by a 'shaft'-type field. The possibility of creating a delay line with a long delay time (units of μsec) and record low losses (units of dB) is shown. In a laboratory model of a delay line at a frequency of 4.2 GHz with a signal attenuation of less than 70 dB, a delay time of more than 800 ns was obtained, which is more than an order of magnitude better than similar values in a uniform field. An important advantage of the delay line on a 'shaft'-type field is a significant distance between the transducers – 5–6 cm or more, which makes it possible to place additional taps and signal processing devices on the wave path.

6. A method for measuring the spatial distribution of an inhomogeneous magnetic field, based on the study of the trajectories of wave beams of the SMSW, has been developed. The possibility of creating a sensor of an inhomogeneous field on an MSW with a high spatial resolution is shown. The laboratory prototype of the sensor made it possible to measure the field with an accuracy of 1–2% at a spatial resolution of 0.5 mm. The possibility of using the SMSW transmission line for determining the orientation of the magnetic field is demonstrated. The high sensitivity of the method was revealed, which makes it possible to register changes in the field orientation by hundredths of a degree, which exceeds the capabilities of the Hall sensor under similar conditions by more than two orders of magnitude.

Bibliography

1. Landau L.D., Lifshitz E.M. On the theory of the dispersion of magnetic permeability in ferromagnetic bodies. // Phys. Zs. der Sowjetunion. 1935. V.8. # 2. P.153.
2. Gurevich A.G. Ferrites at ultrahigh frequencies - Moscow: State Publishing House of Physics and Mathematics Literature, 1960.
3. Gurevich A.G. Magnetic resonance in ferrites and antiferromagnets - Moscow: Nauka, 1973.
4. Gurevich A.G., Melkov G.A. Magnetic vibrations and waves - Moscow: Fizmatlit, 1994.
5. Vonsovsky S.V., Shur Ya.S. Ferromagnetism - M .: OGIZ Gostekhizdat, 1948.
6. Ferrites in nonlinear microwave devices. Sat. articles ed. Gurevich A.G. - M .: IL, 1961.
7. Ferromagnetic resonance and behavior of ferromagnets in alternating magnetic fields. Collection of articles, ed. Vonsovsky S.V. - M .: IL, 1952.
8. Ferromagnetic resonance. Collection of articles, ed. Vonsovsky S.V. - M .: Fizmatgiz, 1961.
9. Lax B., Button K. Superhigh-frequency ferrites and ferrimagnets - M .: Mir, 1965.
10. Monosov Ya.A. Nonlinear ferromagnetic resonance - Moscow: Nauka, 1971.
11. Suhu R. Magnetic thin films - Moscow: Mir, 1967.
12. Vonsovsky S.V. Magnetism - Moscow: Nauka, 1971.
13. Antonov L.I., Mironova G.A., Lukasheva E.V., Malova T.I.
14. Study of the phenomenon of ferromagnetic resonance (FMR) and determination of the nature of the magnetic anisotropy of a ferromagnetic single crystal. Preprint / Moscow State University No. 5/1999 - M .: Moscow: 1999.
15. Kozlov V.I. Investigation of inhomogeneity and anisotropy of magnetic films using gyromagnetic effects. - Abstract dissertation. for a jobDoctor of Physical and Mathematical Sciences - M .: Moscow State University, 1997.
16. Dmitriev V.F., Kalinikos B.A. Excitation of spin waves in perpendicularly magnetized ferrite films // ZhTF. 1987. T. 57. No. 11. P.2212.
17. Kalinikos B.A. Excitation of dipole-exchange spin waves in ferromagnetic films: spin-wave Green's functions // ZhTF. 1984. Vol. 54. No. 9. P.1846.
18. Kalinikos B.A., Miteva S.I. Dispersion of dipole-exchange spin waves in a layered structure // ZhTF. 1981. Vol.51. No. 10. P.2213..
19. Rodrigue G.P. A generation of microwave ferrite devices // Proc. IEEE. 1988. V.76. # 2. P.121. Translation: Glass H.L. Ferrite films for microwave devices // TIIER. 1988. Vol.76. # 2. P.64.
20. Kurushin E.P., Nefedov E.I. Application of thin monocrystalline ferrite films in microwave microelectronic devices // Microelectronics. 1977. Vol.6. No. 6. P.549.
21. Salansky N.M., Erukhimov M.Sh. Physical properties and application of magnetic films - Novosibirsk: Nauka, 1975.
22. Yakovlev Yu.M., Gendelev S.N. Monocrystals of ferrites in radio electronics - M .: Sov. radio, 1975.

23. Tsutaoka T., Ueshima M., Tokunaga T .. No. Akamura T., Hatakeyama K. Frequency dispersion and temperature variation of complex permeability of Ni-Zn ferrite composite materials // JAP. 1995. V.78. No. 6. R.3983.
24. Mikaelyan A. L. Theory and application of ferrites at ultrahigh frequencies - Moscow: Gosenergoizdat, 1963.
25. Rankis G.Zh. Dynamics of magnetization of polycrystalline ferrites - Riga: Zinatne, 1981.
26. Bazhukov K.Yu. Frequency properties of soft magnetic ferrites with different microstructure and shape - Dissertation for thesis. uch. Art. Ph.D. - M .: 1999.
27. Bazhukov K.Yu. Frequency properties of soft magnetic ferrites with different microstructure and shape - Abstract of diss. for a job. uch. Art. Ph.D. - M .: IRE RAN, 1999.
28. Kotov L.N., Bazhukov K.Yu. Calculation of the permeability of polycrystalline ferrite // ZhTF. 1998. Vol. 68. No. 11. P.72.
29. Kotov L.N., Bazhukov K.Yu. Calculation of magnetic spectra of ferrites //RE. 1999. Vol.36. No. 7. P.41.
30. Park D. Magnetic rotation phenomena in a polycrystalline ferrite // PR. 1955. V.97. # 1. P.60.
31. Sparks M. Ferromagnetic resonance porosity linewidth theory in polycrystalline insulators // JAP. 1965. V. 36. No. 5. P.1570.
32. Grimes C.A., Grimes D.M. Permeability and permittivity spectra of granular materials // PR (B). 1991. V.43. P.10780.
33. Ponomarenko V.I., Berzhansky V.N., Zhuravlev S.I., Pershina E.D., Dielectric and magnetic permeability of an artificial dielectric with metallized ferrite particles at ultrahigh frequencies. 1990. Vol. 35. No. 10. P.2208.
34. Ribas R., Labarta A. Magnetic relaxation of a one-domain model for small particle system with dipolar interaction. Monte Carlo simulation // JAP. 1996. V.80. No. 9. P.5192.
35. Bespyatykh Y. I., Tarasenko V. V., Kharitonov V. D. Effective dielectric and magnetic permeability of an ensemble of ferromagnetic particles // RE. 1988. Vol. 33. No. 4. P.872.
36. Ponomarenko V.I., Berzhansky V.N., Zhuravlev S.I. Calculation, design and synthesis of coordinated radio-absorbing structures based on magnetic materials // XI Int. conf. by gyromagnet. electron. and electrodyn.: Abstracts. report Vol. 1. Ukraine, Alushta, 1992.S. 38.
37. Kwon H.J., Shin J.Y., Oh J.H. The microwave absorbing and resonance phenomena of Y-type hexagonal ferrite microwave absorbers // JAP. 1994. V.75. No. 10. P.6109.
38. Pokusin D.N., Chukhlebov E.A., Zalessky M.Yu. Complex magnetic permeability of ferrites in the area of natural ferromagnetic resonance // RE. 1991. Vol.36. No. 11. P.2085.
39. Sparks M., et al., Ferromagnetic resonance in epitaxial garnet thin films // JAP. 1969. V.40. No. 3. P.1518.
40. Laulicht I., Suss J.T., Barak J. The temperature dependence of the ferromagnetic and paramagnetic resonance spectra in thin yttrium-iron garnet films // JAP. 1991. V.70. No. 4. P.2251.
41. Chen H., De Gasperis P., Marcelli R., Pardavi-Horvath M., McMichael R., Wigen P.E. Wide-band linewidth measurements in yttrium iron garnet films // JAP. 1990. V.67. No. 9. P.5530.
42. Pomyalov A.V., Zilberman P.E. Magnetic resonances in small thin-film samples of yttrium iron garnet // RE. 1986. T. 31. # 1. P.94.

43. Archer J.L ,. Bongianni W.L., Collins J.H. Magnetically tunable microwave bandstop filters using epitaxial YIG film resonators //JAP. 1970. V.41. No. 3. P.1360.
44. Hansen P., Krumme J.P. Determination of the local variation of the magnetic properties of liquid-phase epitaxial iron garnet films // JAP. 1973. V.44. No. 6. P.2847.
45. Telesnin R.V., Kozlov V.I., Dudorov V.N. Ferromagnetic resonance in $Y_3Fe_{5-x}Ga_xO_{12}$ epitaxial films. // FTT. 1974. Vol.16. No. 11. P.3532.
46. Algra H.A., Robertson J.M. A FMR study on horizontally dipped LPE grown (La, Ga): YIG films // JAP. 1979. V. 50. No. 3. P.2173.
47. Avaeva I.G., Lisovskiy F.V., Osika V.A., Shcheglov V.I. Investigation of epitaxial films of mixed ferrite-garnets by the ferromagnetic resonance method // FTT. 1975. Vol. 17. No. 10. P.3045.
48. Avaeva I.G., Lisovskiy F.V., Osika V.A., Shcheglov V.I. Ferromagnetic resonance in epitaxial films of mixed ferrite-garnets // RE. 1976. Vol.21. No. 9. P.1894.
49. Avaeva I.G., Lisovskiy F.V., Osika V.A., Shcheglov V.I. Ferromagnetic resonance in epitaxial films of mixed ferrite-garnets with cubic anisotropy // FTT. 1976. Vol. 18. No. 12. P.3694.
50. Hsia L.C., Wigen P.E., De Gasperis P., Borghese C. Enhancement of uniaxial anisotropy constant by introducing oxygen vacancies in Ca-doped YIG // JAP. 1981. V.52. No. 3. P.2261.
51. Temiryazev A.G., Tikhomirova M.P., Zilberman P.E., Maryakhin A.V., He A.S. Microwave properties of Ga, Sc - substituted YIG films with low magnetization // XIV Nationa Seminar "New magnetic materials for microelectronics": Abstracts. report, part 1. M., 1994.S. 57.
52. Silliman S.D., Gualtieri D.M., Stancil D.D. Improvement of FMR linewidth in Bi-substituted luthetium iron garnet thin films for the MSW-optical-mode interaction // JAP. 1993. V.73. No. 10. R.6460.
53. Butler J.C., Kramer J.J., Esman R.D., Craig A.E., Lee J.N., Ryuo T.
54. Microwave and magneto-optic properties of bismuth-substituted yttrium iron garnet thin films // JAP. 1990. V.67. No. 9. R.4938.
55. Grishin A.M., Dellalov V.S .. No.ikolayev E.I., Shkar V.F., Yampolskii S.V. FMR doublet in two-layer iron garnet films (abstract) // JAP. 1994. V.76. No. 10. R.6560.
56. Barak J., Ruppin R. Ferromagnetic resonance of double yttrium-iron-garnet films: perpendicular field // JAP. 1990. V.67. No. 5. P.2549.
57. Glushchenko A.G., Kurushin E.P., Koshkin L.I. Microwave microdevices on thin monocrystalline films of Mg-Mn ferrite // Izv. Univ. Radioelektron. 1975. Vol. 18. No. 11. P.93.
58. Glushchenko A.G., Kurushin E.P., Koshkin L.I. The use of thin monocrystalline films of Mg-Mn ferrite in microstrip microwave transmission lines // RE. 1974.T.19. No. 11. P.2397.
59. Puzyrev V.A. Thin ferromagnetic films in radio engineering circuits - M .: Sov. glad., 1974.
60. Kohmoto O., Munakata M. Ferromagnetic resonance in evaporated Co-CoO films // JAP. 1995. V.77. No. 12. P.6712.
61. Krishnan R., Lassri H., Seddat M., Fessier M., Guruswamy S., Sahay S., Magnetic and ferromagnetic resonance studies in Co-Cu composite films // JAP. 1994. V.75. No. 10. P.6607.
62. Zherikhov S.P., Bochkarev V.F., Rusov G.I., Torba G.F. Features of ferromagnetic resonance in amorphous gadolinium-cobalt films // In the book: Physics of magnetic films. - Irkutsk: Ed. IGPI, 1976, p. 63.
63. Pogorely A.N., Podyalovsky D.I. Spin-wave resonance in amorphous Tb-Fe films at

200-400 K // FTT. 1992. Vol. 34. No. 3. P.972.
64. Acharya B.R., Prasad S., Venkataramani N., Kaabouchi M., Krishnan R., Sella C. Ferromagnetic resonance studies of Fe / Ni and Fe / CoNbZr multilayers: Modes and experiments // JAP. 1995. V.78. No. 6. P.3992.
65. Dupuis V., Perez JP, Tuaillon J., Paillard V., Melinon P., Perez A., Barbara B., Thomas L., Fayeulle S., Gay M. Magnetic properties of nanostructured thin films of transition metal obtained by low energy cluster beam deposition // JAP. 1994. V.76. No. 10. P.6676.
66. Zhou I., Zhang Z., Wigen P.E., Ounadjela K. Interlayer exchange coupling versus ferromagnetic layer thickness in asymmetric Co / Ru / Co trilayer films // JAP. 1994. V.76. No. 10. P.7078.
67. Wigen P.E., Zhang Z., Zhou I., Ye M., Cowen J.A. The dispersion relation in antiparallel coupled ferromagnetic films // JAP. 1993. V.73. No. 10. P.6338.
68. Rezende S.M., Moura J.A.S., de Aguiar F.M., dos Santos C.A., Schreiner W.R., Feixeira S.R. Ferromagnetic resonance in Ag coupled Ni films // JAP. 1993. V.73. No. 10. P.6341.
69. Kordecki R., Meckenstock R., Pelzl J .. No. ikitov S. Investigations of surface and bulk spin wave modes in as-prepared and annealed FeNi multilayers by spin wave resonance // JAP. 1993. V.73. No. 10. P.6359.
70. Artman J.O., DeSmet D.J., Shao X., Cates J.C., Alexander C., Parker M.R., Lacey E.T., Lord D.G., Grundy P.J. Ferromagnetic resonance studies of anisotropy in PtCo multilayers // JAP. 1991. V.70. No. 10. P.6038.
71. Guslienko K. Yu., Lesnik N.A., Mitsek A.J., Vozniuk B.P. FMR in ferromagnetic films with coupled layers // JAP. 1991. V.69. No. 8. P.5316.
72. Voznyuk B.P., Guslienko K.Yu., Kozlov V.I., Lesnik N.A., Mitsek A.I. Influence of interaction of layers on FMR in two-layer ferromagnetic films // FTT. 1991. Vol.33. # 2. P.438.
73. Guslienko K.Yu. Spin-wave modes in exchange-coupled multilayer magnetic films // Phys. 1995. Vol. 37. No. 6. P.1603.
74. Vysotsky S.L., Kazakov G.T., Katz M.L., Filimonov Yu.A. Influence of surface spin pinning on the spin-wave resonance spectrum of a structure with two exchange-coupled films // Phys. 1993. Vol. 35. No. 5. P.1190.
75. Krebs J.J., Lind D.M., Berry S.D. // Ferromagnetic resonance and spin anisotropy in iron oxide thin films and iron oxide / nickel oxide superlattices //JAP. 1993. V.73. No. 10. P.6457.
76. Erukhimov M.Sh., Erukhimov G.M., Berenshtein B.E. Spin wave spectrum and magnetization of ferromagnetic superlattices //FTT. 1994. Vol.36. No. 6. P.1621.
77. Wang W.N., Jiang Z.S., Du Y.W. Ferromagnetic resonance study on Fe-SiO2 granular films // JAP. 1995. V.78. No. 11. P.6679.
78. Jiang Z.S., Ge X., Ji J.T., Sang H., Guo G., Du Y.W., Zhang S.Y. Magnetic properties of the FeSi-SiO2 granular films // JAP. 1994. V.76. No. 10. P.6490.
79. Rubinstein M., Das B.N., Koon N.C., Chrisey D.B., Horwitz J. Granular giant magnetoresistive materials and their ferromagnetic resonances // JAP. 1994. V.76. No. 10. P.6823.
80. Chubing P., Haiying C., Guozhong L., Daosheng D. Magnetic properties of the Fex-Cu1-x granular alloy films // JAP. 1994. V.76. No. 10. P.7102.
81. Rubinstein M., Das B., Chrisey D.B., Horwitz J., Koon N.C. Ferromagnetic resonance studies of granular materials // JAP. 1994. V.75. No. 10. P.6622.
82. Kordecki R., Meckenstock R., Pelzl J., Mühlbauer H., Dumpich G .. No. ikitov S. Ferromagnetic resonance investigations of spin waves in Fe-Ni multilayers // JAP.

1991. V.70. No. 10. P.6418.
83. Berzhansky V.N., Petrov V.E., Kokoz V.L. Spin-wave resonance in ion-implanted ferrite-garnet structures // FTT. 1991. Vol.33. No. 11. P.3372.
84. Zyuzin A.M., Bazhanov A.G. Transformation of spin-wave resonance spectra in multilayer films upon passing through the Curie point of the pinning layer // ZhETF. 1997. Vol. 112. No. 4 (10). P.1430.
85. Svistov L.Ye., Safonov V.L., Khachevatskaya K.R. Spin-wave resonances in non-uniformly deformed $FeBO_3$ plates // ZhETF. 1997. Vol. 112. No. 2 (8). P.564.
86. Miroshnikov Yu.F., Khramov B.V. Perpendicular anisotropy in biaxial films // Izv. Univ. Physics. 1974. No. 11 (150). P.119.
87. Gavrilin V.P., Berezin D.G., Miroshnikov Yu.F. Ferromagnetic resonance and magnetic crystallographic anisotropy of single-crystal films of lithium ferrite // Izv. Univ. Physics. 1973. No. 9 (136). P.86.
88. Meckenstock R., von Geisau O., Peizl J., Wolf J.A. Conventional and photothermally modulated ferromagnetic resonance investigations of anisotropy fields in an epitaxial Fe (001) film // JAP. 1995. V.77. No. 12. P.6439.
89. Komenou K., Zebrowski J., Wilts C.H. Ferromagnetic resonance study of the anisotropy field and nonmagnetic regions in implanted layers of bubble garnet films // JAP. 1979. V. 50. No. 8. P.5442.
90. ilts C.H., Zebrowski J., Komenou K. Ferromagnetic resonance study of the anisotropy profile in implanted bubble garnets // JAP. 1979. V. 50. No. 9. P.5878.
91. Cregg P.J., Crothers D.S.F., Wickstead A.W. An approximate formula for the relaxation time of a single domain ferromagnetic particle with uniaxial anisotropy and collinear field // JAP. 1994. V.76. No. 8. P.4900.
92. Dorsey P., Sokoloff J.B., Vittoria C. Ferrimagnetic resonance lineshape asymmetry due to Suhl instabilities // JAP. 1993. V.74. No. 3. P.1938.
93. Zyuzin A.M., Vankov V.N. Angular dependence of the intensity of the FMR line in anisotropic magnetic films // Phys. 1990. T. 32. No. 7. From 2015.
94. Khalvashi, E., Width of the magnetic resonance lines in the vicinity of the phase transition, FTT. 1994. Vol.36. No. 4. P.1175.
95. Zyuzin A.M., Radaykin V.V. Influence of the dispersion of orthorhombic anisotropy fields on the width of the ferromagnetic resonance line in ferrite garnet films // Zh. 1997. Vol. 67. No. 8. P.131.
96. Hoekstra B., van Stapele R. P., Robertson J. M. Spin-wave resonance spectra of inhomogeneous bubble garnet films // JAP. 1977. V.48. # 1. P.382.
97. Robson M.C., Kwon C., Kim K.-C., Sharma R.P., Venkatesan T., Lofland S.E., Bhagat S.M., Ramesh R., Dominguez M., Tyagi S.D. Characterization of epitaxial $La0.7Ba0.3MnO3$ structures using ferromagnetic resonance // JAP. 1996. V.80. No. 4. P.2334.
98. Belyaev B.A., Zhuravlev V.A., Kirichenko V.I., Suslyaev V.I., Tyurnev V.V. Investigation of the electromagnetic parameters of bicomplex microwave media using an irregular microstrip resonator //Preprint No. 735F. Krasnoyarsk: IF SO RAN, 1994.
99. Belyaev B.A., Ivanenko A.A., Leksikov A.A., Makievsky I.Ya., Pashkevich A.Z., Tyurnev V.V. Ferromagnetic resonance spectrometer of local sections of thin magnetic films // Preprint №761F. Krasnoyarsk: IF SO RAN, 1995.
100. Morgenthaler F.R. Electromagnetic signal processor forming localized regions of magnetic wave energy in gyro-magnetic material // Patent USA №4,152,676. Int. Cl.: H01P 1/34. May 1, 1979.
101. Bass F.G., Kagalovsky V.A., Konotop V.V. Inhomogeneous ferromagnetic resonance in magnets with fluctuating parameters // FTT. 1990. T. 32. # 1. P.77.

102. Averkiev N.S., Vikhnin V.S., Sablina N.I. A new shape of resonance lines broadened by dislocations in thin films // Phys. 1996. Vol.38. # 1. P.138.
103. Baryakhtar V.G., Eremenko V.V., Zvyagin S.A., Pashkevich Yu.G., Pishno V.V., Sobolev V.L., Shakhov V.V. Line width of the exchange modes of magnetic resonance in a four-sublattice rhombic antiferromagnet // ZhETF. 1991. Vol. 100. No. 6 (12). P.1893.
104. McMichael R.D. Method for determining both magnetostriction and elastic modulus by ferromagnetic resonance // JAP. 1994. V.75. No. 10. P.5650.
105. Oliver S.A., Harris V.G., Vittoria C. Magnetostriction measurements on thin films by a slot-line ferromagnetic resonance technique // JAP. 1990. V.67. No. 9. P.5019.
106. Dikshtein I.E., Maltsev O.A. Ferromagnetic resonance and magnetostatic waves in inhomogeneously deformed films of ferrite-garnets // RE. 1992. T. 37. No. 11. S.2003.
107. Smith A.B., Jones R.V. Magnetostriction constants from ferromagnetic resonance // JAP. 1963. V. 34. No. 5. P.1283. ...
108. Morgenthaler F.R. Two-dimensional magnetostatic resonances in a thin film disk containing a magnetic bubble // JAP. 1979. V. 50. No. 3. P.2209.
109. Sürig C., Hempel K.A. Interaction effects in particulate recording media studied by ferromagnetic resonance // JAP. 1996. V.80. No. 6. P.3426.
110. Orth Th., Pelzl J., Chantrell R.W., Veitch R., Jakusch H. Ferromagnetic resonance and transverse susceptibility measurements on particulate recording media // JAP. 1993. V.73. No. 10. P.6738.
111. Yu Y., Harrell J.W., Doyle W.D. Ferromagnetic resonance spectra of oriented barium ferrite tapes // JAP. 1994. V.75. No. 10. P.5550.
112. Gobov Yu.L., Shmatov G.A. Reading information based on ferromagnetic resonance in magnetic films placed in an inhomogeneous magnetic field // FMM. 1995. Vol. 79. No. 6. P.68.
113. Kabychenkov A. Some effects at the resonant interaction of a discrete spectrum light field with microwave ferrite // Proc. XIV Int. Conf. on Microwave Ferrites (Gyromagn. Electron. and Electrodyn.). Hungary. Eger. 1998. V.1. P.117.
114. Bahlmann N., Gerhardt R., Wallenhorst M., Dötsch H. Determination of the ferrimagnetic precession cone of in-plane magnetized garnet films using optical modulation technique // JAP. 1996. V.80. No. 7. P.3977.
115. Mendik M., Frait Z., von Känel H., Onda N. Ferromagnetic resonance and Brillouin light scattering from epitaxial $FexSi1-x$ films on Si (111) // JAP. 1994. V.76. No. 10. P.6897.
116. Meckenstock R., Schreiber F., von Geisau O., Peizl J. Photothermally modulated ferromagnetic resonance investigations of epitaxially grown thin films // JAP. 1994. V.75. No. 10. P.6508.
117. Walker L. Magnetostatic types of precession at ferromagnetic resonance // In the book: Ferrites in nonlinear microwave devices [6] - Moscow: IL, 1961. P. 470.
118. Damon R.W., Eshbach J.R. Magnetostatic modes of a ferromagnet slab // J. Phys. Chem. Solids. 1961. V.19. No. 3/4. P.308.
119. LeCrow R., Spencer E., Porter K. The width of the ferromagnetic resonance curve in single crystals of yttrium ferrite with a garnet structure // In: Ferrites in nonlinear microwave devices [6] - Moscow: IL, 1961. P. 433.
120. Dillon J., Nielsen J. Influence of rare-earth impurities on ferromagnetic resonance in yttrium garnet // In the book: Ferrites in nonlinear microwave devices [6] - Moscow: IL, 1961. [6]. P. 442.
121. Fletcher P., Bell R. Magnetostatic types of precession in ferrimagnetic spheres // In

the book: Ferrites in nonlinear microwave devices [6] - Moscow: IL, 1961. P.497.
122. Schlömann E., Joseph R.I. Instability of spin waves and magnetostatic waves in a microwave field applied parallel to the dc field // JAP. 1961. V.32. No. 6. P.1006.
123. Auld B.A. Walker modes in large ferrite samples // JAP. 1960. V.31. No. 9. P.1642.
124. Fletcher P.C., Kittel C. Consideration on the propagation and generation of magnetostatic waves and spin waves // PR. 1960. V. 120. No. 6. P.2004.
125. Damon R. W., Van de Vaart H. Propagation of magnetostatic spin waves at microwave frequencies. II. Rods // JAP. 1966. V. 37. No. 6. P.2445.
126. Damon R. W., Van de Vaart H. Propagation of magnetostatic spin waves at microwave frequencies in a normally-magnetized disk // JAP. 1965. V. 36. No. 11. P.3453.
127. Damon R.W., Eshbach J.R. Magnetostatic modes of a ferromagnet slab // JAP. 1960. V.31. No. 5. P.104S.
128. Eshbach J.R., Damon R.W. Surface magnetostatic modes and surface spin waves // PR. 1960. V. 118. No. 5. P.1208.
129. Joseph R. I., Schlömann E. Theory of magnetostatic modes in long, axially magnetized cylinders // JAP. 1961. V.32. No. 6. P.1001.
130. Seshadri S.R. Magnetostatic surface waves in ferrite slab // Proc. IEEE. 1970. V. 58. No. 3. P.506. - Translation: Seshadri. Surface magnetostatic waves in a ferrite plate // TIIER. 1970. T. 58. No. 4. P.105.
131. Van de Vaart, H. Influence of metal plate on surface magnetostatic modes of magnetic slab, El. Lett. 1970. V.6. No. 19. P.601.
132. Akhiezer A.I., Bar'yakhtar V.G., Peletminsky S.V. Spin waves - Moscow: Nauka, 1967.
133. Sodha M.S., Srivastava N.S. Microwave propagation in ferromagnetics - New York: Plenum Press, 1981.
134. Vashkovsky A.V., Stalmakhov V.S., Sharaevsky Yu.P. Magnetostatic waves in microwave electronics. Saratov: Saratov University Press, 1993.
135. Vendik O.G., Kalinikos B.A. Wave processes in laminated ferrite film structures - physical foundations of spin-wave electronics // Izv. Univ. Physics. 1988. Vol.31. No. 11. C.3.
136. Gulyaev Yu.V., Zilberman P.E. Magnetoelastic waves in plates and films of ferromagnets // Izv. Univ. Physics. 1988. Vol.31. No. 11. C.6.
137. Dmitriev V.F., Kalinikos B.A. Excitation of propagating waves of magnetization by microstrip antennas // Izv. Univ.. Physics. 1988. Vol.31. No. 11. P.24.
138. Preobrazhensky V.L., Fetisov Yu.K. Magnetostatic waves in a nonstationary medium // Izv. Univ. Physics. 1988. Vol.31. No. 11. P.54.
139. 139. Vashkovsky A.V., Stalmakhov A.V., Shakhnazaryan D.G. Formation, reflection and refraction of wave beams of magnetostatic waves // Izv. Univ. Physics. 1988. Vol.31. No. 11. P.67.
140. Voronenko A.V., Gerus S.V., Kharitonov V.D. Diffraction of surface magnetostatic waves by magnetic gratings in the Bragg regime // Izv. Univ. Physics. 1988. Vol.31. No. 11. P.76.
141. Ignatiev A.A., Stalmakhov V.S. Experimental studies of magnetostatic waves in the millimeter range // Izv. Univ. Physics. 1988. Vol.31. No. 11. P.86.
142. Kudinov E.V., Beregov A.S. Using the principle of decomposition in the construction of mathematical models of spin-wave devices // Izv. Univ. Physics. 1988. Vol.31. No. 11. P.106.
143. Morgenthaler F.R. An overview of electromagnetic and spin angular momentum mechanical waves in ferrite media // Proc. IEEE. 1988. V.76. # 2. P.138. 1988. Vol.76. # 2. P.50.

144. White R., Salt I. Multiplet ferromagnetic resonance in ferrite spheres // In the book: Ferrites in nonlinear microwave devices [6] - Moscow: IL, 1961. P. 529.
145. Eshbach J.R. Spin-wave propagation and the magnetoelastic interaction in yttrium iron garnet // JAP. 1963. V. 34. No. 4. P.1298.
146. Eggers F. G., Strauss W. A UHF delay line using single-crystal yttrium iron garnet // JAP. 1963. V. 34. No. 4. P.1180.
147. Pizzarello F.A., Collins J.H., Coerver L.B. Magnetic steering of magnetostatic waves in epitaxial YIG films // JAP. 1970. V.41. No. 3. P.1016.
148. Huben P., McKinstry K.D., Kabos P. Coupled edge-guided magnetostatic waves // JAP. 1993. V.73. No. 10. P.7015.
149. O'Keeffe T.W., Patterson R.W. Magnetostatic surface wave propagation in finite samples // JAP. 1978. V.49. No. 12. P.4886.
150. Parekh J.P. Theory for magnetostatic forward volume excitation // JAP. 1979. V. 50. No. 3. P.2452.
151. Nikitov S.A., Gavrilin S.N. Dispersion of surface magnetostatic waves propagating over a rough surface. 1988. Vol. 33. No. 11. P.2415.
152. Rukhadze A.A., Chogovadze M.E. Propagation of surface spin waves in isotropic media // Physics and Technology. 1991. Vol.33. No. 4. P.1055.
153. 153. Gribkova Yu.V., Kaganov M.I. Magnetostatic spectrum of a transversely magnetized plate (qualitative analysis) // FTT. 1991. Vol.33. # 2. P.508.
154. Zaiko Yu.N., Lazerson A.G. Propagation of impulses of surface magnetostatic waves // RE. 1988. Vol. 33. # 2. P.409.
155. Prokushkin V.N., Sharaevsky Yu.P. Surface magnetostatic waves in a ferrite structure with impedance boundaries. 1987. Vol. 32. No. 8. P.1750.
156. Suchkov S.G. Variational method for calculating the characteristics of magnetostatic waves in obliquely magnetized two-dimensionally limited ferromagnetic films // RE. 1985.T.30. No. 6. P.1080.
157. Shchuchinsky A.G. Surface waves in a tangentially magnetized ferrite layer. 1984. T. 29. No. 9. P.1700.
158. V. V. Gann. Inhomogeneous resonance in a ferromagnetic plate //FTT. 1966. Vol. 8. No. 11. P.3167.
159. Mikhailovskaya L.V., Khlebopros R.G. Magnetostatic spectrum of a ferromagnetic film // FTT. 1969. Vol.11. No. 10. P.2854.
160. Fetisov Yu.K. Influence of constant field orientation on the dispersion of magnetostatic waves in a ferrite film // RE. 1988. Vol. 33. No. 10. P.2217.
161. Dunaev S.N., Fetisov Yu.K. Influence of pulsed optical heating on the propagation of magnetostatic waves in a ferrite film // RE. 1996. Vol.41. # 1. P.89.
162. Esikov O.S., Toloknov N.A., Fetisov Yu.K. Microwave power transmission spectra on surface magnetostatic waves in IZhG plates // RE. 1980. Vol. 25. # 1. P.128.
163. Kostylov V.N., Mitlina L.A., Yankovskaya T.V. Propagation of magnetostatic waves in films of magnesium-manganese ferrite // XIV Vses. shk-sem. "New magnetic materials for microelectronics" Abstracts. report Part 1. Moscow: Moscow State University, 1994.S. 55.
164. Beregov A.S. Magnetostatic waves in a structure with an arbitrarily magnetized film of a cubic ferromagnet //Izv. Univ. Radio electronics. 1984. Vol.27. No. 10. C.9.
165. 165. Kazakov G.T., Sukharev A.G., Filimonov Yu.A., Shein I.V. Influence of cubic anisotropy on the spectrum of spin waves of an arbitrarily magnetized YIG film with the (111) plane // ZhTF. 1989. Vol.59. # 2. P.186.
166. Reimann H., Hsia L. C., Wigen P. E. Magnetostatic mode spectra of uniaxial ferromagnets // JAP. 1981. V.52. No. 3. P.2264.

167. Dudko G.M., Kazakov G.T., Sukharev A.G., Filimonov Yu.A., Shein I.V. Magnetostatic waves in skew-magnetized layers of anisotropic ferrite // RE. 1990. Vol. 35. No. 5. P.966.
168. Bobkov V.B., Zavislyak I.V., Romanyuk V.F. Magnetostatic waves in ferrite films with trigonal anisotropy // FTT. 1993. Vol. 35. # 2. P.431.
169. Kaganov M. I., Pustylnik N.B. Surface magnetic oscillations in a uniaxial antiferromagnet // ZhETF. 1995. Vol. 107. No. 4. P.1298.
170. Vysotsky S.L., Kazakov G.G., Maryakhin A.V., Nam B.P., Sukharev A.G., Filimonov Yu.A., He A.S. Magnetostatic waves in weakly anisotropic Ga, Sc-substituted films of yttrium iron garnet // RE. 1992. T. 37. No. 6. P.1086.
171. Medved A.V., Nikitin I.P., Filimonova L.M. Influence of crystallographic anisotropy on the spectrum of magnetostatic waves in yttrium iron garnet films // RE. 1987. Vol. 32. No. 7. P.1557.
172. LeCraw R.C., Walker L.R. Temperature dependence of the spin-wave spectrum of yttrium iron garnet // JAP. 1961. V.32. No. 3. P.167S.
173. Kabychenkov A.F. Influence of the light field on the dispersion of magnetic dipole waves in ferromagnets // ZhTF. 1994. Vol. 64. No. 8. P.159.
174. Lugovskoy A.V., Shcheglov V.I. Spectrum of exchange and exchangeless spin-wave excitations in ferrite-garnet films // RE. 1982. Vol.27. No. 3. P.518.
175. VN Dudorov, VV Randoshkin. Spin-wave resonance in yttrium-iron garnet films // Physics and Technology. 1994. Vol.36. No. 6. P.1790.
176. Adam J.D., O'Keeffe T.W., Patterson R.W. Magnetostatic wave to exchange resonance coupling // JAP. 1979. V. 50. No. 3. P.2446.
177. Schiltz W. Experimental evidence for coupling between dipole and exchange dominated spin waves in epitaxial YIG // Solid State Comm. 1972. V.11. No. 3. P.615.
178. Schiltz W. Spin-wave propagation in epitaxial YIG-films // Phil. Res. Rep. 1973. V.28. # 1. P.50.
179. Wolfram T., de Wames R.E. Effects of exchange on the magnetic surface states of yttrium iron garnet films // Solid State Comm. 1970. V.8. # 1. P.191-194.
180. De Wames R.E., Wolfram T. Dipole-exchange spin waves in ferromagnetic films // JAP. 1970. V.41. No. 4. P.987.
181. Wolfram T., de Wames R.E. Magnetoexchange branches and spin wave resonance in conducting and insulating films - perpendicular resonance // PR (B). 1971. V.4. No. 9. P.3125.
182. Dmitriev V.F., Kalinikos B.A., Kovshikov N.G. Observation of exchange oscillations of radiation upon excitation of spin waves in normally magnetized films of yttrium iron garnet // ZhTF. 1985. Vol.55. No. 10. P.2051.
183. Chen M., Patton C.E. New perspective on the Green's function dipole-exchange spin waves theory for thin films // JAP. 1994. V.75. No. 10. P.6084.
184. Chernakova A.K., Cash A., Peruyero J., Stancil D.D. Orientation dependence of dipole gaps in the magnetostatic wave spectrum of Bi-substituted iron garnets // JAP. 1994. V.75. No. 10. P.6606.
185. Kazakov G.T., Sukharev A.G., Filimonov Yu.L. Radiation exchange losses of Damon-Eshbach magnetostatic waves in films of yttrium iron garnet // Physics and Technology. 1990. T. 32. No. 12. P.3571.
186. Andreev A.S., Gulyaev Yu.V., Zilberman P.E., Kravchenko V.B., Lugovskoy A.V., Ogrin Yu.F., Temiryazev A.G., Filimonova L.M. Propagation of magnetostatic waves in films of yttrium iron garnets of submicron thickness // ZhETF. 1984. T. 86. No. 3. P.1005.
187. Zilberman P.E., Temiryazev A.G., Tikhomirova M.P. Excitation and propagation of

exchanged spin waves in films of yttrium iron garnet // ZhETF. 1995. T. 108. No. 1 (7). P.281.
188. Zilberman P.E., Lugovskoy A.V., Sharafatdinov A.A. Spin-wave resonance and the distribution of exchanged spin waves in ferrite films inhomogeneous in thickness // Phys. 1995. Vol. 37. No. 7. 2010.
189. Temiryasev A.G., Tikhomirova M.P., Zil'berman P.E. "Exchange" spin waves in nonuniform yttrium iron garnet films // JAP. 1994. V.76. No. 9. P.5586.
190. Gulyaev Yu.V., Temiryazev A.G., Tikhomirova M.P., Zil'berman P.E. Magnetoelastic interaction in yttrium iron garnet films with magnetic inhomogeneities through the film thickness // JAP. 1994. V.75. No. 10. P.5619.
191. Kaibichev I.A., Shavrov V.G. Surface magnetostatic waves in an inhomogeneous ferromagnet with a deviation of magnetization from the normal to the surface. 1998. Vol.43. # 1. P.90.
192. Morgenthaler F.R. Magnetostatic surface modes in nonuniform thin films with in-plane bias fields // JAP. 1981. V.52. No. 3. P.2267.
193. Zilberman P.E., Temiryazev A.G., Tikhomirova M.P. Exchange spin waves in inhomogeneous ferrite layers: excitation, propagation, application prospects // I Obed. conf. on magnetoelectronics. Abstracts. Dokl., 1995. M .: IRE RAN. P.128.
194. Zilberman P.E., Temiryazev A.G., Tikhomirova M.P. Propagation of pulses of spin and acoustic waves over the thickness of inhomogeneous YIG films // I Obed. conf. on magnetoelectronics. Abstracts. Dokl., 1995. M .: IRE RAN. P.204.
195. Bespyatykh Yu.I., Vasilevsky V., Kharitonov V.D. Attenuation of surface magnetostatic waves in a randomly inhomogeneous anisotropic medium. 1998. Vol.43. No. 5. P.622.Gaynovich I.Yu., Golovach G.P., Zavislyak I.V., Romanyuk V.F.
196. Dipole and exchange spin excitations in inhomogeneous ferrite films // FTT. 1992. Vol. 34. No. 6. P.1680.
197. Gaynovich I.Yu., Zavislyak I.V., Romanyuk V.F. Backward bulk magnetostatic waves in inhomogeneous ferrite {111} films of yttrium iron garnet // RE. 1994. Vol. 39. # 1. P.33.
198. Mason W.P. (ed.). Physical acoustics. T.3B. Lattice dynamics. M .: Mir. 1968.
199. Kingner C., Heil J., Lethi B. Surface acoustic wave propagation in paramagnets and ferromagnets // JAP. 1981. V.52. No. 3. P.2270.
200. Gulyaev Yu.V., Nikitov S.A. Diffraction of an external electromagnetic wave on sound in ferrite // RE. 1985.T.30. No. 5. P.1027.
201. Zavislyak I.V., Kotsarenko N.Ya., Rapoport Yu.G. Conversion of magnetostatic waves into acoustic waves with a large decrease in frequency in a ferromagnetic plate // RE. 1985.T.30. No. 5. P.1028.
202. Kindyak A.S. Nonlinear surface magnetostatic waves in a ferrite-semiconductor structure. // ZhTF. 1999. Vol. 69. No. 6. P.119.
203. Strauss W. Elastic and magnetoelastic waves in yttrium iron garnet // Proc. IEEE, 1965. V.53. No. 10. P.1485. - Translation: Strauss. Elastic and magnetoelastic waves in yttrium iron garnet // TIIER. 1965. Vol.53. No. 10. P.1673.
204. Comstock R.L. Magnetoelastic coupling constants of the ferrites and garnets // Proc. IEEE. 1965. V.53. No. 10. P.1508. -
205. Andreev A.S., Gulyaev Yu.V., Zilberman P.E., Kravchenko V.B., Ogrin Yu.F., Temiryazev A.G., Filimonova L.M. Magnetoelastic effects in tangentially magnetized films of yttrium iron garnet // RE. 1985.T.30. No. 10. S. 1992.
206. Bugaev A.S., Gorsky V.B. Magnetoelastic effects in ferrite films with the participation of higher thickness exchange magnetostatic modes // RE. 1992. T. 37. No. 6. P.961.

Bibliography

207. Tarasenko S.V. Magnetoelastic mechanism of formation of a traveling surface spin wave // Physics and Technology. 1991. Vol.33. No. 10. P.3021.
208. Tarasenko S.V. Anomalous character of diffraction of spin waves under conditions of strong linear magnon-phonon interaction // Phys. 1991. Vol.33. No. 8. P.2394.
209. Kryshtal R.G., Medved A.V. Devices for processing microwave signals based on the phenomenon of scattering of magnetostatic waves by surface acoustic waves // RE. 1991. Vol.36. No. 8. P.1571.
210. Kryshtal R.G., Medved A.V., Popkov A.F. Influence of the attenuation of surface magnetostatic waves on the parameters of their scattering by a surface acoustic wave in yttrium iron garnet films. 1994. Vol. 39. No. 4. P.647.
211. Uchastkin V.I. Surface magnetoelastic waves // FTT. 1971.V.13. No. 8. P.2499.
212. Kittel C. Interaction of spin waves and ultrasonic waves in ferromagnetic crystals // PR. 1958. V. 110. No. 5. P.836.
213. Seavey M.H. Phonon generation by magnetic films // Proc. IEEE. 1965. V.53. No. 10. P.1387.
214. Adam J.D. Analog signal processing with microwave magnetics // Proc. IEEE. 1988. V.76. # 2. P.159.
215. Ishak W.S. Magnetostatic wave technology: a review // Proc. IEEE. 1988. V.76. # 2. P.171.
216. Schlömann E.F. Circulators for microwave millimeter wave integrated circuits // Proc. IEEE. 1988. V.76. # 2. P.188.
217. Vashkovsky A. V., Stalmakhov A. The. Internal magnetostatic waves at the boundary of the magnetization jump. 1984. T. 29. No. 12. P.2409.
218. V. I. Zubkov, V. A. Epanechnikov. Surface magnetostatic waves in two-layer ferromagnetic films // PZhTF. 1985. T.11. No. 23. P.1419.
219. V. I. Zubkov, V. A. Epanechnikov. Spectrum of surface magnetostatic waves in a planar structure with three ferromagnetic components separated by dielectric layers // RE. 1986. T. 31. No. 4. P.656.
220. V. I. Zubkov, V. A. Epanechnikov. Spectrum of surface magnetostatic waves in three-layer ferromagnetic films // ZhTF. 1987. T. 57. No. 4. P.625.
221. V.I. Zubkov, V. A. Epanechnikov. Influence of metal planes on the spectrum of surface magnetostatic waves in two-layer ferromagnetic films // ZhTF. 1989. Vol.59. No. 9. P.53.
222. Novikov G.M., Borisov S.A., Dubovitsky S.A., Petrunkin E.Z. Investigation of the dispersion characteristics of magnetostatic waves in composite ferrite-dielectric-metal structures. 1983. Vol.28. # 1. P.121.
223. IV Vasiliev, G.S. Makeeva. Propagation of magnetostatic waves in a metallized ferrite structure of finite dimensions. 1984. T. 29. No. 3. P.419.
224. Babichev R.K., Zubkov V.I. Effect of the screen on the characteristics of magnetostatic waves in a layered structure with a ferrite film with arbitrary magnetization // RE. 1989. Vol. 34. No. 5. P.973.
225. Babichev R.K., Zubkov V.I. Influence of two screens on the characteristics of magnetostatic waves in a layered structure with a ferrite film with arbitrary magnetization // RE. 1989.T.34. No. 10. S.2074.
226. Zilberman P.E., Umansky A.V. Magnetostatic waves in a periodically inhomogeneous ferrite plate // RE. 1990. Vol. 35. No. 8. P.1623.
227. Kaibichev I.A., Shavrov V.G. Slot magnetostatic waves in ferromagnets with opposite direction of magnetizations // RE. 1993. Vol. 38. No. 10. P.1816.
228. Buris N.E. Crystalline anisotropy and loss effects on magnetostatic waves in layered yttrium iron garnet films // JAP. 1991. V.68. No. 12. P.6442.

229. Vysotsky S.L., Kazakov G.T., Filimonov Yu.L., Shein I.V., He A.S. Magnetostatic waves in a tangentially magnetized structure with two ferrite layers of orientation (111) // RE. 1990. Vol. 35. No. 5. P.959.
230. Vysotsky S.L., Kazakov G.T., Maryakhin A.V., Filimonov Yu.A., He A.S. Surface magnetostatic waves in exchange-coupled ferrite films // FTT. 1996. Vol.38. # 2. P.407.
231. Dmitriev V.F., Kalinikos B.A. Self-consistent theory of excitation of spin waves by a microstrip antenna in a tangentially magnetized layered structure. 1988. Vol. 33. No. 11. P.2248.
232. Balashova E.V., Kazakov G.T., Filimonov Yu.A. Influence of an Ideally Conducting Metal on Resonant Interaction of Magnetostatic Waves with Converted Elastic Waves in a Thin-Film Ferrite Structure. 1985.T.30. No. 10. P. 1930.
233. Schlömann E. Amplification of magnetostatic surface waves by interaction with drifting charge carriers in crossed electric and magnetic fields // JAP. 1969. V.40. No. 3. P.1422.
234. Kudryashkin I.G., Fetisov Yu.K. Interaction of surface magnetostatic waves with conduction electrons in a ferrite-semiconductor thin-film structure // PZhTF. 1989. Vol. 15. No. 8. P.47.
235. Kindyak A.S. Propagation of a surface magnetostatic wave in a ferrite-semiconductor photosensitive structure // RE. 1996. Vol.41. No. 5. P.526.
236. Vysotsky S.L., Kazakov G.T., Sukharev A.G., Filimonov Yu.A. Investigation of the effect of generation of static EMF by traveling surface magnetostatic waves in a thin-film structure of yttrium iron garnet-n-InSb // RE. 1986. T. 31. # 2. P.411.
237. Kazakov G.T., Filimonov Yu.A. Surface magnetostatic waves in ferrite-semiconductor structures of finite width // RE. 1987. Vol. 32. No. 5. P.1105.
238. Bugaev A.S., Filimonov Yu.A. Deformation mechanism of interaction of magnetostatic waves with conduction electrons. 1984. T. 29. # 1. P.141.
239. Gulyaev Yu.V., Zilberman P.E. Interaction of microwave spin waves and electrons in layered semiconductor-ferrite structures (review) // RE. 1978. Vol. 23. No. 5. P.897.
240. Bespyatykh Yu.I., Vashkovsky A.V., Zubkov V.I. Instability of surface magnetostatic waves in a semiconductor-ferromagnet-dielectric-metal structure. 1975. T.20. No. 5. P.1003.
241. Vashkovsky A.V., Zubkov V.I., Kildishev V.N. Investigation of the amplification of surface waves in a layered ferrite-semiconductor structure. "Physics of magnetic phenomena", Materials of the VII All-Union School-Seminar on Gyromagnetic Electronics and Electrodynamics. Ashgabat, 1973, p. 249.
242. Zilberman P.E. Landau's mechanism of electronic absorption and amplification of spin waves // FTT. 1977.T.19. No. 10. P.2986.
243. Vashkovsky A.V., Zubkov V.I., Kildishev V.N., Mansvetova E.G., Salyganov V.I. Magnetoelectric resonance in a layered ferrite-semiconductor structure // ZhETF. 1975. Vol. 68. No. 3. P.1066.
244. Vashkovsky A.V., Zubkov V.I., Kildishev V.N., Murmuzhev B.A. Interaction of surface magnetostatic waves with charge carriers at the ferrite-semiconductor interface // PZhETF. 1972. Vol.16. # 1. C.4.
245. Zilberman P.E., Polzikova N.I., Raevsky A.O. Resonant spectrum rearrangement and nonresonant amplification of spin waves in ferromagnetic semiconductors in an alternating electric field // Phys. 1990. T. 32. No. 3. P.756.
246. Solin N.I., Auslender M.I., Shumilov I.Yu., Samokhvalov A.A. A new mechanism of Cherenkov amplification of spin waves by drifting charge carriers in a ferromagnetic semiconductor // FTT. 1990. T. 32. No. 8. P.2240.

247. Kostylev V.A., Samokhvalov A.A., Chebotarev N.M. Influence of the electric field on the attenuation of magnetostatic waves in the ferromagnetic semiconductor HgCr2Se4 // FTT. 1992. Vol. 34. No. 8. P.2619.
248. Solin N.I., Filippov B.N., Shumilov I.Yu., Samokhvalov A.A. Spectrum and damping of magnetostatic waves in a ferromagnetic semiconductor HgCr2Se4 when magnons are heated by an electric field. 1993. Vol. 35. No. 6. P.1613.
249. Auslender M. I., Falkovskaya L. D. Magnetoelectric damping of low-frequency spin waves in an antiferromagnetic semiconductor of the "easy plane" type. Possibility of Cherenkov amplification by drifting charge carriers in EuTe // FTT. 1991. Vol.33. No. 3. P.840.
250. Kostylev V.A., Samokhvalov A.A., Viglin N.A., Chebotarev N.M. Amplification of magnetostatic waves in a ferromagnetic semiconductor HgCr2Se4 in an electric field // FTT. 1991. Vol.33. No. 5. P.1494.
251. Anfinogenov VB, Verbitskaya TN, Gulyaev Yu.V., Zilberman P.E., Meriakri SV, Ogrin Yu.F., Tikhonov VV. Hybrid electromagnetic-spin waves in contacting layers of ferrite and ferroelectric. I. Theory // RE. 1989. Vol. 34. No. 3. P.494.
252. Anfinogenov VB, Verbitskaya TN, Gulyaev Yu.V., Zilberman PE, Meriakri SV, Ogrin Yu.F., Tikhonov VV. Hybrid electromagnetic-spin waves in contacting layers of ferroelectric and ferrite. II. Experiment // RE. 1990. Vol. 35. No. 2. P.320.
253. Wolsky A.M., Giz R.F., Daniels E.D. New superconductors: application prospects // In the world of science, 1989. No. 4. P.37.
254. Gorkov L. P., Kopnin N.B. High-temperature superconductors from the experimental point of view // UFN, 1988. V.156. No. 1. P.117.
255. Bednorts I.G., Mueller K.A. Perovskite oxides - a new approach to high-temperature superconductivity // Phys. 1988. Vol. 156. No. 2. P.323.
256. Bespyatykh Yu.I., Vasilevsky V., Gaidek M., Simonov A.D., Kharitonov V.D. Dispersion and damping of surface magnetostatic waves in the structure of a ferromagnet-superconductor of the second kind // FTT. 1993. Vol. 35. No. 11. P.2983.
257. Bespyatykh Yu.I., Vasilevsky V., Kharitonov V.D. Influence of pinning of Abrikosov vortices on the propagation of surface magnetostatic waves in a ferromagnet-superconductor structure // Phys. 1998. Vol. 40. No. 1. P.32.
258. Gulyaev Yu.V., Ogrin Yu.F., Polzikova N.I., Raevsky A.O. Observation of absorption of bulk spin waves in a magnetic-superconductor structure // Phys. 1997. Vol. 39. No. 9. P.1628.
259. Vashkovsky A.V., Zubkov V.I., Locke E.G. Propagation of magnetostatic waves in a ferrite-high-temperature superconductor structure in the presence of a transport current in a superconductor // FTT. 1997. Vol. 39. No. 12. P.2195.
260. Bespyatykh Yu.I., Vasilevsky V., Gaidek M., Kharitonov V.D. Influence of elastic interaction of Abrikosov vortices on the dispersion and damping of surface magnetostatic waves in the structure of a ferromagnet-superconductor of the second kind // Phys. 1995. Vol. 37. No. 10. P.3049.
261. Buzdin A.I., Vuyichich B., Kupriyanov M.Yu. Superconductor-ferromagnet structures // ZhETF. 1992. T.101. No. 1. P.231.
262. Polzikova N.I., Raevsky A.O. Peculiarities of the dispersion laws of surface spin waves in structures containing a superconductor // FTT. 1996. Vol.38. No. 10. P.2937.
263. Bespyatykh Yu.I., Vasilevsky V., Gaidek M., Simonov A.D., Kharitonov V.D. Excitation of surface spin waves in a tangentially magnetized structure of a ferromagnet-superconductor of the second kind // FTT. 1991. Vol.33. No. 5. P.1545.

264. Genkin G.M., Skuzovatkin V.V., Tokman I.D. Magnetization of ferromagnet-superconductor structures // FTT. 1993. Vol. 35. No. 3. P.736.
265. Lebed B.M., Nikiforov A.V., Yakovlev S.V., Yakovlev I.A. Scattering of spin waves by a lattice of magnetic vortices in a high-temperature superconductor-ferrite film structure // FTT. 1992. Vol. 34. No. 2. P.656.
266. N.I. Polzikova. Resonant interaction of magnetostatic waves with a lattice of vortices of the magnetic flux of a superconductor // PZhTF. 1993.T.19. No. 22. P.28.
267. Vugalter G. A., Khvoshcheva N.S. Reflection of magnetostatic waves by a segment of a microstrip line // RE. 1988. Vol. 33. No. 10. S.2055.
268. Vugalter G.A., Makhalin V.N. Reflection and excitation of direct volume magnetostatic waves by a metal strip // RE. 1984. T. 29. No. 7. P.1252.
269. Ivanov V.N., Babichev R.K., Zubkov V.I. Radiation resistance and inductance of a microstrip line containing a longitudinally magnetized ferrite layer with a screen. 1997. Vol. 42. # 1. P.38.
270. Ivanov V.N., Zubkov V.I., Babicheva E.R. Impedance of a one-sided coplanar line located above a ferrite film // Electromagnetic waves and electronic systems. 2010.T.15. No. 6. P.56.
271. Babicheva E.R., Babicheva G.V., Babichev R.K. Excitation of surface magnetostatic waves by a coplanar line // Proceedings of the XXIII All-Russian conference "Electromagnetic field and materials". M .: INFRA-M, 2015.S. 81.
272. Nistratov N.P., Vyatkina S.A., Babichev R.K. Experimental study of a MSSW beam excited by a finite-length transducer // Proceedings of the XXI International Conference "Electromagnetic Field and Materials". M .: NIU MEI, 2013.S. 249.
273. Babicheva E.R., Babicheva G.V., Babichev R.K. Influence of parameters of a microstrip converter, exciting surface magnetostatic waves, on its linear impedance // Proceedings of the XXI International Conference "Electromagnetic field and materials". M .: NIU MEI, 2013.S. 254.
274. Auld B.A. Geometrical optics of magnetoelastic wave propagation in a nonuniform magnetic field // Bell Syst. Tech. J. 1965. V.44. No. 3. P.495.
275. Bespyatykh Yu.I., Zubkov V.I., Tarasenko V.V. Propagation of surface magnetostatic waves in a ferromagnetic plate // ZhTF. 1980. T. 50. No. 1. P.140.
276. Valyavsky A.B., Vashkovsky A.V., Stalmakhov A.V., Tyulyukin V.A. Anisotropic properties of wave beams of exchangeless spin waves // ZhTF. 1989. Vol.59. No. 6. P.51.
277. Valyavsky A.B., Vashkovsky A.V., Grechushkin K.V., Stalmakhov A.V. Angular spectrum and spectrum of directions of group velocities of magnetostatic waves // RE. 1988. Vol. 33. No. 9. P.1830.
278. Grechushkin K.V., Stalmakhov A.V., Tyulyukin V.A. Energy flux of backward volume magnetostatic waves // RE. 1986. T. 31. No. 8. P.1487.
279. Grechushkin K.V., Stalmakhov A.V., Tyulyukin V.A. Spatial-frequency dependences of the energy flux of reverse bulk magnetostatic waves // IV Vses. shk-sem. "Problems of improving devices and methods for receiving, transmitting and processing information." Abstracts. report M. 1986. Part 2. P.38.
280. Valyavsky A.B., Stalmakhov A.V., Tyulyukin V.A. Anisotropic properties of wave beams of magnetostatic waves // V All. shk-sem. "Problems of improving devices and methods for receiving, transmitting and processing information." Abstracts. report M. 1988. S. 45-46.
281. Vashkovsky A.V., Stalmakhov A.V., Tyulyukin V.A. Limited beams of magnetostatic waves // The 4th Joint MMM-Intermag Conference. Vancouver. British Columbia. Canada. Book of Abstr. 1988. P. HC-07.

282. Vashkovsky A.V., Grechushkin K.V., Stalmakhov A.V., Tyulyukin V.A. Propagation of a limited wave beam of a surface magnetostatic wave // RE. 1988. Vol. 33. No. 4. P.876.
283. Vashkovskiy A.V., Grechushkin K.V., Stalmakhov A.V. Spatial-frequency dependences of the energy flux of a surface magnetostatic wave. 1985.T.30. No. 12. P.2422.
284. Fetisov Yu.K., Preobrazhensky V.L. Anisotropic propagation of magnetostatic waves in tangentially magnetized ferrite films // ZhTF. 1987. T. 57. No. 3. P.564.
285. Slavin A.N., Fetisov Yu.K. Influence of constant magnetic field orientation on dispersion characteristics of magnetization waves in yttrium iron garnet films // ZhTF. 1988. T. 58. No. 11. P.2210.
286. Baryshev D.A., Vashkovsky A.V., Stalmakhov A.V. Peculiarities of propagation of radio pulses of a surface magnetostatic wave. 1990. Vol. 35. No. 10. P.2224.
287. Baryshev D.A., Valyavsky A.B., Vashkovsky A.V., Stalmakhov A.V. Propagation of a packet of magnetostatic waves in a ferrite film // RE. 1990. Vol. 35. No. 10. P.2164.
288. Beregov A.S. Control of the spectrum and group velocity of magnetostatic waves // RE. 1983. Vol.28. No. 1. P.127.
289. L. V. Lutsev, I. L. Berezin. Thermal stability of parameters of magnetostatic waves propagating in films with an arbitrary direction of magnetization // Electronic engineering. Ser. Microwave electronics. 1989. No. 6 (120). C.3.
290. F.V. Lisovsky Using the method of constructing isoenergy surfaces in k-space to analyze the processes of scattering of spin waves. 1969.T.14. No. 8. P.1511.
291. Locke E.G. Properties of isofrequency dependences and the laws of geometric optics // Phys. 2008. T.178. No. 4. P.397.
292. Locke E.G. Magnetostatic waves in ferrite films and structures based on them - Dissertation for the degree of Doctor of Physics and Mathematics. M. IRE RAS. 2007.
293. Locke E.G. Criteria for positive and negative reflection and refraction for three-dimensional anisotropic geometries // RE. 2009. Vol.54. # 2. P. 166.
294. Vashkovsky A.V., Stalmakhov A.V. Dispersion of magnetostatic waves in a two-layer ferrite-ferrite structure. 1984. T. 29. No. 5. P.901.
295. K. V. Grechushkin, A. V. Stalmakhov. Behavior of the energy flux of surface MSWs in coupled ferrite structures. 1986. T. 31. No. 2. P.397.
296. Locke E.G. Properties of non-collinear magnetostatic waves in magnetic films // Proceedings of the XX International School-Seminar "New Magnetic Materials of Microelectronics". M .: Moscow State University, 2006.S. 642.
297. Locke E.G. Propagation of surface magnetostatic waves in magnetic films with one-dimensional inhomogeneities - Dissertation for the degree of candidate of physical and mathematical sciences. M. IRE AN USSR - 1992.
298. P. A. Kolodin Dipole-exchange spin waves in multilayer structures based on ferromagnetic films -
299. Dissertation for the degree of candidate of physical and mathematical sciences. Leningrad. LETI. 1989.
300. Bespyatykh Yu.I., Dikshtein I.E., Simonov A.D. Propagation of surface magnetostatic waves in a ferrite plate with a metal half-plane // RE. 1989. Vol. 34. No. 8. P.1618.
301. Bespyatykh Yu.I., Dikshtein IE, Simonov A.D. Scattering of magnetostatic waves by a perfectly conducting half-plane //
302. ZhTF. 1989. Vol.59. No. 2. P.10.
303. Bespyatykh Yu.I., Dikshtein I.E., Simonov A.D. Propagation of surface magnetostatic waves in a ferrite plate with a metal half-plane // RE. 1985.T.30. No. 5. P.1028.
304. Bespyatykh Yu.I., Dikshtein I.E., Simonov A.D. Reflection of surface magnetostatic

waves from roughness of the surface of a ferrite film // RE. 1989. Vol. 34. No. 1. P.41.
305. Vashkovsky A. V., Shakhnazaryan D. G. Reflection of a surface magnetostatic wave from the edge of the film // RE. 1987. Vol. 32. No. 4. P.719.
306. Gorobets Yu.I., Reshetnyak S.A. Reflection and refraction of spin waves in uniaxial magnets in the geometric optics approximation // ZhTF. 1998. Vol. 68. No. 2. P.60.
307. Vashkovsky A. V., Locke E. G. The emergence of a negative refractive index during the propagation of a surface magnetostatic wave through the interface between ferrite - ferrite - dielectric - metal // Phys. 2004. Vol. 174. No. 6. P.657.
308. Vashkovsky A.V., Grechushkin K.V., Stalmakhov A.V., Tyulyukin V.A. Focusing transducer of surface magnetostatic waves // RE. 1986. T. 31. No. 4. P.838.
309. Vashkovsky A.V., Stalmakhov A.V., Tyulyukin V.A. Microwave filters based on focusing transducers of magnetostatic waves // XIV Conf. on microwave ferrite technology. Abstracts. report L. 1987.S. 128.
310. V. A. Tyulyukin Formation and focusing of beams of magnetostatic waves in films of magneto-dielectrics - Dissertation for thesis. uch. step. Ph.D. M. IRE AS USSR. 1990.
311. Vashkovsky A.V., Stalmakhov A.V., Tyulyukin V.A., Shakhnazaryan D.G. On the possibility of applying the methods of geometric optics to the creation of devices on magnetostatic waves. 1990. Vol. 35. No. 12. P.2606.
312. Vashkovsky A. V., Grechushkin K. V., Stalmakhov A. V., Tyulyukin V. A. Focusing of surface and backward volume magnetostatic waves // II Vses. shk-sem. "Spinwave microwave electronics". Abstracts. report Ashgabat. 1985.S. 29.
313. Parekh J.P., Tuan H.S. Beams steering and diffraction of magnetostatic backward volume waves // JAP. 1981. V.52. No. 3. P.2279.
314. Vashkovsky A.V., Grechushkin K.V., Stalmakhov A.V., Tyulyukin V.A. Diffraction phenomena in the propagation of bulk magnetostatic waves // II Sem. on functional magnetoelectronics. Abstracts. report Krasnoyarsk. 1986.S. 17.
315. Valyavsky A.B., Vashkovsky A.V., Stalmakhov A.V., Tyulyukin V.A. Features of the behavior of backward volume magnetostatic waves in the ferrite-dielectric-metal structure // X Vses. shk-sem. "New magnetic materials for microelectronics". Abstracts. report Riga. 1986 Part 1. P.178.
316. Vashkovsky A. V., Grechushkin K. V., Stalmakhov A. V., Tyulyukin V. A. Diffraction phenomena in the propagation of limited wave beams of bulk magnetostatic waves // RE. 1987. Vol. 32. No. 11. P.2295.
317. Vashkovsky A. V., Valyavsky A. B., Stalmakhov A. V., Tyulyukin V. A. Features of the behavior of bulk magnetostatic waves in the ferrite-dielectric-metal structure. 1987. Vol. 32. No. 11. P.2450.
318. Grechushkin K.V., Stalmakhov A.V., Tyulyukin V.A. Diffraction phenomena during the propagation of surface magnetostatic waves // XI Vses. conf. "Microwave electronics". Abstracts. Dokl., Ordzhonikidze, 1996, Part 2. P.55.
319. Valyavsky A.B., Vashkovsky A.V., Stalmakhov A.V., Tyulyukin V.A. A limited wave beam of a surface magnetostatic wave in a ferrite-dielectric-metal structure // Reg. conf. "Spin-wave phenomena of microwave electronics". Abstracts. report Krasnodar. 1987. p. 27.
320. Valyavsky A.B., Vashkovsky A.V., Stalmakhov A.V., Tyulyukin V.A. Limited wave beam of a surface magnetostatic wave in a ferrite-dielectric-metal structure. 1988. Vol. 33. No. 9. P.1820.
321. D. A. Baryshev, A. V. Stalmakhov. The effect of channeling the energy of a surface magnetostatic wave during metallization of a ferrite film // PZhTF. 1990. Vol.16.

No. 15. P.73.
322. K. V. Grechushkin, A. V. Stalmakhov, V. A. Tyulyukin. Propagation of magnetostatic waves in ferrite waveguides // RE. 1990. Vol. 35. No. 5. P.977.
323. Vashkovsky A.V., Stalmakhov A.V., Tyulyukin V.A. Wave beams of magnetostatic waves in inhomogeneous magnetic fields // III Sem. on functional magnetoelectronics. Abstracts. report Krasnoyarsk. 1988.S. 216.
324. Vashkovsky A.V., Stalmakhov A.V., Tyulyukin V.A. Wave beams of magnetostatic waves in inhomogeneous magnetic fields // PZhTF. 1988. Vol.14. No. 14. P.1294.
325. Grechushkin K.V., Stalmakhov A.V., Tyulyukin V.A. Spatial structure of beams of satellite waves of a nonlinear surface magnetostatic wave. 1991. Vol.36. No. 11. S.2078.
326. Locke E.G. Angular beam width in diffraction by a slit of waves with non-collinear group and phase velocities // Phys. 2012.T.182. No. 12. P.1327.
327. Locke E.G. Universal formula for the angular width of a diffraction beam in anisotropic media // Proceedings of the XXI International Conference "Electromagnetic Field and Materials". M .: NIU MEI, 2013.S. 107.
328. Locke E.G. On the distribution of the amplitude of the magnetic potential of diffraction beams arising as a result of the incidence of a surface spin wave on a slit in an opaque screen // Proceedings of the XXIII All-Russian conference "Electromagnetic field and materials". M .: INFRA-M, 2015.S. 115.
329. Locke E.G. On the angular width of the beam of a backward spin wave propagating in a ferrite plate // Proceedings of the XXII International Conference "Electromagnetic Field and Materials", Moscow: NIU MPEI, 2014. P.93.
330. Locke E.G. Angular width of a wave beam of a backward spin wave excited by a linear transducer in a ferrite plate // RE. 2015.Vol. 60. # 1. P.102.
331. Annenkov A.Yu., Gerus S.V., Locke E.G. Determination of the angular width of a beam of spin waves on the basis of measuring their spatial distribution (theory and experiment) // Proceedings of the XXIII All-Russian conference "Electromagnetic field and materials". M .: INFRA-M, 2015.S. 122.
332. Rodrigue G.P. Microwave solid-state delay lines // Proc. IEEE. 1965. V.53. No. 10. P.1428. - Translation: Rodrigue. Solid-state microwave delay lines // TIIER. 1965. Vol.53. No. 10. P.1613.
333. Schlömann E. Generation of spin waves in nonuniform magnetic fields. I. Conversion of electromagnetic power into spin-wave power and vice versa // JAP. 1964. V. 35. No. 1. P.159.
334. Schlömann E., Joseph R.I. Generation of spin waves in nonuniform dc magnetic fields. II. Calculation of the coupling length // JAP. 1964. V. 35. No. 1. P.167.
335. Schlömann E., Joseph R.I. Generation of spin waves in nonuniform magnetic fields. III. Magneto-elastic interaction // JAP. 1964. V. 35. No. 8. P.2382.
336. Schlömann E., Joseph R.I., Kohane T. Generation of spin waves in nonuniform magnetic fields, with application to magnetic delay line // Proc. IEEE. 1965. V.53. No. 10. P.1495. - Translation: Schleman, Joseph, Cohain. Excitation of spin waves in inhomogeneous magnetic fields and their application in magnetic delay lines // TIIER. 1965. Vol.53. No. 10. P.1685.
337. Desormiere B., Le Gall H. Magnetostatic mode excitation in YIG rods containing a turning point // JAP. 1969. V.40. No. 3. P.1191.
338. Esikov O.S., Fetisov Yu.K., Tsarkov A.G. Influence of an inhomogeneous magnetic field on the propagation of surface magnetostatic waves in a ferrite plate // ZhTF. 1982. Vol. 52. No. 4. P.719.
339. Kohane T., Schlömann E., Joseph R.I. Microwave magneto-elastic resonances in a

nonuniform magnetic fields // JAP. 1965. V. 36. No. 3. P.1267.
340. Korn G., Korn T. Handbook of mathematics for scientists and engineers. Moscow: Nauka, 1973.
341. Stancil D.D., Morgenthaler F.R. Guiding magnetostatic surface waves with nonuniform in-plane fields // JAP. 1983. V. 54. No. 3. P.1613.
342. Morgenthaler F.R. Control of magnetostatic waves in thin films by means of spatially nonuniform bias fields // Circuits, systems and signal processing, 1985. V.4. No. 1-2. P.63.
343. Poston T.D., Stancil D.D. A new microwave ring resonators using guided magnetostatic surface waves // JAP. 1984. V. 55. No. 6 (pt.2B). P.2521.
344. Shmatov G.A. Backward bulk magnetostatic waves and their excitation in an inhomogeneous magnetic field // FMM, 1995. Vol. 80. No. 3. P.18.
345. Shmatov G.A., Gobov Yu.L. Excitation of magnetostatic waves in a ferromagnetic film placed in an inhomogeneous magnetic field // PZhTF. 1996. Vol.22. No. 1. P. 39. ...
346. G. N. Burlak Magnetostatic waves in ferromagnetic films in an inhomogeneous magnetic field // PZhTF. 1986. Vol.12. No. 24. P.1476.
347. Burlak G.N., Grimalsky V.V., Kotsarenko N.Ya. Magnetostatic waves in ferromagnetic films in an inhomogeneous field // ZhTF. 1989. Vol.59. No. 8. P.32.
348. Vyzulin S.A., Korotkov V.V., Rozenson A.E. Trajectory and amplitude of a monochromatic magnetostatic wave in a ferrite film magnetized by an inhomogeneous field. 1991. Vol.36. No. 10. S.2024.
349. Tsutsumi M. The effect of inhomogeneous bias field on the delay characteristics of magnetostatic forward volume waves // JAP. 1979. V. 34. No. 1. P.204.
350. Morgenthaler F.R. Nondispersive magnetostatic forward volume waves under field gradient control // JAP. 1982. V. 53. No. 3. P.2652.
351. Burlak G.N., Grimalsky V.V. Magnetostatic waves and reconstruction of the profile of a radially symmetric inhomogeneous field. 1994. Vol. 39. No. 1. P.49.
352. Zeskind D.A., Morgenthaler F.R. Localized high-Q ferromagnetic resonance in nonuniform magnetic fields // IEEE Trans. on Magn. 1977. V.13. No. 5. P.1249.
353. Morgenthaler F.R. Magnetostatic waves bound to a DC field gradients // IEEE Trans. on Magn. 1977. V.13. No. 5. P.1252.
354. Morgenthaler F.R. Bound magnetostatic waves controlled by field gradients in YIG single crystal and epitaxial films // IEEE Trans. on Magn. 1978. V.14. No. 5. P.806.
355. Burlak G.N., Grimalsky V.V. Transformation of magnetostatic waves upon reflection from a field inhomogeneity. 1993. Vol. 38. No. 8. P.1407.
356. Stancil D.D. Magnetostatic waves in nonuniform bias field includingexchange effects // IEEE Trans. on Magn. 1980. V.16. No. 5. P.1153.
357. Morgenthaler F.R., Bhattacharjee T. Numerical solution of the integral equations for MSSW // IEEE Trans. on Magn. 1982. V.18. No. 6. P.1636.
358. 358. Annenkov A.Yu., Gerus S.V. Propagation of magnetostatic waves in two coupled channels formed by a magnetic field // RE. 1996. Vol.41. No. 2. P.216.
359. Annenkov A.Yu., Gerus S.V., Sotnikov I.V. Propagation of magnetostatic waves in a stationary spatially periodic magnetic field // RE. 1992. T. 37. No. 8. P.1371.
360. Annenkov A.Yu., Gerus S.V., Kovalev S.I. Numerical modeling of quasi-surface magnetostatic waves in a ferrite film with two magnetic channels // ZhTF. 1998. Vol. 68. No. 2. P.91.
361. Lyashenko N.I., Talalaevsky V.M. Influence of inhomogeneity of magnetic fields on the dispersion characteristics of surface magnetostatic waves in films of yttrium iron garnet // RE. 1996. Vol.41. No. 12. P.1438.

362. Lyashenko N.I., Talalaevsky V.M., Chevnyuk L.V. Influence of temperature and elastic stresses on the dispersion characteristics of surface magnetostatic waves // RE. 1994. Vol. 39. No. 7. P.1164.
363. A. V. Voronenko, S. V. Gerus. Interaction of surface magnetostatic waves with a spatially periodic magnetic field // PZhTF. 1984. T.10. No. 12. P.746.
364. Annenkov A.Yu., Gerus S.V., Sotnikov I.V. Propagation of magnetostatic waves in a stationary spatially periodic magnetic field // RE. 1992. T. 37. No. 8. P.1371.
365. Annenkov A.Yu., Vinogradov AP, Gerus SV, Ryzhikov IA, Shishkov SA, Inoue M. Study of magnetostatic waves in photonic crystals // Izvestiya RAN. Series Physical. 2007. Vol. 71. No. 11. P.1612.
366. Annenkov A.Yu., Gerus S.V. 2-D distribution of magnetostatic waves in the vicinity of the non-transmission zone of a magnonic crystal // Proceedings of the XXI International Conference "Electromagnetic field and materials". M .: NIU MEI. 2013.S. 113.
367. Gerus S.V. Magnetostatic waves in spatially periodic and two-dimensionally inhomogeneous magnetic fields - Dissertation for the degree of Doctor of Physical and Mathematical Sciences M. IRE RAS. 2012.
368. Voronenko A.V., Gerus S.V., Krasnozhen L.A. Method for measuring the parameters of gyromagnetic films // Microelectronics. 1989. Vol. 18. # 1. P.61.
369. Annenkov A.Yu., Vasiliev I.A., Gerus S.V., Kovalev S.I. Modes of surface magnetostatic waves in a channel created by an inhomogeneous magnetic field // ZhTF. 1995. Vol. 65. No. 1. P.71.
370. Annenkov A.Yu., Gerus S.V., Kovalev S.I. Transformation of surface magnetostatic waves channeled by a stepped magnetization field // ZhTF. 2002. Vol. 72. No. 6. P.85.
371. Annenkov A.Yu., Gerus S.V., Kovalev S.I. Bulk and surface-volume magnetostatic waves in waveguides created by a stepped bias field // ZhTF. 2004. Vol. 74. # 2. P.98.
372. Annenkov A.Yu., Gerus S.V. Interaction of bulk and surface magnetostatic waves in a channel created by an inhomogeneous magnetic field // Proceedings of the XIX International Conference "Electromagnetic Field and Materials". M .: NIU MEI. 2011.S. 90.
373. Annenkov A.Yu., Gerus S.V. Study of the distribution of surface magnetostatic waves by scanning the surface of a ferrite plate // RE. 2012.T.57. No. 5. P.572.
374. Annenkov A.Yu., Gerus S.V. Formation of a beam of surface magnetostatic waves in the ferrite-dielectric-ferrite structure // Proceedings of the XX International Conference "Electromagnetic Field and Materials". M .: NIU MEI. 2012.S. 289.
375. Annenkov A.Yu., Gerus S.V. Dispersion properties of a magnonic crystal with non-reciprocity // Proceedings of the XXIII All-Russian conference "Electromagnetic field and materials". M .: INFRA-M, 2015.S. 242.
376. Parekh J.P., Tuan H.S. Reflection of magnetostatic surface wave at a shallow groove on a YIG film // Appl. Phys. Lett. 1977. V.30. No. 10. P.709.
377. Kalinikos B.A. Dipole-Exchange Spin Waves in Ferromagnetic Films - Dissertation for the degree of Doctor of Physical and Mathematical Sciences. SPb. LETI. 1985.
378. N.Yu. Grigorieva Dipole-exchange spin waves in periodic structures based on thin ferromagnetic films. Dissertation for the degree of candidate of physical and mathematical sciences. SPb. LETI. 2009.
379. Grigorieva N.Yu., Kalinikos B.A. The theory of spin waves in multilayer ferromagnetic film structures. SPb .: Technolit, 2008.
380. Demokritov S.O., Serga A.A., Andre A., Demidov V.E., Kostilev M.P., Hillebrands B., Slavin A.N. Tunneling of dipolar spin waves through a region of inhomogeneous

magnetic field // Phys. Rev. Lett. 2004. V.93. P.047201.
381. Vysotsky S.L., Nikitov S.A., Filimonov Yu.A. Magnetostatic spin waves in two-dimensional periodic structures - magnetophotonic crystals // ZhETF. 2005. Vol. 128. No. 3 (9). P.636.
382. Vashkovsky A. V., Locke E. G. Surface electromagnetic wave in a ferrite-dielectric structure with negative dielectric constant // Proceedings of the XVIII International School-Seminar "New Magnetic Materials of Microelectronics". M .: Moscow State University. 2002. p. 781.
383. Vashkovsky A. V., Locke E. G. Backward surface electromagnetic waves in composite structures using ferrites // RE. 2003. T.48. No. 2. P.169.
384. Vashkovsky A. V., Locke E. G. Surface magnetostatic waves in a ferrite-dielectric structure surrounded by half-spaces with negative dielectric constant // RE. 2002. Vol.47. # 1. P.97.
385. Pendry J.B., Holden A.J., Stewart W.J., Youngs I. Extremely low frequency plasmons in metallic mesostructures // Phys. Rev. Lett. 1996. V.76. No. 25. P.4773.
386. Locke E.G. Dispersion of magnetostatic waves in a ferrite-lattice composite structure of metal strips. 2003. T.48. No. 12. P.1484.
387. Locke E.G. Propagation of surface magnetostatic waves in a ferrite-lattice composite structure of metal strips. 2005. Vol. 50. # 1. P.74.
388. V. I. Zubkov, Ya. A. Monosov, V. I. Shcheglov. Mandelstam-Brillouin spin effect // PZhETF. 1971.V.13. No. 5. P.229.
389. Popkov A.F. Collinear scattering of spin waves in a plate under acoustic pumping // RE. 1982. Vol.27. No. 7. P.1366.
390. Mednikov A.M., Popkov A.F., Anisimkin V.I., Nam B.P., Petrov A.A., Spivakov D.D., He A.S. Inelastic scattering of a surface spin wave in a thin YIG film by a surface acoustic wave // PZhETF. 1981. Vol. 33. No. 12. P.646.
391. Kiryukhin N.N., Lisovskiy F.V. Spin waves in a medium with spatio-temporal periodicity // FTT. 1968. Vol.10. No. 3. P.709.
392. Kryshtal R.G., Medved A.V. Experimental study of the scattering of nonlinear surface magnetostatic waves by a surface acoustic wave // Physics and Technology. 1992. Vol. 34. No. 1. P.333.
393. Kryshtal R.G., Medved A.V., Osipenko V.A., Popkov A.F. Transformation of modes of magnetostatic waves when scattered by a surface acoustic wave in YIG films // Zh. 1988. T. 58. No. 12. P.2315.
394. Kryshtal R.G., Medved A.V. Scattering of surface magnetostatic waves by a surface acoustic wave in a nonreciprocal structure of a YIG-metal film // ZhTF. 1989. Vol.59. No. 6. P.82.
395. Kryshtal R.G., Medved A.V. Influence of the electrical conductivity of a metal film on the scattering of surface magnetostatic waves by a surface acoustic wave in the structure of a GGG-YIG metal film // ZhTF. 1991. Vol. 61. No. 4. P.105.
396. Vashkovsky A.V., Zubkov V.I., Locke E.G., Shcheglov V.I. Surface magnetostatic waves in ferrite films magnetized by a transversely inhomogeneous field // Proceedings of the XI International Conference on Gyromagnetic Electronics and Electrodynamics. Ukraine, Alushta: MPEI, 1992. V.1. P.104.
397. Vashkovsky A.V., Zubkov V.I., Locke E.G., Shcheglov V.I. Propagation of surface magnetostatic waves in transversely inhomogeneous fields of magnetization // RE. 1993. Vol. 38. No. 5. P.818.
398. Vashkovsky A. V., Locke E. G. Directional patterns of radiation arising from the conversion of surface magnetostatic waves into electromagnetic waves. 1995. Vol. 40. No. 7. P.1030.

399. Vashkovsky A. V., Locke E. G. On the parameters of the radiation patterns arising from the transformation of a surface magnetostatic wave into an electromagnetic one. 2004. Vol.49. No. 8. P.966.
400. Vashkovsky A. V., Locke E. G. On the radiation diagrams arising from the transformation of magnetostatic waves into electromagnetic waves // Proceedings of the XX International Conference "Electromagnetic field and materials". M .: NIU MEI, 2012.S. 181.
401. Vashkovsky A.V., Locke E.G. On the mechanism of transformation of a surface magnetostatic wave into an electromagnetic one. 2009. Vol.54. No. 4. P.476.
402. V. Shevchenko. Smooth transitions in open waveguides - Moscow: Nauka, 1969.
403. Malozemov A., Slonzuski J. Domain walls in materials with cylindrical magnetic domains - Moscow: Mir, 1982.
404. Vilkov E.A. Magnetostatic surface waves on a moving domain boundary of a ferrogarnet crystal // PZhTF. 2000. Vol.26. No. 20. P.28.
405. Vilkov E.A., Shavrov V.G., Shevyakhov N.S. On the influence of the motion of the domain wall on the spectral properties of confined magnetostatic waves // Journal of Radioelectronics (electronic journal). 2000. No. 8 (http://jre.cplire.ru/win/aug00/2/tex).
406. Limaye P.S., Parker A.A., Stroud D.G., Wigen P.E. Low field microwave excitations of magnetic domain walls // JAP. 1979. V. 50. No. 3. P.2027.
407. Shimokhin I.A. The spectrum of bending vibrations of a 180-degree domain wall in yttrium ferro-garnet // PZhETF. 1996. Vol. 63. No. 10. P.797.
408. Alekseev A.M., Detch H., Kulagin N.E., Popkov A.F., Synogach V.T. Microwave excitations of a domain wall in a cubic magnet with induced anisotropy // ZhTF. 1999. Vol. 69. No. 6. P.55.
409. Dikshtein I.E., Kryshtal R.G., Medved A.V. Critical scattering of electromagnetic waves on spin fluctuations in nonsaturated magnetic films under acoustic pump // JAP. 1994. V.76. No. 10. P.6880.
410. Stancil D.D. A magnetostatic wave model for domain-wall collective excitations // JAP. 1984. V.56. No. 6. P.1775.
411. Kilshakbaeva Zh.A., Rusinov A.A., Kandaurova G.S. Phase diagrams of dynamic systems of magnetic domains. Single-domain state // I Combined. conf. on magnetoelectronics. Abstracts. report 1995, M .: IRE RAN. P. 39.
412. G.S. Kandaurova. Chaos and order in a dynamic system of magnetic domains. Anger condition // I United. conf. on magnetoelectronics. Abstracts. report 1995. M .: IRE RAN. P.41.
413. G.S. Kandaurova. New phenomena in the low-frequency dynamics of a team of magnetic domains // UFN, 2002. Vol. 172. No. 10. P.1165.
414. Kandaurova G.S., Svidersky A.E. Self-organization processes in a multidomain magnetic medium and the formation of stable dynamic structures // ZhETF. 1990. Vol. 97. No. 4. P. 1218.
415. Dikshtein I.E., Lisovskiy F.V., Mansvetova E.G., Chizhik E.S. Formation of a reflective domain structure during unipolar and cyclic magnetization reversal of a uniaxial magnet // ZhETF. 1991. Vol. 100. No. 5. P.1606.
416. Lisovsky F.V., Mansvetova E.G., Nikolaeva E.P., Nikolaev A.V. Dynamic self-organization and symmetry of the magnetic moment distribution in thin films // ZhETF. 1993. Vol. 103. # 2. P.213.
417. M.V. Logunov, N.V. Moiseev. Formation of a honeycomb domain structure in magnetic films // PZhTF. 1997. T. 23. No. 9. P.46.
418. M.V. Logunov, M.V. Gerasimov. Formation and evolution of giant dynamic domains

in a harmonic magnetic field // Physics and Technology. 2003. Vol.45. No. 6. P.1031.
419. Cherechukin A.A., Koledov L.V., Koledov V.V., Mitroshin V.A., Shavrov V.G., Tulaikova A.A., Yasrebdust H.H. Unidirectional stripe domain motion in the highly stressed single crystal YIG mechan plates // I conf. on magnetoelectronics. Abstracts. report 1995. M .: IRE RAN. P.45.
420. Bi S.Y. Stripe domain lattice ferromagnetic resonance in garnet thin films // JAP. 1990. V.67. No. 6. P.3179.
421. Rachford F. J., Lubitz P., Vittoria C. Magnetic multi-domain resonance in single crystal ferrite platelets // JAP. 1981. V.52. No. 3. P.2259.
422. Artman J.O., Charap S.H. Domain mode ferromagnetic resonance in materials with K1 and Kau // JAP. 1979. V. 50. No. 3. P.2024.
423. Gulyaev Yu.V., Zilberman P.E., Karmazin S.V. Observation of traveling magnetostatic waves in YIG films at incomplete saturation // PZhTF. 1983. Vol. 9. No. 3. C.11.
424. Kostenko V.I., Sigal M.A. Magnetostatic waves in a thin uniaxial platelet with stripe domain magnetized along easy axis // Phys. Stat. Sol. (b). 1992. V. 170. No. 3. P.569.
425. Galkin O.L., Zilberman P.E., Karmazin S.V. Statistical magnetization and magnetostatic waves in ferromagnetic films with low uniaxial anisotropy at incomplete saturation in a normal field. 1985.T.30. No. 4. P.735.
426. Charap S.H., Artman J.O. Magnetostatic modes of stripe domain structures // JAP. 1978. V.49. No. 3. P.1585.
427. E. Beginin. Excitation of electromagnetic waves in layered structures based on highly anisotropic ferrites in mono- and multi-domain modes // I Obed. conf. on magnetoelectronics. Abstracts. report 1995. M .: IRE RAN. P.167.
428. Zilberman P.E., Kazakov G.T., Tikhonov V.V. Magnetostatic and fast magnetoelastic waves in films of yttrium iron garnet with an irregular domain structure. 1987. Vol. 32. No. 4. P.710.
429. Zilberman P.E., Kazakov G.T., Kulikov V.M., Tikhonov V.V. Influence of weak magnetizing fields on the propagation of magnetostatic waves in films of yttrium iron garnet of submicron thickness. 1988. Vol. 33. No. 2. P.347.
430. Zilberman P.E., Kulikov V.M., Tikhonov V.V., Shein I.V. Magnetostatic waves in yttrium iron garnet films with weak magnetization // RE. 1990. Vol. 35. No. 5. P.986.
431. Smit J., Beljers H. G. Ferromagnetic resonance absorption in BaFe12O19 a highly anisotropic crystal // Philips Res. Rep. 1955. V.10. No. 2. P.113.
432. Kazakov G.T., Sukharev A.G., Filimonov Yu.A. Resonant interaction of magnetostatic waves (MSW) in Ga, Sc: YIG films with a stripe domain structure (PDS) // V All. shk-sem. "Spin-wave microwave electronics". Abstracts. report Zvenigorod: IRE RAS. 1991. p.83.
433. Shamsutdinov M.A., Farztdinov M.M., Ekomasov E.G. Dynamic bevel of magnetic sublattices in a magnetic field and spin waves in rare-earth orthoferrites with a domain structure // Phys. 1990. T. 32. No. 4. P.1133.
434. Zilberman P.E., Kulikov V.M., Tikhonov V.V., Shein I.V. Nonlinear effects in the propagation of surface magnetostatic waves in yttrium iron garnet films in weak magnetic fields // ZhETF. 1991. Vol. 99. No. 5. P.1566.
435. Vashkovsky A. V., Locke E. G. Influence of dielectric substrate and magnetic losses on dispersion and properties of a surface magnetostatic wave // RE. 2001. Vol. 46. No. 6. P.729.
436. Vashkovsky A. V., Locke E. G. Forward and backward noncollinear waves in magnetic films // Phys. 2006. Vol. 176. No. 5. P.557.
437. Vashkovsky A. V., Locke E. G. Properties of backward electromagnetic waves and

the appearance of negative reflection in ferrite films // Phys. 2006. Vol. 176. No. 4. P.403.
438. Vashkovsky A. V., Locke E. G. Dispersion equations for magnetostatic waves in ferrite structures with a conducting plane and a "magnetic wall" based on Maxwell's equations // Proceedings of the XX International Conference "Electromagnetic Field and Materials". M .: NIU MEI, 2012.S. 177.
439. Locke E.G. Spin waves in a dielectric-ferrite-dielectric structure with "magnetic walls" or ideal conductors (based on Maxwell's equations) // RE. 2014.T.59. No. 7. P.711.
440. Locke E.G. On the configuration of the lines of the high-frequency field of a surface spin wave // Proceedings of the XXII International Conference "Electromagnetic Field and Materials". Moscow: NIU MEI, 2014, p. 87.
441. Locke E.G. Vector lines of a surface spin wave in tangentially magnetized ferrite structures with a metal screen and a magnetic wall // Proceedings of the XXIII All-Russian Conference "Electromagnetic Field and Materials". M .: INFRA-M. 2015. S. 96.
442. Locke E.G. The structure of high-frequency fields of a surface spin wave in a tangentially magnetized ferrite plate. 2015.Vol. 60. No. 12. P.1149.
443. Locke E.G. Isofrequency dependences of electromagnetic waves in an unlimited ferrite medium // Proceedings of the XXIII All-Russian conference "Electromagnetic field and materials". M .: INFRA-M, 2015.S. 103.
444. Vashkovsky A. V., Locke E. G. Characteristics of a surface magnetostatic wave in a ferrite-dielectric structure placed in a slowly varying inhomogeneous magnetic field. 2001. Vol. 46. No. 10. P.1257.
445. Vashkovsky A. V., Locke E. G. On the properties of the backward magnetostatic wave // Proceedings of the XIX International Conference "Electromagnetic Field and Materials", Moscow: NIU MPEI, 2011. P.75.
446. Vashkovsky A. V., Locke E. G. On the relationship between the energy and dispersion characteristics of magnetostatic waves in ferrite structures // Phys. 2011.T.181. No. 3. P.293.
447. Vashkovsky A. V., Locke E. G. On the physical properties of a backward magnetostatic wave in its description based on Maxwell's equations // RE. 2012.T.57. No. 5. P.541.
448. Morgenthaler F.R. Novel devices based upon field gradient control of magnetostatic modes and waves // Ferrites: Proc. of the Int. Conf. 1980. Japan. P.839.
449. Castera J.P. State of the art in design and technology of MSW devices // JAP. 1984. V. 55. No. 6 (pt.2B). P.2506.
450. Magnetostatic waves and applications in signal processing (Thematic issue) // Circuits, systems and signal processing. 1985. V.4. No. 1-2. P.1.
451. Parekh J.P., Chang K.W., Tuan H.S. Propagation characteristics of magnetostatic waves // Circuits, systems and signal processing. 1985. V.4. No. 1-2. P.9.
452. Ishak W., Reese E., Huijer E. Magnetostatic wave devices for UHF band applications // Circuits, systems and signal processing. 1985. V.4. No. 1-2. P.285.
453. Adam J.D. MSW devices for EW receiver applications // Circuits, systems and signal processing. 1985. V.4. No. 1-2. P.301.
454. Owens J.M., Collins J.H., Carter R.L. System applications of magnetostatic wave devices // Circuits, systems and signal processing. 1985. V.4. No. 1-2. P.317.
455. Taub J. Applications of magnetostatic wave technology to EW systems - an assessment // Circuits, systems and signal processing. 1985. V.4. No. 1-2. P.351.
456. Rodrigue G.P. A generation of microwave ferrite devices // Proc. IEEE. 1988. V.76.

2. P.121. - Translation: Rodrigue G.P. Development stages of microwave ferrite equipment // TIIER. 1988. Vol.76. # 2. P.29.
457. Le Gall H. Magnetoelectronics and nanorecording // I Ed. conf. on magnetoelectronics. Abstracts. Dokl., 1995, M .: IRE RAN. P.27.
458. Shone M. The technology of YIG film growth // Circuits, systems and signal processing. 1985. V.4. No. 1-2. P.89.
459. Adam J.D. YIG film characterization for MSW devices // Circuits, systems and signal processing. 1985. V.4. No. 1-2. P.105.
460. Lysenko V.A. Transfer function of microwave devices on magnetostatic waves // RE. 1986. T. 31. No. 8. P.1627.
461. V.N. Prokushkin, Yu.P. Sharaevsky. Influence of reactive impedance load on the characteristics of magnetostatic waves //RE. 1993. Vol. 38. No. 9. P.1551.
462. Paladiy N.V. Methods for calculating the main parameters of delay lines on magnetostatic waves. 1994. Vol. 39. No. 6. P.1011.
463. Beregov A.S., Kudinov E.V., Ereshchenko I.N. Improving the thermal stability of devices on magnetostatic waves // Electronic engineering. Ser. Microwave electronics. 1987. No. 1 (395). P.19.
464. Kurushin E.P., Nefedov E.I. Electrodynamics of anisotropic wave-guiding structures - Moscow: Nauka, 1983.
465. El-Sharawy E., Jackson R. W. Coplanar waveguide and slot line on magnetic substrates: analysis and experiment // IEEE Trans. on MTT. 1988. V.36. No. 6. P.1071.
466. Kwan P., Vittoria C. Propagation characteristics of a ferrite image guide // JAP. 1993. V.73. No. 10. P.6466.
467. Novikov G.M., Petrunkin E.Z. Experimental study of the propagation of MSW in film waveguides. 1984. T. 29. No. 9. P.1691.
468. Zilberman P.E., Kazakov G.T., Tikhonov V.V. Separate measurement of useful signal and pickup parameters in magnetostatic wave transmission lines. 1985.T.30. No. 6. P.1164.
469. Kwan P., Vittoria C. Nonreciprocal coupling structure of dielectric wave guides // JAP. 1993. V.73. No. 10. P.7012.
470. Grechushkin K.V., Sharaevsky Yu.P. Interaction of high-frequency current in a transmission line with resonant modes of a magnetostatic wave. 1994. Vol. 39. No. 6. P.1017.
471. S.N.Dunaev Influence of pulsed optical heating on the propagation of magnetostatic waves in a ferrite film // I Obed. conf. on magnetoelectronics. Abstracts. report M .: IRE RAN, 1995.S. 159.
472. Romanov IV, Studenov VB, Fetisov Yu.K. Laser control of microwave signal parameters in a transmission line on magnetostatic waves // I Objed. conf. on magnetoelectronics. Abstracts. report M .: IRE RAN, 1995.S. 238.
473. Sethares J.C., Weinberg I.J. Theory of MSW transducers // Circuits, systems and signal processing. 1985. V.4. No. 1-2. P.41.
474. Vugalter G.A. Excitation of short surface magnetostatic waves by a metal strip. 1989. Vol. 34. No. 5. P.965.
475. Vugalter G.A., Gilinsky I.A. Excitation and reception of surface magnetostatic waves by a segment of a microstrip line // RE. 1987. Vol. 32. No. 3. P.465.
476. Smelov M.V. Experimental study of the excitation of magnetostatic waves by a meander-type decelerating system. 1988. Vol. 33. No. 2. P.440.
477. Smelov M.V. Experimental study of single-mode excitation of magnetostatic waves by a loop transformer // RE. 1989. Vol. 34. No. 10. P.2222.
478. Smelov M.V. Experimental study of the influence of the forms of the ferrite wave-

guide and exciters on the propagation of magnetostatic waves // RE. 1987. Vol. 32. No. 11. P.2290.
479. IV Krutsenko, GA Melkov, SA Ukhanov. Influence of a magnetic pin on the excitation of magnetostatic waves in thin ferromagnetic films // RE. 1987. Vol. 32. No. 9. P. 1976.
480. Sethares J.C., Weinberg I.J. Apodization of variable coupling MSSW transducers // JAP. 1979. V. 50. No. 3. P.2458.
481. Goldberg L.B. Calculation of spatial-frequency characteristics of curvilinear microstrip emitters of magnetostatic waves // RE. 1990. Vol. 35. No. 6. P. 1212.
482. De Gasperis P., Miccoli G., Di Gregorio C., Roveda R. Slowly dispersive, short time delay line based on very thick liquid phase epitaxially grown yttrium iron garnet // JAP. 1984. V. 55. No. 6 (pt.2B). P.2512.
483. Daniel M.R., Adam J.D., Emtage P.R. Dispersive delay at gigahertz frequencies using magnetostatic waves // Circuits, systems and signal processing. 1985. V.4. No. 1-2. P.115.
484. Adkins L.R. Dispersion control in magnetostatic wave delay lines // Circuits, systems and signal processing. 1985. V.4. No. 1-2. P.137.
485. Kindyak A.S., Kolosov V.A., Demchenko A.I., Makutina L.I. Dispersion delay lines on magnetostatic waves with preset parameters // RE. 1992. T. 37. No. 11. S.2063.
486. Adkins L.R., Glass H.L., Stearns F.S., Carter R.L., Chang K.W., Owens J.M. Electronically variable time delays using cascaded magnetostatic delay lines // JAP. 1984. V. 55. No. 6 (pt.2B). P.2518.
487. Bajpai S.N., Carter R.L., Owens J.M. Insertion loss of magnetostatic surface wave delay lines // IEEE Trans. on MTT. 1988. V.36. No. 1. P.132.
488. Wu H. J., Smith C. V., Owens J. M. Bandpass filtering and input impedance characterization for driven multielement transducer pair-delay line magnetostatic wave devices // JAP. 1979. V. 50. No. 3. P.2455.
489. Kindyak A.S., Kolosov V.A., Demchenko A.I., Makutina L.N. On the matching of dispersive delay lines on magnetostatic waves // RE. 1992. T. 37. No. 8. P.1520.
490. Adam J.D., Stitzer S.N. A magnetostatic wave signal-to-noise enhancer // Appl. Phys. Lett. 1980. V.36. No. 3. P.485.
491. Sethares J.C., Stiglitz M.R., Weinberg I.J. Magnetostatic wave oscillator frequencies // JAP. 1981. V.52. No. 3. P.2273.
492. Dunaev S.N., Fetisov Yu.K., Tokarev V.G., Vasiliev V.P. Experimental studies of the phase noise of generators on magnetostatic waves // RE. 1992. T. 37. No. 7. P.1274.
493. Dunaev S.N., Fetisov Yu.K. Multifrequency generation and mode locking in an oscillator with a delay line on magnetostatic waves. 1992. T. 37. No. 2. P.290.
494. Tuan H.S., Parekh J.P. Q-factor for an MSW delay-line based oscillator // Circuits, systems and signal processing. 1985. V.4. No. 1-2. P.221.
495. Makovkin A. V., Fetisov Yu.K. Modulation of the amplitude of magnetostatic waves in a yttrium iron garnet film by optical pulses // I Obed. conf. on magnetoelectronics. Abstracts. report M .: IRE RAN, 1995.S. 198.
496. Dunaev S.N., Gryaznykh I.V., Myasoedov A.N., Rybakov V.P., Fetisov Yu.K. Modulator of the frequency of radio pulses on magnetostatic waves. 1990. Vol. 35. No. 11. P.2453.
497. Vlaskin S.V., Novikov G.M., Petrunkin E.Z., Popov S.N. Tunable band-pass filters on spin waves // Electronic Engineering. Ser. Microwave electronics. 1990. No. 1 (425). C.8.
498. AV Vashkovsky, VI Zubkov, BM Lebed, GM Novikov. Narrowband filtering of microwave signals upon excitation of magnetostatic waves in yttrium iron garnet films.

1985.T.30. No. 8. P.1513.
499. Altman A.B., Lebed B.M. Calculation model of the filter on magnetostatic waves // RE. 1997. Vol. 42. No. 6. P.719.
500. Paladiy N.V., Kharchikov A.F. Research of intermodulation components in filters on magnetostatic waves // RE. 1991. Vol.36. No. 8. P.1604.
501. Rogozin, AN Averin, SV Zagryadskiy. Calculation of multi-link band-stop filters taking into account the near fields of ferrites // RE. 1985.T.30. No. 11. P.2270.
502. Reed K.W., Owens J.M., Carter R.L. Current status of magnetostatic reflective array filters // Circuits, systems and signal processing. 1985. V.4. No. 1-2. P.157.
503. Smelov M.V. Formation of phase and amplitude characteristics of microwave devices on magnetostatic waves // RE. 1988. Vol. 33. No. 11. P.2432.
504. Dmitriev V.F. Synthesis of frequency characteristics of spin-wave devices based on curvilinear antenna systems. 1989. Vol. 34. No. 10. P.2233.
505. Castera J.P., Hartemann P. Magnetostatic wave resonators and oscillators // Circuits, systems and signal processing. 1985. V.4. No. 1-2. P.181.
506. Chang K. W., Ishak W. Magnetostatic surface wave straight edge resonators // Circuits, systems and signal processing. 1985. V.4. No. 1-2. P.201.
507. Koike T .. No. Agano T. Digital control method for resonant frequencies of straight-edge MSW resonators // I Ed. conf. on magnetoelectronics. Abstracts. report M .: IRE RAN, 1995.S. 232.
508. Zhilinskas M.A., Zubovsky I.I., Ivashka V.P. Magnetostatic oscillations in ferrite resonators of rectangular cross section // RE. 1993. Vol. 38. No. 7. S. 1215.
509. Brinlee W.R., Owens J.M., Smith C.V., Carter R.L. "Two-port" magnetostatic wave resonators utilizing periodic metal reflective arrays // JAP. 1981. V.52. No. 3. P.2276.
510. Gulyaev Yu.V., Nikitov S.A., Plessky V.P. Bragg reflection of surface magnetostatic waves from a periodic system of thin conducting strips // ZhTF. 1982. Vol. 52. No. 4. P.799.
511. Anfinogenov VB, Afanasyev AI, Zilberman PE, Maltsev VA, Rudy Yu.B., Finkelstein Yu.Kh. Resonant attenuation of electromagnetic waves in the structure of a yttrium iron garnet film - a slowing down system of the "meander" type at a weak slowdown // RE. 1990. Vol. 35. No. 12. P.2481.
512. Pomyalov A.V., Andreev A.S. Waves of coupled oscillations in the periodic system of thin-film ferrite resonators // RE. 1986. T. 31. No. 9. P.1739.
513. Blank A.Ya., Sharshanov A.Ya. Resonant excitation of surface magnetostatic waves by electromagnetic radiation in periodic magnetic structures // RE. 1991. Vol.36. No. 8. P.1468.
514. Kryshtal R.G., Maltsev O.A., Medved A.V. Investigation of absorbers of surface magnetostatic waves in films of yttrium iron garnet // RE. 1986. T. 31. No. 1. P.200.
515. Sethares J.C., Floyd R. MSW applications for phased array antennas // Circuits, systems and signal processing. 1985. V.4. No. 1-2. P.335.
516. How H., Rainville P., Harackiewicz F., Vittoria C. Radiation frequencies for magnetically nonsaturated ferrite patch antennas // JAP. 1993. V.73. No. 10. P.6469.
517. Goldie H., Stitzer S.N. Broadband circuit for microwave s / n enhancers - Patent USA N4,292,607. Int. Cl.: H01P 5/00. Sep. 29, 1981.
518. Stitzer S.N., Emtage P.R. Nonlinear microwave signal-processing devices using thin ferrimagnetic films // Circuits, systems and signal processing. 1985. V.4. No. 1-2. P.227.
519. Parekh J.P., Tuan H.S., Chang K.W. Magnetostatic wave convolvers // Circuits, systems and signal processing. 1985. V.4. No. 1-2. P.253.
520. Fisher A.D. Optical signal processing with magnetostatic waves // Circuits, systems

and signal processing. 1985. V.4. No. 1-2. P.265.
521. Wilber W.D., Wetling W., Kabos P., Patton C.E., Jantz W. A wave-vector selective light scattering magnon spectrometer // JAP. 1984. V. 55. No. 6 (pt.2B). P.2533.
522. Takahashi H., Mitsuoka K., Komuro M., Sugita Y. Ferromagnetic resonance studies of Fe16N2 films with a giant magnetic moment // JAP. 1993. V.73. No. 10. P.6060.
523. Soohoo R.F., Morrish A.H. FMR measurement of anisotropy dispersion in amorphous GdFe films // JAP. 1979. V. 50. No. 3. P.1639.
524. Vyzulin S.A., Rozenson A.E., Shekh S.A. On the spectrum of surface magnetostatic waves in a ferrite film with losses. 1991. Vol.36. No. 1. P.164.
525. Vyzulin S.A. Generalized plane waves in problems of electrodynamics of magnetogyrotropic media - Dissertation. to apply for an account. Art. Doctor of Physical and Mathematical Sciences Krasnodar. 2000.
526. Annenkov A.Yu., Gerus S.V. Attenuation of noncollinear surface magnetostatic waves // Book of Abstracts of Moscow International Symposium on Magnetism (MISM). Published by "Publishing House of Phys. Fac. Moscow State University ". Moscow. 2014. P.863.
527. Annenkov A.Yu., Gerus S.V. Indicatrixes of surface magnetostatic waves // Proceedings of the XXII International Conference "Electromagnetic Field and Materials". Moscow: NRU MEI. 2014.S. 99.
528. Makarov P.A., Shavrov V.G., Shcheglov V.I. Influence of dissipation on the properties of surface magnetostatic waves in a tangentially magnetized ferrite plate // Electronic Journal of Radioelectronics. 2014. No. 7. (31). http://jre.cplire.ru/jre/jul14/8/text.html.
529. Makarov P.A., Shavrov V.G., Shcheglov V.I. Surface magnetostatic waves in a ferrite plate with dissipation // Proceedings of the XXII International Conference "Electromagnetic Field and Materials". Moscow: NRU MEI. 2014.S. 221.
530. Keller Yu.I., Makarov P.A., Shavrov V.G., Shcheglov V.I. Surface magnetostatic waves in a ferrite plate with dissipation. Part 1. Dispersion relations // Electronic "Journal of radio electronics", 2016. No. 2, (46). http://jre.cplire.ru/jre/feb16/2/text.html.
531. Keller Yu.I., Makarov P.A., Shavrov V.G., Shcheglov V.I. Surface magnetostatic waves in a ferrite plate with dissipation. Part 2. Wave propagation perpendicular to the field direction // Electronic "Journal of Radioelectronics", 2016. No. 3, (57). http://jre.cplire.ru/jre/mar16/1/text.html.
532. Keller Yu.I., Makarov P.A., Shavrov V.G., Shcheglov V.I. Surface magnetostatic waves in a ferrite plate with dissipation. Part 3. Wave propagation in an arbitrary direction relative to the field // Electronic Journal of Radioelectronics, 2016. No. 3, (49). http://jre.cplire.ru/jre/mar16/2/text.html.
533. Busse F., Mansurova M., Lenk B., Von der Ehe M., Münzenberg M. A scenario for magnonic spin-wave traps // Scientific Reports / 5: 12824 / DOI: 10.1038 / srep12824. (www.nature.com/scientificreports). 2015.
534. Obry B., Vasyuchka V.I., Chumak A.V., Serga A.A., Hillebrands B. Spin-wave propagation and transformation in a thermal gradient // Appl. Phys. Lett. 2012. V.101. P.192406.
535. Kolokoltsev O., Qureshi N., Mejla-Uriarte E., Ordonez Romero C.I. Hot spin-wave resonators and scatterers // JAP. 2012. V.112. P.013902.
536. Satoh T., Terui Yu., Moriya R., Ivanov B.A., Ando K., Saitoh E., Shimura T., Kuroda K. Directional control of spin-wave emission by spatially shaped light // Nature Photonics. 2012. V.6. No. 10. P.662.
537. Pezeril T., Gusev V., Mounier D., Chigarev N., Ruello P. Surface motion induced

by laser action on opaque anisotropic crystals // J. Phys. D: Appl. Phys. 2005. V.38. No. 9. P.1421.
538. Temnov V. Ultrafast acousto-magneto-plasmonics Nature Photonics. 2012. V.6. No. 10. P.728.
539. Kovalenko O., Pezeril T., Temnov V. New concept for magnetization switching by ultrafast acoustic pulses // Phys. Rev. Lett. 2013. V.110. No. 26. P.2666602 (5).
540. Linnik T.L., Scherbakov A.V., Yakovlev D.R., Liu X., Furdyna J.K., Bayer M. Theory of magnetization precession induced by a picosecond strain pulse in ferromagnetic semiconductor (Ga, Mn) As // PR (B). 2011. V.84. P.214432 (11).
541. Chernov A.I., Kozhaev M.A., Vetoshko P.M., Dodonov D.V., Prokopov A.R., Shumilov A.G., Shaposhnikov A.N., Berzhansky V.N., A.K. Zvezdin, V.I. Belotelov Local probing of magnetic films using optical excitation of magnetostatic waves // FTT. 2016.T.58. No. 6. P.1093.
542. S. P. Strelkov. Mechanics - Moscow: Nauka, 1965.
543. V.G. Levich. A course in theoretical physics. Vol. 1 - Moscow: Nauka, 1969.
544. V. I. Zubkov, V. I. Scheglov. Dispersion properties of magnetostatic waves in a relatively magnetized metal-dielectric-ferrite-ferrite-dielectric-metal structure // XVII International School-Seminar "New Magnetic Materials of Microelectronics". Collection of works. M .: Moscow State University. 2000.S. 328.
545. G.S. Landsberg. Optics - Moscow: Nauka, 1976.
546. II Olkhovsky. Theoretical Mechanics Course for Physicists -
547. Tikhonov A.N., Samarskii A.A., Equations of mathematical physics. Moscow: Nauka, 1972.
548. Antonets IV, Kotov LN, Shavrov VG, Shcheglov VI. Algorithm for determining the amplitudes of external and internal waves at the opposite incidence of two one-dimensional waves on a multilayer structure. Proceedings of the XIX International Conference "Electromagnetic Field and Materials". M .: NIU MEI. 2011.S. 154.
549. Antonets IV, Kotov LN, Shavrov VG, Shcheglov VI. Algorithm for determining the amplitudes of reflected and transmitted waves at symmetric incidence of two counterpropagating waves on a multilayer inhomogeneous structure. 2012.T.57. No. 1. P.67.
550. Antonets IV, Kotov LN, Shavrov VG, Shcheglov VI. Algorithm for determining the amplitudes of external and internal waves at the incidence of counterpropagating waves on a multilayer structure with stepped inhomogeneity. 2013.T.58. No. 1. P.16.
551. Antonets I.V., Kotov L.N., Shavrov V.G., Shcheglov V.I. Two simple algorithms for determining the amplitudes of external and internal oscillations during the propagation of counterpropagating waves in a multilayer inhomogeneous structure // Electronic "Journal of Radioelectronics", 2013. No.1. (26). http://jre.cplire.ru/jre/jan 13/11 / text.html.
552. Antonets I. V., Scheglov V. I. Propagation of waves through multilayer structures. Part five. Algorithmic methods - Syktyvkar: IPO SyktGU, 2014.
553. Antonets IV, Shavrov VG, Shcheglov VI. Application of the elimination method for analyzing wave propagation in multilayer media with a random distribution of layer parameters. 2013.T.58. No. 12. P.1149.
554. Antonets I. V., Scheglov V. I. Propagation of waves through multilayer structures. Part four. Specific methods - Syktyvkar: IPO SyktGU, 2013.
555. Antonets IV, Kotov LN, Shavrov VG, Shcheglov VI. Exclusion method for calculating wave propagation through media with stepwise inhomogeneities // Electronic Journal of Radioelectronics. 2013. No.4. (38). http://jre.cplire.ru/jre/apr13/6/text.html.

556. Antonets I. V., Scheglov V. I. Propagation of waves through multilayer structures. Part six. Exclusion method - Syktyvkar: IPO SyktSU, 2015.
557. Antonets IV, Shavrov VG, Shcheglov VI. Investigation by the method of elimination of wave propagation through a multilayer medium with barrier inhomogeneities // Electronic "Journal of Radio Electronics". 2015. No.1, (38). http://jre.cplire.ru/jre/jan15/22/text.html.
558. Antonets IV, Shavrov VG, Shcheglov VI. Spatial distribution of the amplitudes of external and internal waves in a limited multilayer structure with periodic inhomogeneity. Part 1. One-dimensional wave // Electronic "Journal of radio electronics". 2015. No.2, (46). http://jre.cplire.ru/jre/feb15/1/text.html.
559. Antonets IV, Shavrov VG, Shcheglov VI. Spatial distribution of the amplitudes of external and internal waves in a limited multilayer structure with periodic inhomogeneity. Part 2. Electromagnetic wave // Electronic "Journal of Radio Electronics". 2015. No.2, (40). http://jre.cplire.ru/jre/feb15/2/text.html.
560. Antonets IV, Shavrov VG, Shcheglov VI. Spatial distribution of the amplitudes of external and internal waves in a limited multilayer structure with periodic inhomogeneity. Part 3. Electromagnetic wave in a magnetic medium // Electronic "Journal of Radio Electronics". 2015. No.3. (51). http://jre.cplire.ru/jre/mar15/13/text.html.
561. Antonets I.V., Shavrov V.G., Shcheglov V.I. Spatial distribution of the amplitudes of external and internal waves in a limited multilayer structure with periodic inhomogeneity. Part 4. Wave in a dissipative medium // Electronic "Journal of Radio Electronics". 2015. No.3. (56). http://jre.cplire.ru/jre/mar15/14/text.html.
562. Schwarzburg A.B. Video pulses and non-periodic waves in dispersive media (exactly solvable models) // UFN, 1998. V.168. No. 1. P.85.
563. Schwarzburg A.B. Dispersion of electromagnetic waves in layered and nonstationary media (exactly solvable models) // UFN, 2000. Vol. 170. No. 12. P.1297.
564. V. I. Zubkov, V. I. Scheglov. Forward and backward surface magnetostatic waves in a ferrite-dielectric-metal structure magnetized by a transversely inhomogeneous field // Proceedings of the XIX International School-Seminar "New magnetic materials of microelectronics" M .: Moscow State University. 2004.S. 287.
565. V. I. Zubkov, E. G. Locke, V. I. Shcheglov. Passage of surface magnetostatic waves under a metal strip located above the surface of a ferrite film. 1989. Vol. 34. No. 7. P.1381.
566. Vashkovsky A.V., Zubkov V.I., Locke E.G., Shcheglov V.I. Refraction of surface magnetostatic waves at the interface between ferrite and ferrite-dielectric-metal // RE. 1991. Vol.36. No. 10. P. 1959.
567. Vashkovsky A.V., Zubkov V.I., Locke E.G., Shcheglov V.I. Refraction of surface magnetostatic waves by a metal strip located on the surface of a ferrite film. 1991. Vol.36. No. 12. P.2345.
568. Zubkov V.I., Locke E.G., Nam B.P., He A.S., Shcheglov V.I. Dispersion of surface magnetostatic waves in two-layer ferrite films // ZhTF. 1989. Vol.59. No. 12. P.115.
569. Zubkov V.I., Locke E.G., He A.S., Shcheglov V.I. Surface magnetostatic waves and spin-wave resonances in two-layer ferrite films // ZhTF. 1993. Vol. 63. No. 5. P. 166.
570. V. I. Zubkov, V. A. Epanechnikov, V. I. Shcheglov. Dispersion characteristics of surface magnetostatic waves in a two-layer ferromagnetic film // RE. 2007. Vol.52, v.2. P.192.
571. V. I. Zubkov, E. G. Locke, V. I. Shcheglov. Propagation of surface magnetostatic waves in the structure of a ferrite film - a metal strip // II All-Union School-Seminar "Interaction of Electromagnetic Waves with Semiconductors and Semiconductor-Dielectric Structures". Abstracts. report Saratov. 1988. Part 3. P.35.

572. Vashkovsky A.V., Zubkov V.I., Locke E.G., Shcheglov V.I. Refraction of surface magnetostatic waves at the interface between ferrite and ferrite-dielectric-metal // XII All-Union School-Seminar "New Magnetic Materials of Microelectronics". Abstracts. report Novgorod. 1990. Part 1. P.109.
573. Vashkovsky A.V., Lock E.H., Shcheglov V.I., Zubkov V.I. Magnetostatic surface waves refraction at a boundary dividing ferrite and ferrite-dielectric-metal media // European magnetic materials and applications conference. Dresden. Digest booklet. 1991. P.197.
574. Vashkovsky A.V., Zubkov V.I., Locke E.G., Shcheglov V.I. Propagation of surface magnetostatic waves in transversely inhomogeneous magnetizing fields // V All-Union School-Seminar on Spin-Wave Microwave Electronics. Abstracts. report Zvenigorod: IRE RAN, 1991, p. 13.
575. Bespyatykh Yu.I., Dikshtein I.E., Zubkov V.I., Locke E.G., Shcheglov V.I. Reflection of surface magnetostatic waves from a metal strip // V All-Union School-Seminar on Spin-Wave Microwave Electronics. Abstracts. report Zvenigorod: IRE RAN, 1991, p. 119.
576. V. I. Zubkov, V. I. Scheglov. The spectrum of magnetostatic waves propagating in an arbitrary direction in the tangentially magnetized metal-dielectric-ferrite-ferrite-dielectric-metal structure // IX International Conference on Spin-Wave Electronics. Sat. tr. M .: IRE RAN, 2000.S. 211.
577. Shcheglov V.I., Zubkov V.I. The magnetostatic waves in metal-dielectric-ferrite-ferrite-dielectric-metal structure // IX International Conference on Spin Wave Electronics. Sat. tr, Appendix. M .: IRE RAN, 2000.S. 460.
578. V. I. Zubkov, V. I. Scheglov. Magnetostatic waves in a metal-dielectric-ferrite-ferrite-dielectric-metal medium // Baikal International Scientific and Practical Conference "Magnetic Materials". Abstracts. report Irkutsk, 2001, p. 51.
579. V. I. Zubkov, V. I. Scheglov. Magnetostatic waves in the metal-dielectric-ferrite-ferrite-dielectric-metal structure // International conference "Physics and technical applications of wave processes". Abstracts. report T.2. Samara. 2001.S. 17.
580. Vashkovsky A.V., Zubkov V.I., Locke E.G., Shcheglov V.I. Influence of inhomogeneity of a constant magnetic field on the trajectories of surface magnetostatic waves // Technical Physics Letters. 1989. Vol. 15. No. 4. C.1.
581. Vashkovsky A.V., Zubkov V.I., Locke E.G., Shcheglov V.I. Trajectories of surface magnetostatic waves in inhomogeneously magnetized ferrite films // Technical Physics Letters. 1989. Vol. 15. No. 4. C.5.
582. Vashkovsky A.V., Lock E.H., Shcheglov V.I., Zubkov V.I. Passage of surface magnetostatic waves through magnetic "valley" and "ridge" // IEEE Trans. on Magn. 1990. V.26. No. 5.P.1480.
583. Vashkovsky A.V., Zubkov V.I., Locke E.G., Shcheglov V.I. Propagation of surface magnetostatic waves in an inhomogeneous constant magnetic field such as an extended well // ZhTF. 1990.Vol. 60. No. 7. P.138.
584. V. I. Zubkov, E. G. Locke, V. I. Shcheglov. Propagation of surface magnetostatic waves in an inhomogeneous constant magnetic field with a roll-type profile. 1990. Vol. 35. No. 8. P.1617.
585. Vashkovsky A.V., Zubkov V.I., Locke E.G., Shcheglov V.I. Surface magnetostatic waves in linearly inhomogeneous magnetic fields // RE. 1991. Vol.36. No. 1. P.18.
586. Vashkovsky A.V., Zubkov V.I., Locke E.G., Shcheglov V.I. Surface magnetostatic waves in an inhomogeneous magnetizing field of the "valley" type // RE. 1994. Vol. 39. No. 2. P.217.
587. Vashkovsky A.V., Zubkov V.I., Locke E.G., Shcheglov V.I. Surface magnetostatic

waves in an inhomogeneous magnetizing field of the "shaft" type // RE. 1995. Vol. 40. No. 2. P.313.
588. Vashkovsky A.V., Zubkov V.I., Locke E.G., Shcheglov V.I. Ray trajectories of a surface magnetostatic wave in monotonic inhomogeneous magnetic fields // RE. 1995. Vol. 40. No. 6. P.950.
589. Vashkovsky A.V., Lock E.H., Shcheglov V.I., Zubkov V.I .. The influence of parabolical non-uniformity of bias field on wave beam paths of MSW // Proc. of Int. Conf. on Electronics Components and Materials. Beijing. China. 1989. P.430.
590. Vashkovsky A.V., Zubkov V.I., Locke E.G., Shcheglov V.I. Influence of inhomogeneity of the bias field on the trajectories of wave beams of surface magnetostatic waves // XI All-Union School-Seminar "New Magnetic Materials for Microelectronics". Abstracts. report Tashkent, 1988. P. 177.
591. Vashkovsky A.V., Zubkov V.I., Locke E.G., Shcheglov V.I. Propagation of surface magnetostatic waves in transversely inhomogeneous magnetizing fields // V All-Union School-Seminar on Spin-Wave Microwave Electronics. Abstracts. report Zvenigorod: IRE RAN, 1991, p. 13.
592. Vashkovsky A.V., Zubkov V.I., Locke E.G., Shcheglov V.I. Trajectories of rays of a surface magnetostatic wave in monotonic inhomogeneous magnetic fields // XIV All-Russian School-Seminar "New Magnetic Materials of Microelectronics". Abstracts. report Part 1. Moscow: Moscow State University, 1994. p. 89.
593. Vashkovsky A.V., Zubkov V.I., Locke E.G., Shcheglov V.I. Propagation of direct surface magnetostatic waves in a ferrite-dielectric-metal structure magnetized by a linearly inhomogeneous magnetic field // ZhTF. 1995. Vol. 65. No. 8. P.78.
594. Vashkovsky A.V., Zubkov V.I., Shcheglov V.I. Direct surface magnetostatic waves in a ferrite-dielectric-metal structure magnetized by an inhomogeneous magnetic field of the "valley" type // RE. 1996. Vol.41. No. 6. P.670.
595. Vashkovsky A.V., Zubkov V.I., Shcheglov V.I. Direct surface magnetostatic waves in a ferrite-dielectric-metal structure magnetized by an inhomogeneous field of the "shaft" type. 1996. Vol.41. No. 12. P.1413.
596. Vashkovsky A.V., Zubkov V.I., Locke E.G., Shcheglov V.I. Influence of an inhomogeneous magnetizing field on the propagation of surface magnetostatic waves in a ferrite-dielectric-metal structure // First Joint Conference on Magnetoelectronics. Abstracts. report M .: IRE RAN, 1995.P. 133.
597. Vashkovsky A.V., Zubkov V.I., Locke E.G., Shcheglov V.I. Direct surface magnetostatic waves in a ferrite-dielectric-metal structure magnetized by a linearly inhomogeneous magnetizing field // First Joint Conference on Magnetoelectronics, Tez. report M .: IRE RAN, 1995.S. 135.
598. Vashkovsky A.V., Zubkov V.I., Shcheglov V.I. Direct surface magnetostatic waves in a ferrite-dielectric-metal structure magnetized by an inhomogeneous magnetizing field of the "valley" // First Joint Conference on Magnetoelectronics. Abstracts. report M .: IRE RAN, 1995.S. 137.
599. Vashkovsky A.V., Zubkov V.I., Shcheglov V.I. Direct surface magnetostatic waves in a ferrite-dielectric-metal structure magnetized by an inhomogeneous magnetizing field of the "shaft" type // First Joint Conference on Magnetoelectronics. Abstracts. report M .: IRE RAN, 1995.S. 139.
600. V. I. Zubkov, V. I. Shcheglov. Backward surface magnetostatic waves in the ferrite-dielectric-metal structure. 1997. Vol. 42. No. 9. P.1114.
601. V. I. Zubkov, V. I. Shcheglov. Conditions for the existence of backward surface magnetostatic waves in the ferrite-dielectric-metal structure // Technical Physics Letters. 1998. Vol.24. No. 13. C.1.

602. V. I. Zubkov, V. I. Scheglov. Propagation of backward surface magnetostatic waves in a ferrite-dielectric-metal structure magnetized by a linearly inhomogeneous magnetic field // ZhTF. 1999. Vol. 69. No. 2. P.70.
603. V. I. Zubkov, V. I. Scheglov. Trajectories of backward surface magnetostatic waves in a ferrite-dielectric-metal structure magnetized by an inhomogeneous field of the "shaft" type // Letters to ZhTF. 1999. Vol. 25. No. 23. P.61.
604. V. I. Zubkov, V. I. Shcheglov. Backward surface magnetostatic waves in a ferrite-dielectric-metal structure magnetized by an inhomogeneous field of the "valley" type. 2000. Vol.45. No. 1. P.116.
605. Zubkov V.I., Locke E.G., Shcheglov V.I. Propagation of forward and backward surface magnetostatic waves in a non-uniformly magnetized ferrite-dielectric-metal structure // XV All-Russian School-Seminar "New Magnetic Materials of Microelectronics". Abstracts. report M .: Moscow State University, 1996.S. 265.
606. V. I. Zubkov, V. I. Scheglov. Backward surface magnetostatic waves in a ferrite-dielectric-metal structure magnetized by an inhomogeneous field of the "valley" type // XVI International School-Seminar "New Magnetic Materials of Microelectronics". Abstracts. report Part 1. Moscow: Moscow State University, 1998, p. 89.
607. Shcheglov V.I., Zubkov V.I. The propagation of magnetostatic surface waves in ferrite-dielectric-metal structure magnetized by nonuniform magnetic field in the shape of "ridge" // Moscow International Symposium on Magnetism. Proceedings of MISM-99. Part.2. Moscow, 1999. P.206.
608. V. I. Zubkov, V. I. Scheglov. Phase incursion of surface magnetostatic waves propagating in nonuniformly magnetized ferrite films and ferrite-metal structures // Technical Physics Letters. 1999. Vol. 25. No. 4. P.79.
609. V. I. Zubkov, V. I. Scheglov. Phase incursion of surface magnetostatic waves in a ferrite film with free and metallized surfaces, magnetized by an inhomogeneous field of the "valley" type. 2000. Vol.45. No. 11. P.1369.
610. V. I. Zubkov, V. I. Scheglov. Phase incursion of surface magnetostatic waves in a ferrite film with free and metallized surfaces, magnetized by an inhomogeneous field of the "shaft" type. 2001. Vol. 46. No. 1. P.98.
611. V. I. Zubkov, V. I. Scheglov. Delay time of surface magnetostatic waves in ferrite films and ferrite-metal structures magnetized by inhomogeneous fields. 2003. T.48. No. 8. P.1012.
612. V. I. Shcheglov. Phase characteristics of surface magnetostatic waves in nonuniformly magnetized ferrite films and ferrite-metal structures. XVI International School-Seminar "New Magnetic Materials of Microelectronics". Abstracts. report Part 1. M .: Moscow State University, 1998.S. 125.
613. Shcheglov V.I., Zubkov V.I. The giant phase shift of magnetostatic surface waves propagating in ferrite film with metalized surface magnetized by nonuniform magnetic field in the shape of the valley // IX International Conference on Spin-Wave Electronics. Sat. tr. Application. Moscow: IRE RAN, 2000.S. 462.
614. V. I. Zubkov, V. I. Scheglov. Group delay time of surface magnetostatic waves propagating in nonuniformly magnetized ferrite films and ferrite-metal structures. Proc. XI Intern. Conference on Spin-Electronics and Gyrovector Electrodynamics. Moscow (Firsanovka). Publ.UNC-1 MPEI (TU), 2002. P.162.
615. V. I. Zubkov, V. I. Scheglov. Application of the method of partial wave beams for the analysis of amplitude-frequency and phase-frequency characteristics of devices on magnetostatic waves // RE. 1999. Vol. 44. No. 3. P.287.
616. V. I. Zubkov, V. I. Scheglov. Calculation of the amplitude-frequency characteristics of a surface magnetostatic wave filter with two linear transducers. 1999. Vol. 44.

No. 6. P.758.
617. V. I. Zubkov, V. I. Scheglov. Calculation of the amplitude-frequency characteristics of devices based on surface magnetostatic waves propagating in a ferrite film magnetized by a linearly inhomogeneous field. 2000. Vol.45. No. 7. P.865.
618. V. I. Zubkov, V. I. Scheglov. Calculation of the amplitude-frequency characteristics of the filter on magnetostatic waves in the field of the "valley" type // RE. 2001. Vol. 46. No. 6. P.739.
619. V. I. Zubkov, V. I. Scheglov. Amplitude-frequency characteristics of the waveguide channel of magnetostatic waves formed by an inhomogeneous field of the "shaft" type // RE. 2001. Vol. 46. No. 9. P.1121.
620. V. I. Zubkov, V. I. Scheglov. Amplitude-frequency characteristics of the filter on magnetostatic waves propagating in an inhomogeneous field of the "shaft" type // RE. 2002. Vol.47. No. 5. P.625.
621. V. I. Zubkov, V. I. Scheglov. Amplitude-frequency characteristics of the filter on magnetostatic waves propagating arbitrarily relative to the magnetizing field. 2002. Vol.47. No. 7. P.886.
622. V. I. Zubkov, V. I. Scheglov. Calculation of the amplitude and phase-frequency characteristics of devices based on magnetostatic waves by the method of partial wave beams // XVI International School-Seminar "New Magnetic Materials of Microelectronics". Abstracts. report Part 1. Moscow: Moscow State University, 1998.S. 99.
623. V. I. Zubkov, V. I. Scheglov. Amplitude-frequency characteristics of the filter based on magnetostatic waves propagating in a field of the "valley" type // The second joint conference on magnetoelectronics (international). Abstracts. report Ekaterinburg, 2000.S. 22.
624. V. I. Zubkov, V. I. Scheglov. Dependence of the amplitude-frequency characteristics of the filter on magnetostatic waves propagating in an inhomogeneous field of the "shaft" type on the apertures of the input and output converters // X International Conference on Spin Electronics and Gyrovector Electrodynamics. Sat. tr. M .: UNTs-1 (MPEI), 2001.S. 128.
625. V. I. Zubkov, V. I. Scheglov. Analysis of the amplitude-frequency characteristics of a filter based on magnetostatic waves for an arbitrary orientation of the magnetizing field by the method of partial wave beams // X International Conference on Spin Electronics and Gyrovector Electrodynamics. Sat. tr. M .: UNTs-1 (MPEI), 2001.S. 147.
626. V. I. Zubkov, V. I. Scheglov. Amplitude-frequency characteristics of the transmission line on surface magnetostatic waves with an arbitrary orientation of the magnetizing field // Proceedings of the XVIII International School-Seminar "New Magnetic Materials of Microelectronics". M .: MGU, 2002.S. 823.
627. V. I. Zubkov, V. I. Scheglov. Influence of the phase incursion of surface magnetostatic waves on the characteristics of their transmission lines. 2005. Vol. 50. No. 4. P.464.
628. V. I. Zubkov, V. I. Scheglov. Wavefront of surface magnetostatic waves in inhomogeneously magnetized ferrite films // Proceedings of the XX International School-Seminar "New Magnetic Materials of Microelectronics". M .: Moscow State University, 2006.S. 674.
629. V. I. Zubkov, V. I. Scheglov. Deformation of the plane wavefront of surface magnetostatic waves propagating in ferrite films magnetized by a linearly inhomogeneous field // Proceedings of the XVI International Conference "Radiolocation and Radio Communication". M .: NIU MEI, 2008, p. 286.
630. Dwight G.B. Tables of integrals and other mathematical formulas - Moscow: Nauka,

1973.
631. G. Fikhtengolts Differential and integral calculus course. Volume 1 - M., L .: State. ed. technical-theor. lit., 1951.
632. Annenkov A.Yu., Vashkovsky A.V., Zubkov V.I., Shcheglov V.I. The use of a package of nonuniformly magnetized ferrite films in a multichannel microwave frequency separation filter // RE. 1995. Vol. 40. No. 7. P.1146.
633. V. I. Zubkov, V. I. Scheglov. Radiation of electromagnetic waves due to the acceleration of magnetostatic waves in an inhomogeneously magnetized ferrite film. 2001. Vol. 46. No. 4. P.433.
634. Vashkovsky A.V., Zubkov V.I., Locke E.G., Shcheglov V.I., Yakovlev S.V. Calculation of the shape of transducers of surface magnetostatic waves in devices with a flat magnet // Electronic engineering. Series 1 "Microwave technology". 1992. No. 1 (445). P.18.
635. V. I. Zubkov, V. I. Scheglov. Group delay time of surface magnetostatic waves propagating in a tangentially magnetized ferrite-dielectric-metal structure // Electrodynamics and microwave and EHF technique. 1999. Vol.7. No. 3 (24). P.35.
636. Vashkovsky A. V., Zubkov V. I., Locke E. G., Shcheglov V. I. Influence of inhomogeneity of the magnetizing field on the delay of a pulsed microwave signal in a ferrite film // Proc. of 10-th Int. Conf. on Microwave Ferrites. Part 1. Szczyrk. Poland, 1990. P.220.
637. Annenkov A.Yu., Vashkovsky A.V., Zubkov V.I., Shcheglov V.I. Frequency-separating filter on a package of non-uniformly magnetized ferrite films // Proceedings of the XI International conference on gyromagnetic electronics and electrodynamics. Vol. 1. Ukraine. Alushta, 1992, p. 86.
638. V. I. Zubkov, E. G. Locke, V. I. Shcheglov. Branched waveguide structures for surface magnetostatic waves on film magnets // Proceedings of the XI International Conference on Gyromagnetic Electronics and Electrodynamics. T1. Ukraine. Alushta, 1992, p. 89.
639. AV Vashkovsky, VI Zubkov, EG Locke, VI Shcheglov, SV Yakovlev. Calculation of the shape of transducers of surface magnetostatic waves in devices with an inhomogeneous magnetic field. // Proceedings of the XI International Conference on Gyromagnetic Electronics and Electrodynamics. Vol. 1. Ukraine. Alushta, 1992.S. 93.
640. Vashkovsky A.V., Zubkov V.I., Locke E.G., Shcheglov V.I. Delay of a pulsed microwave signal propagating in a non-uniformly magnetized ferrite film. // IV All-Union School-Seminar "Spinwave Microwave Electronics". Abstracts. report Lvov, 1989.S. 52.
641. Vashkovsky A.V., Zubkov V.I., Locke E.G., Shcheglov V.I. Multichannel microwave frequency separation filter on surface magnetostatic waves // Ferrite microwave devices and materials. Materials of the XV All-Union Scientific and Technical Conference on Microwave Ferrite Technology. T.2. Moscow. P.61.
642. Vashkovsky A. V., Zubkov V. I., Locke E. G., Shcheglov V. I. Tunable band-pass and band-stop filters on surface magnetostatic waves // Ferrite microwave devices and materials. Materials of the XV All-Union Scientific and Technical Conference on Microwave Ferrite Technology. T.2. Moscow. 63.
643. Vashkovsky A. V., Zubkov V. I., Locke E. G., Shcheglov V. I. Tunable microwave delay line on surface magnetostatic waves // Ferrite microwave devices and materials. Materials of the XV All-Union Scientific and Technical Conference on Microwave Ferrite Technology. T.2. Moscow, 1990, p. 65.
644. V. I. Zubkov, E. G. Locke, V. I. Shcheglov, S. V. Yakovlev. Optimization of the shape of transducers of surface magnetostatic waves for a non-uniformly magnetized fer-

rite film // V All-Union School-Seminar on Spin-Wave Microwave Electronics. Abstracts. report Zvenigorod: IRE RAN, 1991, p. 121.
645. Annenkov A.Yu., Vashkovsky A.V., Zubkov V.I., Shcheglov V.I. Multichannel frequency separating filter on stacked ferrite structures // V All-Union School-Seminar on Spin-Wave Microwave Electronics. Abstracts. report Zvenigorod: IRE RAN, 1991, p. 123.
646. Vashkovsky A. V., Locke E. G., Shcheglov V. I. Comparative analysis of the efficiency of microwave and magneto-optical methods for measuring the parameters of YIG films with a complex nature of anisotropy // XVI International School-Seminar "New Magnetic Materials of Microelectronics". Abstracts. report Part 2. Moscow: Moscow State University, 1998.S. 505.
647. Lock E.H., Shcheglov V.I. The measurements of yttrium iron garnet films magnetic parameters // XIV-th International Conference on Microwave Ferrites (Gyromagnetic Electronics and Electrodynamics). Conference Proceedings. Hungary. Eger, 1998. V.1. P.152.
648. V. I. Zubkov, V. I. Scheglov. Investigation of the spatial configuration of an inhomogeneous magnetic field using a surface magnetostatic wave sensor // Second Joint Conference on Magnetoelectronics. Abstracts. report Ekaterinburg, 2000.S. 24.
649. V. I. Zubkov, V. I. Scheglov. Delay of surface magnetostatic waves during propagation in a ferrite-dielectric-metal structure // Second Joint Conference on Magnetoelectronics. Abstracts. report Ekaterinburg, 2000, p. 28.
650. V. I. Zubkov, V. I. Scheglov. The use of ferromagnetic resonance in unsaturated films of yttrium iron garnet to determine their magnetic parameters // The Second Joint Conference on Magnetoelectronics. Abstracts. report Ekaterinburg, 2000.S. 32.
651. Locke E.G., Shcheglov V.I. Measurement of magnetic parameters of yttrium iron garnet films by methods of ferromagnetic resonance and magneto-optics // Second Joint Conference on Magnetoelectronics. Abstracts. report Ekaterinburg, 2000.S. 34.
652. V. I. Zubkov, V. I. Scheglov. Measurement of the spatial distribution of an inhomogeneous magnetic field using a magnetostatic wave sensor // 7th All-Russian Scientific and Technical Conference "State and Problems of Measurements". Abstracts. report Moscow, 2000.S. 113.
653. V. I. Zubkov, V. I. Scheglov. Measurement of parameters of films of yttrium iron garnet in an unsaturated state by the method of ferromagnetic resonance // 7th All-Russian Scientific and Technical Conference "State and Problems of Measurements". Abstracts. report Moscow, 2000.S. 115.
654. V. I. Zubkov, V. I. Scheglov. Wave-conducting properties of a ferrite film magnetized by a "shaft" type field. // Baikal International Scientific and Practical Conference "Magnetic Materials". Abstracts. report Irkutsk, 2001.S. 52.

Index

C

curves
  dispersion curves 17, 19, 140, 141, 142, 143, 145, 157, 178, 180, 181, 182
  isofrequency curves vii, 10, 11, 16, 19, 20, 30, 142, 143, 144, 145, 146, 159, 162, 163, 164, 165, 169, 171, 173, 175, 176, 182, 183

E

easy magnetization axis 4
effect
  Faraday effect 31
  Kerr effect 30, 31
EMA (easy magnetic axis) 4, 13
equation
  Landau–Lifshitz equation 33, 34, 39, 48
  Laplace equation 5, 45, 46, 53, 75
  Maxwell's equation 26
  Walker equation 5, 43, 45, 46, 48, 49, 50, 51, 75, 137

F

ferromagnetic resonance 1, 4
field
  linearly inhomogeneous field 195, 213
  shaft'-type field 205, 206, 207, 225, 231, 232, 241, 247, 252, 277, 283, 294, 295, 342, 343, 363, 364
  valley-type field 197, 203, 210, 218, 220, 223, 231, 342, 349, 350
FMR (ferromagnetic reosnance) 1, 2, 4, 16, 22, 23, 24, 27, 28, 235

G

gyromagnetic constant 307

I

isofrequency 185, 189, 190, 191, 193, 194, 195, 196, 197, 198, 208, 210, 211, 213, 214, 215, 216, 218, 219, 220, 223, 226, 241, 242, 244, 245
isofrequency curves 10, 11, 16, 19, 20, 30

M

magnetization vector 2, 3, 4, 29
magnetostatic waves 341, 343, 349, 359, 362
method
  Euler method 130, 137
  Hamilton–Auld method 113, 114, 127, 129, 130, 137, 185, 189, 191, 193, 198, 208, 213, 220, 221, 236, 242, 249
  method of isofrequency curves 185, 189, 190, 193, 194, 241, 242
  Runge–Kutta method 15, 129, 130, 137, 192
MSW (magnetostatic waves) 1, 5, 7, 8, 9, 10, 11, 12, 13, 14, 15, 16, 17, 20, 21, 23, 24, 27, 28, 29, 30, 31, 32, 341, 342, 358, 364

P

problem
  Damon–Eshbach problem 43, 45, 48, 49, 84, 86, 92, 93, 94, 97, 137

S

SMSW (surface magnetostatic waves) vi, 9, 10, 12–25, 30, 139, 140, 141, 142, 146, 147, 149, 152, 154, 155, 156, 157, 158, 159, 160, 161, 162, 164, 165, 166, 167, 168, 169, 170, 171, 172, 173, 174, 175, 176, 177, 178, 179, 180, 181, 182, 183, 184, 185, 187, 188, 189, 191, 192, 193, 194, 195, 196, 197, 198, 199, 272, 275, 276, 277, 278, 279, 284, 286, 287, 290, 291, 292, 293, 294, 295, 296, 297, 298, 299, 300, 144, 306, 307, 308, 309, 310, 311, 312, 313, 315, 316, 333,
structure
  FDM (ferrite-dielectric-metal) structure 139, 140, 142, 144, 146, 154, 155, 156, 157, 158, 161, 162, 164, 165, 167, 169, 170, 171, 173, _)
  ferrite–metal (FM) structure 208, 189, 209, 210, 211, 212, 218, 219, 222, 223, 226. 234, 235, 243, 244, 245, 246

V

vector
  Poynting vector 25, 26

W

waves
  electromagnetic waves 21, 24, 25, 26, 29
  magnetostatic waves 1, 5, 7, 9, 11, 21, 24, 26, 29, 31

Y

YIG (yttrium–iron garnet) 7, 9, 15, 16, 23, 24, 25, 27, 28, 29, 239, 299, 300, 343, 344, 345, 346, 347, 350, 352, 353, 354, 355, 357, 359, 363

Printed in the United States
by Baker & Taylor Publisher Services